# Urban Migrants in China

Daming Zhou

# Urban Migrants in China

中国财经出版传媒集团
经济科学出版社
Economic Science Press

Springer

Daming Zhou
Sociology and Anthropology Department
Sun Yat-sen University
Guangzhou, China

*Translated by*
Yu Cao
College of Liberal Arts
Jinan University
Guangzhou, China

This book is sponsored by the Chinese Fund for the Humanities and Social Sciences

ISBN 978-981-99-3113-2 ISBN 978-981-99-3114-9 (eBook)
https://doi.org/10.1007/978-981-99-3114-9

Jointly published with Economic Science Press
The print edition is not for sale in Chinese Mainland. Customers from Chinese Mainland please order the print book from: Economic Science Press.



This Springer imprint is published by the registered company Springer Nature Singapore Pte Ltd.
The registered company address is: 152 Beach Road, #21-01/04 Gateway East, Singapore 189721, Singapore

# Preface

As Fei Xiaotong stated in his magnum opus *From the Soil: The Foundations of Chinese Society:* "Chinese so ciety is fundamentally rural", and this argument stays valid until the 1970s. China has a prolonged agrarian stage, and its basic livelihood is a small peasant economy with intensive farming, then nostalgia is an important trait of Chinese culture. Although some unexpected situation caused several major migrations in Chinese history, such calamities as famine, war, and forced evacuation. In an agrarian society, migration was either imposed by exogenous forces, or as a survival strategy to balance insufficient farmland and dense population. Afterall, Chinese are reluctant to migrate, and continue the theme of nostalgia for the agrarian civilization. Even in the early 20th century, dozens of industrial cities rise on China's coastal and riparian lines, the dominance of agrarian society remains unchallenged.

After the founding of the People's Republic of China in 1949, the state enforced the *household registration(Hukou)* system, built up a strict urban and rural dual system, and greatly limits migration, especially, the rural to urban domestic migration is almost stagnant. On the contrary, a few times of mass migration happened during 1949 to 1978 were the inverse of the urbanization.

After the reform and opening up in 1978, on the one hand, the rural reform in early 1980s released many agricultural surplus-labour; on the other hand, the decentralized development of industry in coastal areas caused the rapid rise of small towns, they absorb not only agricultural surplus-labour on the same region but also attracted immigrant labour across the country, a great population migration prelude. In the recent 30 years, started from predominantly rural migrant to today's diverse immigrants' communities from all kinds of resources; started from aiming developed coastal areas to spread the aiming at medium cities as well as the central and western regions; started from the first generation of rural migrant to the emergence of new immigrants who intend to settle in the city today. New problems, new features, and new trends emerged in China's domestic migration endlessly, embody with the continuity, and characteristics of complexity and diversity. The nationwide city-centered population migration from rural areas was an important driving force for China's transformation from agricultural civilization to modern industrial civilization and urban civilization. Since then, migration has replaced nostalgia and has become the majority of China as part of people's daily life.

The huge population migration has brought about drastic changes in China's urban and rural lives. One of the important manifestations is the trend of urban immigration. In the Pearl River Delta, Yangtze River Delta, Beijing-Tianjin-Hebei and other urban regions with the highest concentration of immigrants in China, the number of immigrants is large, diverse in composition, and extremely heterogeneous. In terms of ethnic identity, there are the Han ethnic group, ethnic minority groups, and foreigners; in terms of source, there are people who migrate within the province and the region, and there are also people who migrate across provinces and countries; in terms of occupation, there are migrant labour, professional and technical personnel, entrepreneurs, etc.; in terms of household registration status, there are local residents, domestic immigrants who obtain local household registration through various channels, and immigrants who do not have a local household registration; in terms of migration methods, there are people who move but do not stay, and there are also a large number of new immigrants with settled intentions; etc. They gathered in a very short period of time, which made the immigration degree of these regions extremely high. Moreover, in the past two decades, with the implementation of the national strategy for functional zones, as well as the rapid economic development and urbanization in the central and western regions, this trend of urban immigration is spreading to cities and small-and medium-sized cities in the central and western regions. In addition to domestic migration, China is also becoming a destination for international immigrants. In Beijing, Shanghai, Guangzhou, Shenzhen and other global cities, foreigners' "economic zones" (or economic clusters) that have emerged due to business exchanges are particularly interesting, such as Wangjing in Beijing, Xiaobei Road and Yuanjing Road in Guangzhou, Chouzhou Road and Gongren Road in Yiwu. These ethnic communities often have distinct characteristics of different cultures, which make modern cities present distinctive characteristics of cultural diversity.

The rapid gathering of diverse and heterogeneous immigrants in the cities not only brings great vitality to the cities, but also brings challenges to the harmonious development of the cities. How to reform a series of urban and rural welfare, social resource distribution, land and industry policy, which formed around the household registration system, has often become the focus of discussion from all classes. In the author's opinion, a series of urban-rural dual systems have been closely connected with people's daily lives, and become a social mechanism that builds differences and inequality between people. Therefore, the harmonious development of immigrant cities is not only a question of the construction and improvement at the administration level, but also a question of promoting the social integration of immigrants and developing the habit of multicultural coexistence under the guidance of the administration and based on the perspective of daily life.

Based on the above understanding, this book shows the trends and problems of urban immigration in China from three levels and puts forward policy recommendations on related issues.

The first is to grasp the latest trends in migration, especially to break the limitations of the concept of "rural migrant" and use the concept of new immigrants to accommodate immigrant groups with more diversity, occupations, identities, and educational levels. Our research takes those who are "willing to settle in cities" as the main research subject, grasping the basic status of their migration and development, and accumulating a wealth of first-hand information.

The second is to think about how to bridge the dual structure of urban and rural social settings and promote the harmonious development of immigrant cities from the system level. In the author's view, the urban-rural dual system is not only a series of system designs, but also a social mechanism that distinguishes people and builds differences. Therefore, reforming of the system requires, first of all understanding how the system works as a social structure. The author regards the trend of urban immigration as a practical process for bridging the dual structure of urban and rural social settings, and puts the dual system in this process of practice, focuses on the operation of the system as a social structure, and thinking about how to use the system to improve the bridging of the dual structure of urban and rural.

The third is to analyze how to promote the social integration of urban new immigrants from the level of daily life. Realizing the social integration of new immigrants is an important goal for achieving the harmonious development of immigrant cities. In addition to institutional guidance, social integration also involves a series of adaptations and adjustments in daily life. For new immigrants, from agrarian society to urban society, from acquaintance society to stranger society, their thinking and behavior methods need to be adjusted accordingly to adapt to the external environment. For locals and new immigrants, how to effectively integrate different groups instead of forming opposition or even social rupture, highlighting modernity and rationality, is very important.

This book is the final result of the major research project on philosophy and social sciences of the Ministry of Education, "Issues and Countermeasures for Urban New Immigrants". This project was initiated in early 2008, and the final results were submitted in early 2012. The project has gone through four years. The survey points were set in six cities, Guangzhou, Dongguan, Hangzhou, Zhengzhou, Shenyang and Chengdu, and more than 3,000 questionnaires and 150 interviews were produced. The cooperating faculties on this project include Zhejiang University, Zhengzhou University, Liaoning University, Sichuan University, etc. The main participants include Liu Zhaohui (Zhejiang University), Han Heng (Zhengzhou University), Jiang Fan (Liaoning University), Wang Chuan (Sichuan Normal University), Yang Xiaoliu (Sun Yat-sen University), Wang Xingzhou (Sun Yat-sen University), Yu Chengpu (Sun Yat-sen University) and other scholars. In such a wide-ranging and long-lasting project, the research team members coordinate well and overcome difficulties. While completing the project tasks efficiently, they also formed a joint research force, laying a foundation for further research on related issues in the future. During the implementation of the project, the Department of Social Sciences of the Ministry of Education, the Department of Social Sciences of Sun Yat-sen University, and various cooperative faculties provided strong support from all aspects to ensure the smooth development of the project. The author hereby represents the research team to express great appreciation to all aforementioned faculties.

The reasons and mechanisms for the emergence of urban new immigrants in China are extremely complex, and the relevant countermeasures must also consider various factors. The errors and omissions in the research of this book are inevitable. We hope this book can provide some reference and formation for more in-depth related research in the future.

Daming Zhou

# Main Members of the Research Group

| | | |
|---|---|---|
| Wang Chuan | Wang Xingzhou | Xiang Chunling |
| Liu Zhiyang | Liu Jiaji | Liu Zhaohui |
| Jiang Fan | Yang Xiaoliu | Li Zhigang |
| Li Ruojian | Li Ling | Li Cuiling |
| Yu Chengpu | Zhou Runan | Qin Hongzeng |
| Ji Guangxu | Han Heng | Cheng Yu |

# Contents

# Chapter 1
# Introduction

## 1.1 Background

Since 1978, China has adopted the reform and opening-up policy for over 4 decades and achieved great progress in industrialization. By 2008, China's non-agricultural production shares as high as 88.7% of GDP, and the urbanized population percentage rose from 17.9% in 1978 to 45.7% in 2008, while the proportion of non-agricultural employment exceeded 50%.[1] However, the quality of urbanization in China is still quite low, and the proportion of non-agricultural employment is far lower than the proportion of non-agricultural production. The entire urbanization process is not coordinated with the progress of industrialization. Since the founding of the People's Republic of China in 1949, the state has made hesitations in promoting urbanization progress.[2] In particular cases, after the adoption of reform and opening-up policy, the state has imposed various restrictions on the flow of rural labour into cities, and denied to treat the rural migration population to equal to as urban residents; the urbanization population flows from the pendulum mode of moving without settling down,[3] the typical waves of migration labour have risen and fallen for many times. The state has long pursued economic development as primary task, and neglects the development of livelihood, which caused a long-term "industry focus" urban planning and development model, the supplies of livelihood infrastructure and public services did not keep up with the accelerated speed of urbanization; the consequences of this contradiction mostly fallen on the migration population. In the 1980s, millions of

[1] China's National Bureau of Statistics, *China Statistical Abstract 2009*, Beijing: China Statistics Press, 2009.

[2] Zhou Tianyong, The Hesitation, Missteps and Consequences of Urbanization in China, *Urban Insight*, Vol. 1, 2009.

[3] Zhou Daming, *Longing for Survival: An Anthropology Observation on Rural Labour's Mobility*, Guangzhou: Sun Yat-sen University Press, 2005.



D. Zhou, *Urban Migrants in China*,
https://doi.org/10.1007/978-981-99-3114-9_1

migration labour flooded Guangzhou and neighboured regions, government worried that the influx of people might cause "urban diseases" such as traffic jams, housing shortages and deterioration of public security. Therefore, throughout the 1980s, the population movement was strictly restricted, there were several nationwide migration screening and repatriation actions aimed at rural migrant performed.

During the 1990s, the policies regarding population and migration became more regular, relevant departments successively set up migration population management and coordination institution, and issued a series of orders on household registration(*hukou*) system,[4] employment and social welfare. Since 2003, the "No.1 Central Document"[5] has repeatedly affirmed the status and contribution of migration labour as industrial workers in China's modernization progress, and said it would "push forward the reform of the household registration system in large and medium-sized cities, and ease the conditions for rural migrant to enter cities for employment and residency".[6] The Outline of the Eleventh Five-Year Plan pointed out that "for rural migrant who have stable occupation and residence in cities, conditions should be created to gradually convert them into urban residents". Beijing, Shanghai, Guangzhou, Shenzhen, Hangzhou, Zhengzhou, Wuhan and Shenyang have all abolished their temporary residence permits and replaced them with residence permits and relevant household registration entry policies during the 10th and 11th Five-year Plan periods. Now, China is in the mature stage of industrialization, thus the economic structure is upgrading and adjusting for the future development. Promote urbanization is "a major measure to promote the virtuous circle of national economy and the coordinated development of society". The state "encourages the rural population to settle in small-and medium-sized cities and towns, and the megalopolis should start from the source of industrial structure adjustment and form the mechanism to control the over-rapid growth of population through economic measures".[7]

Since 1996, the national urbanization level has maintained an average annual growth rate of around 1.3%, which indicates that China has been and will continue to be in the acceleration stage of urbanization,[8] and the urbanization process will gradually catch up with the pace of industrialization. Now, China's industrialization and urbanization enter a new growth period of medium-term acceleration. The economic and social structure is expecting undergo profound transformation, and

---

[4] *Hukou* is a system of household registration in China, it is a complex social control method affect Chinese livelihood in many ways, it is necessary to comprehend this concept before reading this book.—*Tr.*

[5] "No.1 Central Document" is issued by the Central Committee of the Communist Party of China(CPC) yearly, usually on the beginning of a year, is a guidance document of the whole year's rural policy.—*Tr.*

[6] Opinions of the Central Committee of the CPC and the State Council Concerning Several Policies on Promoting the Increase of Farmers' Income, issued by Departments and Institutions of the CPC Central Committee, Central Committee of the Communist Party of China, State Council, 2003.

[7] The Outline of the 11th Five-Year Plan for the National Economic and Social Development of the People's Republic of China, issued by National People's Congress (NPC) of the People's Republic of China, 2006.

[8] Northam, Ray M., *Urban Geography*, John Wiley & Sons, 1975, pp. 65–67.

the social structure change will enter the new growth period of breaking down the urban–rural dual structure.[9] According to official prediction, China's urbanization percentage will be between 64 to 75% by 2030, which means 400–500 million people will become permanent urban residents in the next 20 years. However, China has to solve the same problem left in the past, there are already 201 million migrants in cities not been converted to urban resident, yet hundreds of million more to come. If the migration population not being accepted as urban resident by now, a situation of trisection may occur after 30 years, which is composed by 500 million urban residents, 500 million floating migrants who live and work in cities but not cover by cities' welfare, 500 million rural residents,[10] and this situation will jeopardize China's social stability and national competency.

## 1.2 Research Overview

Many developed countries have been through hundreds of years processing industrialization and urbanization, yet developing countries generally facing the "time compression",[11] to complete the same process in a much shorter time period, which brings many social problems; China is experiencing this compression, and formed a special urban–rural dual social structure. Development economics holds that the urban–rural dual transformation is the irresistible trend of economic and social development of any country. The mechanism of urban–rural dual social structure transformation is that urbanization promotes the transformation of agricultural society to industrial society, first by transferring more agricultural labour force to the employment of higher-income non-agricultural industries, and then promotes the transformation of social structure from rural society to urban society. We believe that forming the urban–rural dual economic and social structure is actually the migration process.

### *1.2.1 Rural–urban Migration Study Outside China*

In the early nineteenth century, Britain, ahead of any other country, started its industrialization process and transferred its rural labour to industrial labour in a century of lasting progress. Ernst G. Ravenstein, a British geographer, has written twice about

[9] Li Peilin, Chen Guangjin, Research Group of "Analysis and Forecast of Social Situation Institute" in Chinese Academy of Social Sciences, "New Growth Stage of China's Development: Analysis and Forecast of China's Social Development Situation from 2009 to 2010", *Society of China Analysis and Forecast(2010),* Beijing: Social Sciences Academic Press(China), 2009, pp. 9–10.

[10] Lyu Xueli, China's Population Facing Situation of Trisection, retrieved at May 21, 2020, http://politics.peoplecom.cn/GB/1027/9129924.html

[11] Li Peilin, Modernity and the Chinese Experience, *Chinese Journal of Sociology*, Vol. 3, 2008.

the migration base on the study of county-level migration in Britain.[12] He determined migration absorption or dispersion by the gain and loss of population on county level, thus mapped and summarized the laws of migration. In 1889, Ravenstein used the same method on more than 20 European and North American countries' population statistics and try to prove that "different countries have similar migration patterns under the similar condition".[13] Everett S. Lee summarized in his book *A Theory of Migration*, migration flow is positively correlated with economic prosperity. In the case of economic blooming, migration flow is bound to increase.[14] In the early twentieth century, with the blooming of the United States' economy, a large number of young farmers (mostly aged below 25) were pushed out of countryside and turned to cities for non-agricultural employment,[15] these migrants were usually low-educated (with few college graduates) and low-skilled (or even unskilled).[16]

We believe that domestic migration, especially rural–urban migration, is the most obvious social phenomenon during the rapid industrialization era of a country, regardless of developed or developing countries. However, developed countries may able to "provide the opportunity for rural migrants to pursue a decent life",[17] while the developing country was trapped in the dilemma of "infinite labour supply",[18] which is a large number of rural–urban migrants can improve income of the rural population, but the income maintains at a low level for quite a long time, making the gap between urban and rural and the urban–rural dual structure hard to overcome. Michael P. Todaro believes rural–urban migration in developing countries determined by the rural–urban "expected" income differential and the probability of finding an urban job.[19] He divides the process of migration in developing countries into two stages: first, rural migrants with little skills migrate to cities to find jobs in conventional sectors, and then after a period of effort and adaptation (generation or generations)to find stable modern sector jobs. Half a century later, South Korea, Japan, and other developed countries and regions in East Asia successfully completed the transformation in this way. Rural residents could not only find stable jobs in cities, but also

---

[12] Ravenstein, E. G., "The Birthplace of the People and the Laws of Migration", *The Geographical Magazine*, Vol. 3 (1876), pp. 173–177, 201–206, 229–233. Ravenstein, E. G., "The Laws of Migration", *Journal of the Statistical Society of London*, Vol. 48, No. 2. (June, 1885), pp. 167–235.

[13] Ravenstein, E. G., "The Laws of Migration", *Journal of the Royal Statistical Society*, Vol. 52, No. 2. (June, 1889), pp. 241–305.

[14] Lee, E. S., "A Theory of Migration". *Demography*, Vol. 3, No. 1. (1996), pp. 47–57.

[15] Mercer, Blaine E., "Rural Migration to Urban Settings, Educational and Welfare Problems", *International Migration Digest*, Vol. 2, No. 1 (Spring, 1965), pp. 52–62.

[16] Rose, Arnold, Warshay, Leon, "The Adjustment of Migrants to Cities", *Social Force*, (October, 1957), pp. 72–76.

[17] Mercer, Blaine E., "Rural Migration to Urban Settings, Educational and Welfare Problems", *International Migration Digest*, Vol. 2, No. 1 (Spring, 1965), pp. 52–62.

[18] Lewis, W. Arthur, "Economic Development with Unlimited Supplies of Labour", *Manchester School of Economic and Social Studies*, Vol. 22, No. 2. (1954), pp. 139–191.

[19] Todaro, Michael P., "A Model of Labour Migration and Urban Unemployment in Less Developed Countries", *The American Economic Review*, Vol. 59, No. 1 (1969), pp. 138–148.

choose to return to rural areas for non-agricultural employment.[20] On the contrary, Latin American countries have failed to past the Lewis turning point, where economic and social development has stagnated, where great numbers of migrants are concentrated in urban conventional sectors, and even generations of efforts have failed to achieve upward mobility.[21]

Seymour M. Lipset and Reinhard Bendix studied the impact of rural migration in industrial society on urban structure of occupation, and found that the expansion of urbanization makes rural migrants and urban residents have different motion trend under the same conditions: rural migrants become the bottom level in cities, while the local bottom level has the opportunity to move upward on occupation.[22] Peter M. Blau and Otis D. Duncan have a similar finding in *The American Occupational Structure,* rural background migrants are disadvantaged in social economics status, often found in the lowest occupation category, while the urban resident and migrants from other urban area with good education and skills, stacked on the top of rural migrants' occupations, gained a better occupational status.[23] Hagen Koo summarized this situation as two models with structural linkage, statically underprivilege model and dynamical push-up model.[24] But he found that cases in Asia and Latin America do not support his models. First of all, rural–urban migrants have a strong selectivity,[25] which means that a considerable proportion of rural–urban migrants come from upper-middle-class families in rural society, rather than the imagined low-skilled and low-level groups. They have enough education and resources to compete with urban residents for better occupations. Therefore, the role of rural migration as the bottom of occupation structure is non-existent, and the urban economy in developing countries is usually divided into conventional and modern industrial sectors. Most urban residents are engaged in marginal economic activities, and only a small part of them have relatively good occupations. In other words, the advantages of urban residents are not obvious.

Calvin Goldscheider's research on developing countries also shows that rural migrants are less disadvantaged in cities, especially after a period of migration;

---

20 Speare, Jr., "Alden, A Cost–Benefit Model of Rural to Urban Migration in Taiwan", *Population Studies,* Vol. 25, No. 1 (March, 1971), pp. 117–130.

21 Borges, Manoela Guidorizzi, *Citizenship for the Urban Poor? Inclusion trough Housing Program in Rio De Janeiro,* Brazil, Doctoral Dissertation, The University of Colorado, 2005.

22 Lipset, Seymour M., Bendix, Reinhard, *Social Mobility in Industrial Society,* Berkeley: University of California Press, 1959, pp. 204–216.

23 Blau, Peter M., Duncan, Otis Dudley, *The American Occupational Structure,* New York: Wiley, 1967, p. 269.

24 Koo, Hagen, "Rural–Urban Migration and Social Mobility in Third World Metropolises: A Cross-National Study", *The Sociological Quarterly,* Vol. 19, No. 2 (Spring, 1978), pp. 292–303.

25 In *The American Occupational Structure* (p. 272), Blau and Duncan also discussed similar immigration selectivity. (in any case) "immigrants would gain higher professional status and achieve upward flow more easily than non-immigrants". We see this as an immigration characteristic of the United States as a developed industrial society, different from the immigration selectivity of developing countries.

there are not many systematic differences between them and urban residents.[26] (1) Economic motivation is primary in driven rural migrants to cities (such as finding non-agricultural jobs or looking for better job opportunities). However, migration decisions are based not on actual job vacancies but potential job opportunities and the willingness of migrants and their family strategies. (2) In essence, rural–urban migration digests the surplus rural labour force and meets the demand of urban economic expansion, which is a process of realizing generational and intergenerational occupational mobility and an important mechanism of national economic integration. (3) Rural migrants are selected, who are more of a skilled, clearly motivated group able to compete with urban residents. (4) Education and skills are essentials for rural migrants and urban residents in the formal and informal sector to achieve their occupational attainment, one of the most important factors for rural migrants, is that they not only gathered in the informal sector, but also a lot of them get white-collar and blue-collar jobs. (5) Housing of migrants is related to poverty, especially when new migrants have not yet integrated into the urban economy. In summary, Calvin Goldscheider argues that previous studies have shown that there are no structural factors that prevent new migrants from participating in urban economies; new migrants in cities have the opportunity to gain economic and social benefits thus integrate well into cities.

Before David Goldscheider and Hagen Koo, some scholars found the rural–urban migrants in Latin American countries, whose selectivity would change along with the change of socioeconomic status. On the one hand, the attractive job opportunities in big cities and the population pressure in rural areas as the pull and push factors; on the other hand, the urban development craved for labour force, thus the number of rural migrants is increasing and the selectivity is becoming lower.[27] Low selectivity means rural–urban migration turns from selective pattern to massive pattern, in other words, the selective migration of high-skilled population in rural areas and small towns is expanded to the vast rural areas, including a large number of low-skilled population.[28] This structural change of selective pattern to population pattern has made it difficult for the large numbers of low-skilled migrants to integrate into the city's economic and social structure, and this change partly explains the deteriorating of the socioeconomic status of some Latin American countries.

---

[26] Goldscheider, Calvin, "Migration and Social Structure, Analytic Issues and Comparative Perspectives in Developing Nations", *Sociological Forum*, Vol. 2, No. 4, Special Issue, Demography as an Interdiscipline (Autumn, 1987), pp. 674–696.

[27] Balan, Jorge, "Migrant-native Socioeconomic Differences in Latin American Cities, a Structural Analysis", *Latin American Research Review* Vol. 4, No. 1, (Spring, 1969), pp. 3–29; Browning, Harley L., Feindt, Waltraut, "Selectivity of Migrants to a Metropolis in a Developing Country: A Mexican Case Study", *Demography*, Vol. 6, No. 4 (November, 1969), pp. 347–357; Mangin, William P., "Latin American Squatter Settlements, a Problem and a Solution", *Latin American Research Review*, Vol. 2, No. 3(Summer, 1967), pp. 65–98.

[28] Simmons, Alan B., Cardona G., Ramiro, "Rural–Urban Migration, Who Comes, Who Stays, Who Returns? The Case of Bogotá, Columbia, 1929–1968", *International Migration Review*, Vol. 6, No. 2, Internal Migration in Latin America (Summer, 1972), pp. 166–181.

### 1.2.2 Domestic Migration and Rural Migrant Study in China

Over the past 40 years, China's socioeconomic structure has gradually shifted from an agricultural society to an industrial society. The domestic migration, which being named as "alien population", "floating population" and "temporary resident", is increasing dramatically. In the household registration system, "alien population", "floating population" and "temporary resident" are statistical indexes to non-local population census, yet naming migrants as "floating" or "temporary" is inaccurate, for a large number of these people are de-facto resident.[29] These inaccurate concepts make people focus on the mobility of non-local population and neglect its residency. Therefore, this book uses "non-local population" to refer to the settled migrants who involved in urban economic activities but non-local registered in the household registration system. Academics have been paying attention to the living conditions, adaptation process and social demands of the migrant population in the city, as well as the administrative policy towards the migrant population, especially on migrants' salary, social welfare, working environment, social security, civil rights, identity transformation and other aspects of urban public services. Different disciplinary has different perspective, economy pay attention to the cause and scale of the rural surplus labour transfer, sociology emphasizes the living condition of non-local population, demography studies the age, gender and other demographic characteristics of the migrant population and the law of change, anthropology focuses on the cultural adaptation, social organization, group relationship and ethnic identity of non-local population.[30] This chapter focuses on the classification of non-local population and the characteristics of each sub-group, the stratification and its influence on social strata in cities.

Gu Shengzu and Li Debin categorized China's domestic migration as economic migration and non-economic migration, economic migration refers to the migrants who participate directly in urban economic activities, including "any non-local labourer contracted or temporarily hired in any trade in a city", vendors or peddlers—who buy or sell farm or industrial product, and craftsman from rural area. Non-economic migration refers to the migrants who temporary stay in cities for family visit, sightseeing tour or business trip.[31] In this chapter, we only consider the economic migration as our research object, excluding the non-economic migration. Economic migration population can be segmented into industrial labourer—who participate in industry, construction and logistics; agricultural labourer—who participate in planting, breeding and animal husbandry; businessmen—who participate in business management or trade; service provider—who participate in catering,

---

[29] Li Ruojian, A Preliminary Analysis of the Settlement and Mobility of the Migrant Population in Guangdong Province, *Population Research*, Vol. 6, 2007.

[30] Zhou Daming, Zhou Jianxin, Liu Zhijun, *"Free" Urban Marginal People: Study on Casual Labourer in Southeast Coast of China*, Guangzhou: Sun Yat-sen University Press, 2007, pp. 3–11.

[31] Gu Shengzu, A Study on Population Migration, *Wuhan University Journal(Social Science Edition)*, Vol. 2 1989; Li Debin, The Characteristics and Causes of Migration in Contemporary China, *Sociological Studies*, Vol. 4, 1993.

commerce or repair, and casual labourer who drifting frequently among trades. The high-profile rural migrant is the major part of the non-local labour force, which initially refers to temporary labour contracted to the state-own or collective-own industrial enterprise (mostly in construction, mining and manufacturing sector) engaged in manual labour, on the contrary to the permanent staff under the planned economy.[32]

Early studies regarding rural migrant mostly focus on the rural labour flowing to industrial enterprises,[33] gradually, the concept of "rural migrant" became a label for all migrants from rural areas to cities.[34] Academics use the concept of rural migrant generally refers to the people conducting manual or business activities in cities from rural areas,[35] or even the whole non-local population.[36] In fact, the economic transformation and social structure adjustment along with the rapid influx of rural migration causes expansion and class stratification of migrants. The non-local population in cities is now divided into three distinct social strata, the business owners who possess large capital and employ others, the self-employed individuals who possess small capital and conducting small businesses, the dependent employee who live mostly on salary.[37] As early as 1991, the author has noticed the reality of social stratification among the non-local population and conduct a series of survey on the sub-group of "causal labourer". Causal labourer refers to those who engage in various "freelance" occupations among the rural migrants they have neither individual business certificates nor official employees in various enterprises.[38] Li Peilin's research on urban villages shows that the difference between migrants is not limited to the dimension of presence or absence capital, but also differs by having or not education and skill.[39] Specifically, the non-local population is divided into: (1) self-employed small business owners engaged in various commercial and service industries; (2) the working class who lives entirely on salary, and it can be subdivided into: (a). "white-collars" are generally technicians, marketing personnel, teachers, doctors,

---

32 Editorial, Let Rural Migrant Play a Role in the Construction Industry, *Economic Daily*, June 7, 1984; Gong Yongquan, Recruit Permanent Labour, Replace Rural Labour, Nanjing Nitrogen Fertilizer Factory Tap the Potential of Labour, *People's Daily*, August 17, 1988. Zhu Limin, Pros and Cons of Massive Rural Labour Recruit by Enterprises, *Shanghai Enterprise*, Vol. 8, 1988.

33 Yang Hongshan, An Investigation on Pearl River Delta's "Rural Labour Rush", *Population Research*, Vol. 2, 1995;"Rural Labour" Research Group, The Status of Rural Labour in Pearl River Delta, *Social Science in China*, Vol. 4, 1995.

34 Chen Yingfang, Rural Migrant: System and Identity, *Sociological Studies*, Vol. 3, 2005.

35 Li Qiang, The Occupational Mobility of Rural Migrant in Cities of China's mainland, *Sociological Studies*, Vol. 3, 1999; Wang Dong, Qin Wei, A Study on Intergenerational Difference of Rural Migrant: Comparative Stratification of Rural Migrant in Chengdu, *Population Research*, Vol. 5, 2002.

36 Li Qiang, Tang Zhuang, The Rural–Urban Workers in Cities and the Irregular Employment in Cities, *Sociological Studies*, Vol. 6, 2002.

37 Li Peilin, Rural Migrant's Social Network and Status, *Sociological Studies*, Vol. 4, 1996.

38 Zhou Daming, A Study on Migrant Casual Labour in Guangzhou, *Sociological Studies*, Vol. 4, 1994.

39 Li Peilin, Tremendous Changes, The End of Villages: A Study of Villages in the Center of Guangzhou City, *Social Sciences in China*, Vol. 1, 2002.

taxi drivers, editors, journalists, company clerk and other occupations; (b). "blue-collars" are generally processing and manufacturing workers, construction and decoration industry employees, catering, logistics and other service sector employees, casual labourer. "Rural migrant" is a generalized concept, "it cannot be regarded as a static and intrinsic concept, but only a relative concept, temporary established by its subject-identified by the household registration status-under a specific relationship of recognition and exclusion".[40] The non-local population is all given the "rural" label because they are in a non-resident state, and then labelled as "rural migrant".

From this resident/ non-resident aspect, some scholars proposed the approach of "urbanization of rural migrant". Liu Chuanjiang divided China's urbanization process into two stages: the first is rural labour migrates to cities and become rural migrant, which is the process of "non-agriculturalization of rural labour"; the second is rural migrant change to the urban occupation and identity, which is process of "urbanization of rural migrant".[41] The urbanization of rural migrant includes four aspects: (1) occupation, rural migrant transformed from rural labour in the secondary segment of labour market into non-agricultural workers in the primary segment of labour market; (2) social identity, from peasant to urban resident; (3) improvement of skills, rural migrant gain skill and education under the urbanizing process; (4) urbanization of ideology, lifestyle and behaviour of rural migrant. The classification and strata of the "rural migrant" is very important in the process of urbanization of whom.[42] In summary, relevant research suggests that rural migrant is divided into: (1) entrepreneurs who have large capital and hire others; (2) self-employed business owner who have a small capital; (3) skilled workers; (4) unskilled workers. The academics pay many attentions to "new age rural migrant" or "second generation rural migrant",[43] some research suggest "new age rural migrant" are generally more urban than rural, they are neither willing to return to the countryside nor acclaimed by urban society. Therefore, they are the group most eager to urbanize and need to be assimilated by urban society. Some research emphasizes that first two groups have high income, good living conditions and high social status thus make them easy to complete the process of urbanization. The urbanization of rural migrant is theoretically concerned with the fact that "rural labour become industrial labour", yet the reality is the rural migrant is highly stratified. The urbanization subject is not only rural labour but also non-local private entrepreneurs, non-local self-employed

---

[40] Wang Xiaozhang, From "Existence" to "Recognition": Issues of Peasant Workers in the View of Citizenship, *Sociological Studies*, Vol. 1, 2009.

[41] Liu Chuanjiang, A Study of Urbanization of Rural Migrant in China, *Theory Monthly*, Vol. 10, 2006.

[42] Gu Shengzu, Yi Shance, Zheng Lingyun, Sustainable Development of Industrialization and Urbanization in China: An Approach Based on the Characteristics of Rural Migrant Workers, *Population Research*, Vol. 5, 2006; Hu Chunjuan, Urbanization of Migrant Workers Should Pay Attention to Classified Guidance, *Guangming Daily*, November 13, 2009.

[43] Wang Chunguang, The Social Identity of New Generation Rural Migrant and the Relationship with Rural–urban Integration, *Sociological Studies*, Vol. 3, 2001; Zhang Zhiyong, Intergenerational Urbanization of Rural Migrant: A Comparative Study Based on Multiple Employment of Rural Migrant, *Jianghan Tribune*, Vol. 11, 2009.

businessman and even non-local white-collar worker. This difference shows the self-contradiction of "urbanize + rural migrant" in the formulation of "urbanization of rural migrant", and it also shows the conceptual expansion of the expression of "rural migrant" in the study of migration population.[44]

## 1.3 Two Definition and Approaches on China's Domestic Migration Research

### *1.3.1 The Definition and Approach of "Rural Migrant"*

As Jorge Balan said, in the context of urbanization, "the social composition of the city determined by the economic development model of a country, especially the socioeconomic differences between urban and rural areas"[45]; rural–urban migration studies in other countries show that countries in Europe and North America, countries in East Asia, and Latin American countries have three different urban social structure patterns during the industrialization period: (1) European and North American countries quickly complete their rural–urban migration, the urban economic-social structure completely replicates the rural–urban dual structure, but economic development quickly dispels this difference; (2) post-developed countries in East Asia have obvious selective rural–urban migration, and the migrants in urban structure tends to be elite, therefore there is no obvious systematic difference between rural migrants and urban residents; (3) Latin American rural–urban migration is massive, urban economic-social structure is seriously divided, and economic development cannot eliminate the growing differentiation. In China's planned economy era, the "heavy industry priority" development strategy has caused serious segregation between urban and rural areas. After the reform and opening-up policy, the dislocation development of industrialization and urbanization has resulted in "semi-urbanization", which means rural migrants cannot achieve urbanization.[46] As a result, the academics invented the expression of "rural migrant" in the study of domestic migration. This expression of "rural migrant" emphasizes the recognition that the migrants from rural areas are subject to the institutional restrictions related to their household registered status,[47] which duplicate the rural–urban dual structure based

[44] Wang Yanhua, A Sociological Study on Urbanization of New Generation Rural Migrant, *China Youth Study*, Vol. 5, 2007. The article refers to the case of the new generation, "it is unacceptable to identify themselves as rural migrant, and rural migrant should refer to those who do manual work in building houses at the construction site, but they are vague about their household registration, social security and other issues". This case expresses the tension of the concept.

[45] Balan, Jorge, Migrant-native Socioeconomic Differences in Latin American Cities, a Structural Analysis. *Latin American Research Review*, Vol. 4, No. 1, (Spring, 1969), pp. 3–29.

[46] Wang Chunguang, A Study of Floating People's "Semi-urbanization", *Sociological Studies*, Vol. 5, 2006.

[47] Chen Yingfang, Rural Migrant: System and Identity, *Sociological Studies*, Vol. 3, 2005.

on the household registration system. It contains the following key points: (1) non-local populations mainly migrants from rural areas to cities and towns, especially to metropolises; (2) non-local populations are generally at the bottom of urban society[48]; (3) household registration system and relevant policies are the main barriers for non-local populations to achieve economic and social integration in cities.[49]

Many cases in developing countries show the rural–urban migration is not the only form of urbanization, migration from towns and small cities to major cities, and major cities to metropolis is also important parts of urbanization migration.[50] As Li Peilin wrote "migrants are not all from rural areas, those who migrate from small cities to major cities, from cities in less-developed areas to cities in developed areas, from economically depressed cities to economically blooming cities is becoming common".[51] Otherwise, the stratification of non-local population is a common sense among the academics, beside the migrants at the bottom of social-economic strata, there are also white-collars and small business-owner who want to "make money and then retire to hometown", and those migrants who "already gained social-economic position at middle-class or even higher".[52] In industrial enterprise, "it is easy for non-local labourer compete with local labourer and stand out".[53] Some academics point out that China's domestic migration is an exceptional case, China's unique household registration system makes the migrants cannot convert to permanent residents.[54] Whereas the household registration system exist or not, the variables and mechanisms differ the social-economic status between locals and non-locals still exist, therefore, the author argues that the studies regarding non-local population and related policies based on the origins (rural/urban) or identities (household registration definition) should be questioned.

---

48 Zhu Li, Characteristics and Social Status of the Peasant Worker Class, *Journal of Nanjing University (Philosophy, Humanities and Social Sciences)*, Vol. 6, 2003.

49 Li Ruojian, "Urban White-collar" Group: Analysis on Guangzhou Population Data, *Chinese Journal of Population Science*, Vol.2, 2009.

50 Simmons, Alan B., Cardona G. Ramiro, "Rural–Urban Migration, Who Comes, Who Stays, Who Returns? The Case of Bogotá, Columbia, 1929–1968", *International Migration Review*, Vol. 6, No. 2, *Internal Migration in Latin America* (Summer, 1972), pp. 166–181. Kim Hyung-Kook, "Social Factors of Migration from Rural to Urban Areas with Special Reference to Developing Countries, The Case of Korea", *Social Indicators Research*, Vol. 10, No. 1 (January 1982), pp. 29–74. Kemper, Robert V., "Rural–Urban Migration in Latin America, A Framework for the Comparative Analysis of Geographical and Temporal Patterns, *International Migration Review*", Vol. 5, No. 1 (Spring, 1971), pp. 36–47.

51 Li Peilin, Tremendous Changes the End of Villages: A Study of Villages in the Center of Guangzhou City, *Social Sciences in China*, Vol. 1, 2002.

52 Zhai Zhenwu, Hou Jiawei, Migrant Population Concentrated Communities in Beijing: Patterns and Trends, *Population Research*, Vol. 1, 2010.

53 Li Ruojian, "The Opportunities and Obstacles for Status Attainment: An Analysis of Occupational Structure in Communities with Dense Floating Population", *Chinese Journal of Population Science*, Vol. 5, 2006.

54 Zhu Yu, Overseas Research on Non-permanent Migration and Its Implications for the Issues of Floating Population in China, *Population Research*, Vol. 5, 2004.

From the perspective of urban–rural dual structure, the academics often discuss the follow subjects: shall the rural surplus labour migrate to cities or convert to non-agricultural industries in situ; shall the population growth in metropolises be controlled; and etc. Therefore, numerous studies used quantitative method to calculate "the maximum capacity of cities", "the maximum migration allowance", "the social cost to accommodate rural migrants in cities" and etc., try to ease the fears and concerns of government. The author finds an interesting phenomenon: when academics made an appeal to improve the status quo of the vulnerable groups, even if the central government responds positively and issue guiding policies with high profile, and the corresponding policies of local governments tend to be obviously conservative, such methods as "selective urban admittance", "integrated social welfares", are quite deviate from the academics' intention. These policies and corresponding studies are mainly aimed at improving rural migrant's temporary living conditions in cities, rather than solving the development problem of a permanent population. Reviewing on 30 years of development, if the government want to restrict rural labour from entering the city, they entered the city still; if the government want to control the city size, the city has grown and the population has repeatedly broken through the control indicators. "Urbanization is a result of natural selection and cannot be determined by administrator's will, the level of urban development is a self-determined process of economic development".[55] The author believe that the study of migrant populations requires a broader and deeper perspective, especially focusing on social transformation and urban integration of new migrants.[56]

### *1.3.2 3.2. The Definition and Approach of "Urban New Immigrant".*

In recent years, the academics have paid attention to the perspective beyond the urban–rural dual structure, and started to discuss the social adaptation, integration and even residency of the migrant population. Chen Yingfang has disputed the use of concepts such as "non-local population" and "rural migrant", and tried to describe rural migrant groups as "urban new immigrant", and defined their rights and interests as "urban resident rights".[57] Zhu Li labels migrants as urban new migrants and classifies them as intellectual migrants, entrepreneurial migrants and labour migrants, nonetheless he mainly focuses on the living conditions and social status of labour

---

55 Lin Yifu, Rural Modernization and Urban Development, *China Financial and Economic News*, August 17, 2001.

56 Massey, Douglas S., Joaquin Arango, Graeme Hugo, Ali Kouaouci, Adela Pellegrino, J. Edward Taylor. "Theories of International Migration, A Review and Appraisal", *Population and Development Review* 19, No. 3 (1993), 431–466.

57 Chen, Yingfang, Focus on Urban Immigrant (A Public Speech), *Jiefang Daily*, August 22, 2004.

migrants.[58] Wen Jun also focused on the labour migrant, and called it "new labour immigration", the so-called "new labour immigration" refers to the migrant mainly engaged in simple manual labour in cities, but has obtained fairly stable jobs and fixed residence and subjectively willing to settle in the city for a long time.[59] Zhang Wenhong and Lei Kaichun also regarded the migrant population as urban new immigrants, and they mainly studied the white-collar new immigrants among them.[60] Lu Wei belives that the settlement of new immigrants is not limited to improving the living conditions of rural migrant, but should extend to the ground of urban settlement and liveability, and then enlarge the perspective of promoting China's urbanization, industrialization and modernization.[61]

With the deepening of industrialization, on the one hand, the industry began to transfer in an orderly way, and the industrial development presented a trend of decentralized distribution; on the other hand, metropolis and major cities seek industrial upgrading and develop service economy, with the development of new rural construction, China's urbanization will become increasingly diversified and the occupational differentiation will become more obvious. The social and economic environment nowadays is quite different from the formation background of the expression "rural migrant". First, urbanization is accelerating, and urbanization will be the main driving force for economic development in the future. Second, as we learnt from previous research results, the migrant population in big cities is no longer mainly composed by rural migrants, and the proportion of the migrant population from city background is increasing. Third, the human asset of the migrant population is very diverse, and the employment is very extensive, from engineers, white-collar workers, to private business-owners, industrial labourer, casual labourer and so on. Fourth, the strata distribution of migrant population is relatively scattered as the occupation distribution of them. Fifth, it is a fact that migrants have achieved social mobility in cities.[62] Sixth, the household registration system and the segregated social welfare system will continue to exist. Seventh, market reforms have largely dismantled institutional barriers for migrant and helped to strip away institutional benefits. These are the practical foundations for our expression of "urban new immigrant".

---

58 Zhu Li, How to Recognize Rural Migrant, A Preface, Zhu Li, Chen Ru, ed. *New Urban Immigrant: Report on Immigrant of Nanjing*, Nanjing: Nanjing University Press, 2003, p. 22.

59 Wen Jun, The System Composition and Behavior Preference of Urban New Immigrant in China, *Nanjing Journal of Social Sciences*, Vol.1, 2005; Wen Jun, Are They Migrant or Resident? A Study on Shanghai's New Labour Migrant Since 1980s, The Publicity Department of CPC Shanghai Municipal Committee, ed., *Modern Awareness and Urban Studies*, Shanghai: Shanghai People's Publishing House, 2006, pp. 32–67.

60 Zhang Wenhong, Lei Kaichun, Structural Equation Model of the Social Identifications of the New Urban Immigrant, *Sociological Studies*, Vol.4, 2009; Zhang Wenhong, Lei Kaichun, The Urban New Immigrants' Social Inclusion: Internal structure, present situation and influential factors, *Sociological Studies*, Vol.5, 2008.

61 LU Wei, *Residential Urbanization: A Perspective on Sciences of Human Settlements*, Beijing: Higher Education Press, 2005, p. 140.

62 A large part of the social mobility of the migrant population is a form of transformation within the external system, which cannot be noticed in the context of "rural migrant".

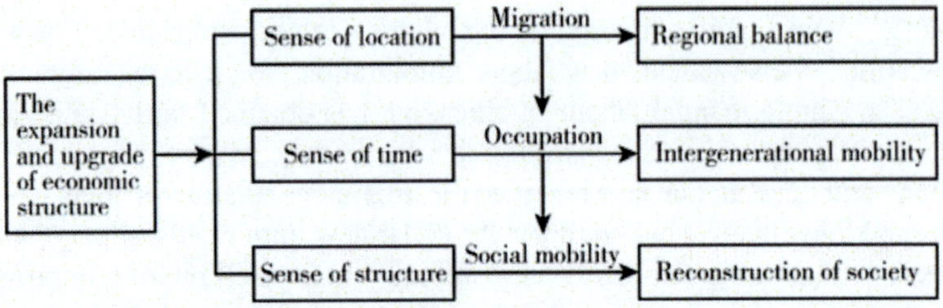


**Fig. 1.1** Diagram of "migration as a mode of development"

The author uses the concept of "urban new immigrant" to describe the study of migrant population, in order to get rid of the urban–rural binary opposition thinking, so that our attention is focused on urban society. On the one hand, "urban new immigrant" cover urban migrants from different sources and different levels. On the other hand, we regard the division of household registration as a result—the fact of social structure. Therefore, we pay more attention to how the urban new immigrants affect the urban social composition and thus "force" the reform of the system.[63] As mentioned above, "the modernization transformation of urban–rural dual economic and social structure is actually the process of migration". In other words, immigration is a way of social development. The expression of "urban new immigrant" provides a sociological interpretation for this development mode (see Fig. 1.1). Firstly, the starting point of this process is the expansion and upgrading of the urban economic structure, which attracts a large number of populations to the city and promotes regional balance in the sense of location, then the urban new immigrants through the labour force reproduction to achieve the intergenerational flow of the occupation in the sense of time. Therefore, the local strata structure accepts urban new immigrant and realize the social mobility in the sense of structure, eventually, urban society was reconstructed.

When this macro process of "migration as a mode of development" is implemented at the city level, our perspective focuses on the "social incorporation" of new immigrants in cities.[64] the expression of "social incorporation" and "new immigrants", refers to urban new immigrant from the same background (such as education) who have the same opportunities as local residents under the same conditions (such as institutional environment, economic environment, etc.) to obtain the equality on employment and social strata. Considering there are 200 million immigrants in China's cities, it is of more practical significance to study the mechanism of immigration and how to promote the integration of these immigrants into urban society. First, the long-term, family and settlement trend of the immigrant in urban society shows that the immigrant has been incorporated into urban society in a certain way and to a certain extent. Secondly, social incorporation is an important way to understand

[63] Li Peilin, Modernity and Chinese Experience, *Chinese Journal of Sociology*, Vol. 3, 2008.

[64] DeWind, Josh; Kasinitz, Philip, "Everything Old is New Again? Processes and Theories of Immigrant Incorporation", *International Migration Review*, Vol. 31, No. 4, Special Issue, Immigrant Adaptation and Native-Born Responses in the Making of Americans (Winter, 1997), pp. 1096–1111.

and solve urban social problems. Therefore, social incorporation is not only a social process but also an ultimate goal. Thirdly, social incorporation includes economic incorporation and structural incorporation. Economic incorporation refers to immigrant have the equal opportunity as the local resident to obtain occupation with the corresponding background, and realize the generational and intergenerational occupational mobility, thus obtain the corresponding economic status. Structural incorporation[65] means that immigrant have sufficient opportunities to have social connections (such as marriage) with local resident of similar social strata after gaining corresponding economic status. For immigrant population, social incorporation enables them to cross the boundary of immigrant communities and eliminate prejudice and discrimination between them and local residents. This process is reflected in the identity proximity and mutual tolerance of cultural customs. For individual immigrant and their families, social incorporation enables immigrants and local residents to cross the social space and enhance communication in schools, workplaces and neighbourhoods. This process is reflected in the reduction of social distance and mutual respect of customs and habits.

Some Chinese academics often compare "psychology", "culture", "identity" and "customs and habits" with "economic incorporation" when studying the social integration of urban immigrant, and pay attention to "the interdependence and interaction between them". According to these studies, relevant foreign theories and experiences emphasize cultural incorporation, as well as "the mutual integration and infiltration of cultures", which puts cultural incorporation in the same position with economic incorporation.[66] Milton M. Gordon's theory of immigrant assimilation, as mentioned by Yang Juhua and Zhang Wenhong: although Gordon has classified social assimilation and acculturation or behavioural assimilation as two processes of integration, he believes that "once structural integration is achieved, all other kinds of integration will occur". In the United States, due to the complex race relations, the structural integration of the whole society did not appear, while the intra-racial cultural integration occurred to a large extent.[67] In fact, the background of relevant foreign studies is somewhat different from that of China. On the one hand, foreign researches on social integration of immigrants are mostly aimed at international migration (especially migrants from developing countries) and cultural differences based on race. However, the social integration of urban immigrant population in the study did not encounter obvious racial and cultural differences, but cultural differences based on

[65] Gordon, Milton M., "Assimilation in America, Theory and Reality, Daedalus", Vol. 90, No. 2, *Ethnic Groups in American Life* (Spring, 1961), pp. 263–285.

[66] Yang Juhua, From Isolation and Selection to Integration, Theoretical Contemplating on Integration of Immigrant, *Population Research*, Vol. 1, 2009; Zhang Wenhong, Lei Kaichun, The Urban New Immigrants' Social Inclusion: Internal Structure, Present Situation and Influential Factors, *Sociological Studies*, Vol. 5, 2008.

[67] Gordon, Milton M., Assimilation in America, Theory and Reality, Daedalus, Vol. 90, No. 2, *Ethnic Groups in American Life* (Spring, 1961), pp. 263–285. Richard Alba and Victor Nee: Rethinking Assimilation Theory for a New Era of Immigration, *International Migration Review*, Vol. 31, No. 4, Special Issue, Immigrant Adaptation and Native-Born Responses in the Making of Americans (Winter, 1997), pp. 826–874.

stratification. On the other hand, the international migration studied abroad takes place in the stage of developed industry, while China is in the period of modernization transformation. This book puts forward the expression of "urban new immigrant" in the context of modernization transformation, and derives the social integration process in accordance with China's actual situation, economic incorporation takes place before structural incorporation, while creates conditions for structural incorporation. In our study, cultural differences, as Pierre Bourdieu views it, is the result of "members of different classes is always under the restriction of the respective class, with their particular class fractions, enter the field of different tastes, and by choosing different ways of life to show their class identity".[68] Cultural differences change with the change of the social integration condition, impact on individuals as well as families and groups simultaneously. To sum up, the expression of "urban new immigrant" not only deepens our understanding of the way of social development, but also derives the perspective of "social incorporation" of urban new immigrants. This theoretical turn attempts to put the changes of urban social composition and the structural relationship between immigrants and local residents at the core of the study, laying the foundation for the study of urban new immigrant. Generally speaking, immigration, as a process, can be divided into three stages: first, middle and last. Correspondingly, the question of urban new immigrant also contains three issues: (a) What has happened? The basic status of urban new immigrant and its impact on the composition of urban society (such as changes in population structure, labour market and class structure); (b) What should happen? How to define the social integration of cities as an ultimate goal; (c) How does urban social integration happen as a process? Immigrants, as the research object, can cut into the above three topics from the macro, meso and micro perspectives. Therefore, in the actual research, there are not only macroscopic argumentation, but also involve the influence of meso and micro variables, which makes the empirical research on the urban new immigrant issue colourful.

## 1.4 Definition of Core Concepts

According to the 2006 national provisional population census, which regards the non-local population who migrate to their destination for more than a year as immigrant, 65.43% immigrant is migrant labour, 13.20% immigrant is engaged in business, 7.49% immigrant is engaged in the service industry, and a small proportion farming in migrate destination.[69] Compared with 1997 census,[70] the share of migrant labour rose

---

[68] Liu Xin, Class Habitus and Taste: Bourdieu's Theory of Social Class, *Sociological Research*, Vol. 6, 2003.

[69] The Public Security administration Bureau of the Ministry of Public Security, *2006 National Census of Provisional Population Statistics*, Beijing: Qunzhong Publishing House, 2006.

[70] The Public Security administration Bureau of the Ministry of Public Security, *1997 National Census of Provisional Population Statistics*, Beijing: Chinese People's Public Security University Publishing House, 1997.

by 15% in 2006, while the share of those engage in business and farming fell sharply, but the share change of those engage in service industry is nearly invariable. In absolute numbers, immigrant in all sectors are growing, but the population of migrant labour is increasing very fast, which is related to the excessive industrialization of China's economic structure. Given the changing economic structure in the future, the proportion of migrant labour is bound to fall, while the proportion of others will rise accordingly. There are three major kinds of capital for urban new immigrant to participate in urban economic activities: manual, intelligence and capital. On the basis of referring to relevant classification of foreign migration research,[71] the author divides urban new immigrant into three categories: labour immigration, intellectual immigration and entrepreneurial immigration, and conduct preliminary exploration respectively.

In this research, "urban new immigrant" is defined by the five elements listed following:

Neither born nor household registered local, relocate for 2–5 years;
Willingness to relocate to cities;
Legal residence;
Legal income;
Mainly categorized into labour immigration, intellectual immigration and entrepreneurial immigration.

According to the different natural of occupation, urban immigrant can be divided into three categories: labour immigration, intellectual immigration and entrepreneurial immigration.

The features of labour immigration are:

Education—usually did not receive higher education;
Occupation—have legal income but low technical content;
Time frame—urban immigrant and their family members who neither born nor household registered local, relocate for 2–5 years;
Residency—located in their migrate destination, and willing to settle down;
Origin—rural emigrant and laid-off urban emigrant, mostly rural.

The features of intellectual immigration are:

Education—usually received higher education;
Occupation—high employment threshold;
Time frame—urban immigrant and their family members who neither born nor household registered local, relocate for 2–5 years;
Residency—located in their migrate destination, and willing to settle down.

Entrepreneurial immigration refers to those who invest or own a business in cities, includes small business, workshop, restaurant, store and etc. who rely on capital or skills to run their business.

[71] Richard Alba, Victor Nee, "Rethinking Assimilation Theory for a New Era of Immigration", *International Migration Review*, Vol. 31, No. 4, Special Issue, Immigrant Adaptation and Native-Born Responses in the Making of Americans (Winter, 1997), pp. 826–874.

## 1.5 Methods

According to the project design, this project is based on the empirical data of quantitative and qualitative research. Based on China's current regional economic development pattern, the distribution pattern of urban new immigration and the new measures taken by various regions in dealing with the rising urban immigration, this book selected cities as the sampling objects respectively in the Pearl River Delta, the Yangtze River Delta, the Old Industrial Base in Northeast China, the urban agglomerations in the central and western regions and their radiation area. The selected cities and reasons are as follows (see Table 1.1).

The survey was conducted by quota sampling, with 550 samples per region, of which 50 were backup samples, proportions are intellectual immigration: labour immigration: entrepreneurial immigration = 150:200:150. From November 2008 to April 2009, urban immigration survey was conducted in Guangzhou, Dongguan, Shenyang, Chengdu, Hangzhou and Zhengzhou. A sum of 3,300 questionnaires were distributed, and 3,234 questionnaires were collected, among which 66 were invalid. Final sampling amount is 3,168, with an effective rate of 96%. The above quota completion is as follows (see Table 1.2).

According to the above mentioned three basic features of immigration, intellectual immigrants refers to those who have received higher education (college degree or above), the sampling criteria mainly are those who have been working in cities for 2–5 years after graduation. The labour immigrants are who have no

**Table 1.1** Selected cities and reasons

| City | Economic area | Status of new immigrant | Related policies |
|---|---|---|---|
| Guangzhou | Pearl River Delta metropolitan area | Concentration of all types of new immigrant | |
| Dongguan | Pearl River Delta metropolitan area | Concentration of new immigrant, especially labour immigrant | New Dongguan people scheme, bold attempt to implement a series of security measures for immigrant |
| Hangzhou | Yangtze River Delta metropolitan area | Concentration of all types of new immigrant | Zhejiang Province began a pilot implementation of the residence permit system |
| Zhengzhou | Central China | Emigration and immigration | |
| Shenyang | Northeast China | Emigration and immigration | |
| Chengdu | West China | Emigration and immigration | |

**Table 1.2** Quantitative quota sampling completion

| Categories | Regions | | | | | | Total |
|---|---|---|---|---|---|---|---|
| | Guangzhou | Dongguan | Shenyang | Chengdu | Hangzhou | Zhengzhou | |
| Intellectual immigration | 115 | 162 | 175 | 174 | 206 | 152 | 984 |
| Labour immigration | 277 | 263 | 204 | 217 | 123 | 236 | 1,320 |
| Entrepreneurial immigration | 146 | 155 | 149 | 161 | 123 | 130 | 864 |
| Total | 538 | 580 | 528 | 552 | 452 | 518 | 3,168 |

higher education (college degree or below) and work in factories, shops, restaurants, and etc. Entrepreneurial immigrants are those who invest or own a business in cities, includes small business, workshop, restaurant, and etc. During our survey, the snowball method is adopted to find the survey object through acquaintances. Each acquaintance can recommend up to three people as the next survey object, which will be rolled successively, and at the same time to avoid the survey object being too similar. For labour immigration, we mainly survey migrant labourers in manufacturing industry, service industry and other industries. For intellectual immigration, we mainly survey the newly employed college students, especially who lives in urban villages, house renting neighbourhood, and enterprise dormitory. For entrepreneurial immigration, we mainly survey small business owners, who vending in various professional wholesale markets, such as manufacturing industry, service industry and other professional market.

We selected three variables including "age", "degree of education" and "monthly disposable income" for homogeneous analysis.[72] It can be seen from Fig. 1.2 that the composition characteristics of urban new immigrants are as follows: (1) most of the labour immigrants are aged 25 or below, with monthly disposable income of less than RMB2,000, and their education level between middle school and high school; (2) most intellectual immigrants are between aged 26 to 30 with monthly disposable income of RMB2,001–4,000; (3) most entrepreneurial immigrants are aged 31 or above. Firstly, unlike the new generation of immigrants discussed in the study of foreign migration,[73] urban new immigrants in China's cities, especially young labour immigrants, are mostly new cohort, which means they are only the first generation, or

[72] Homogeneous analysis is a kind of corresponding analysis, which is a multivariate statistical method to explore the relationship between classification variables, and a graphical technique to analyze the contingency table, to visually present the relationship between multiple variables and the relationship between variable categories. There are two principles to read the graph, first, check the difference between the horizontal and vertical axes in different variables respectively. Second, compare the position relation between the different values of different variables, and the classification points of different variables in roughly the same region are related to each other from the point of graph centre (0,0).

[73] Hirschman, Charles, "Problems and Prospects of Studying Immigrant Adaptation from the 1990 Population Census, From Generational Comparisons to the Process of 'Becoming American'",

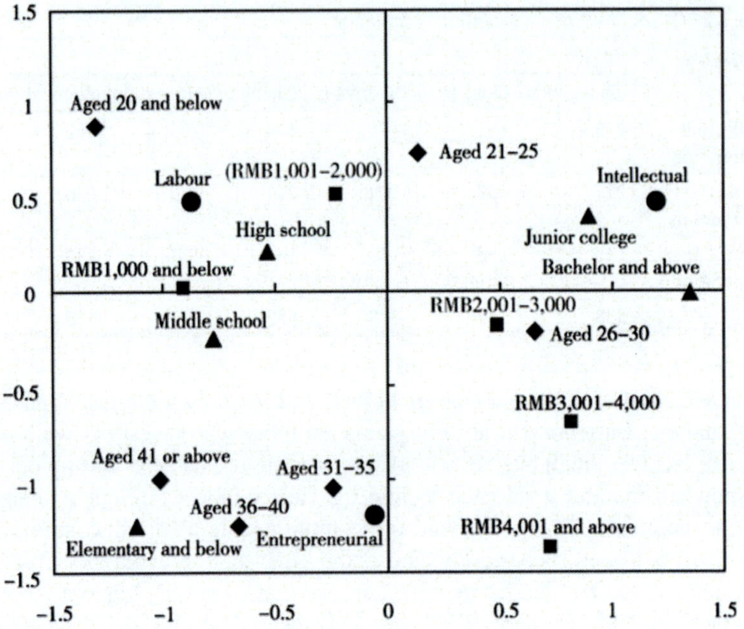


**Fig. 1.2** Corresponding analysis diagram of urban new immigrant

somewhat as 1.5 generation. This fact is the starting point of the immigrant integration process and an important background for our research on urban immigration.[74] Secondly, same as the new cohort, the new immigrants are much higher in education than they were 20 years ago, this would be another transformational advantage after the demographic dividend is gradually reduced.[75] In addition, different from the past condition of low wage, no social security or little social security and no family burden, the tendency of urban new immigrants are increasingly family-oriented, and the intergenerational reproduction of immigrants in cities is an important driving force for urban social integration.[76] Finally, there are some facts about the mutual flow between various types of immigrants: for example, the new generation is the

---

*International Migration Review*, Vol. 28, No. 4, Special Issue, The New Second Generation (Winter, 1994), pp. 690–713.

74 "1.5 generation" is referring from Ruben G. Rumbaut, "The Crucible within, Ethnic Identity, Self-Esteem, and Segmented Assimilation among Children of Immigrants", *International Migration Review*, Vol. 28, No. 4, Special Issue, The New Second Generation (Winter, 1994), pp. 748–794; Zhou Daming, Cheng Luxiao, "A Study of Migrant Workers' Occupational Diversification and Their Children's Education: Taking Youxian Hunan Province as an Example", *Journal of South China Normal University (Social Science Edition)*, Vol. 6, 2009.

75 Cai Fang, "China's Future Demographic Dividend: Digging New Source of Economic Growth", *Chinese Journal of Population Science*, Vol. 1, 2009.

76 Lieberson, Stanley, "Generational Differences among Blacks in the North", *American Journal of Sociology*, Vol. 79, No. 3 (November, 1973), pp. 550–565.

**Table 1.3** Distribution of individual interviews

| Categories | Shenyang | Hangzhou | Zhengzhou | Chengdu | Guangzhou and Dongguan | Total |
|---|---|---|---|---|---|---|
| Intellectual immigration | 10 | 17 | 10 | 10 | 2 | 49 |
| Labour immigration | 10 | 12 | 10 | 11 | 5 | 48 |
| Entrepreneurial immigration | 11 | 8 | 10 | 9 | 11 | 49 |
| Total | 31 | 37 | 30 | 30 | 18 | 146 |

*Note* For qualitative information of each city, please refer to *The Case Collection of Qualitative Investigation on "Urban Immigration and its Countermeasures" (2013)*

majority of both labour immigrants and intellectual immigrants. Some labour immigrants can acquire knowledge and skill capital through a period of efforts and then transform into intellectual immigrants. The entrepreneurial immigrants, on the other hand, are largely from labour immigrants and partly from intellectual immigrants, although the transition will take time, and they are obviously older.

After the quantitative survey, the research group conducted 146 individual interviews in six cities. The interview adopts semi-structured outline, which mainly includes the interview background, migration process, current situation and changes of social network, participation in urban society, social security situation and future planning, etc., which provides abundant and substantial data for the study of urban new immigration and integration process. Distribution of individual interviews listed as follows (see Table 1.3).

# Chapter 2
# Labour Immigrants

With the development of economy and the acceleration of urban construction, tens of millions of labour immigrant flooded cities in China, from inland to coastal areas, from rural to urban areas, they emerge in all economic sectors and support the development and expansion of cities. In this chapter, we will sketch the living conditions of the labour immigrants based on the survey data, including their basic characteristics, livelihood and financial balance, living conditions, working conditions, social welfare and public security, filial education, recreation and psychological conditions.

## 2.1 Basic Characteristics

In terms of the demographic characteristics of the labour immigrant, the study focuses on the indicators of gender, age, ethnic group, educational level and marital status. As for gender difference, the surveyed ratio of gender was determined by the actual gender ratio of labour immigrant in their region respectively. In addition, except the imbalance gender ratio in Dongguan, the overall ratio of male to female is close to 1:1.

According to the survey, labour immigrant's marital status is mostly unmarried, with an average age of 27.5, age distribution concentrated around 19–28. In Shenyang and Zhengzhou, the average age is smaller, the proportion of unmarried is also obviously high. As for education the labour immigrant is concentrated between high school to junior college. Guangzhou and Shenyang's immigrant are weighted on junior college, with Dongguan, Chengdu, Hangzhou and Zhengzhou's immigrant weighted on high school, especially in Zhengzhou, where the labour immigrant with


D. Zhou, *Urban Migrants in China*,
https://doi.org/10.1007/978-981-99-3114-9_2

junior college degrees only account for about one-third of the high school or above education group. As for ethnic group, the Han ethnic group in each region (except Shenyang) is absolutely dominant, reaching over 90%. The Han ethnic group in Shenyang also reached nearly 80%, but the Manchu ethnic group also takes 15.7% in Shenyang, where the number of ethnic groups was significantly higher than that in other cities, this is determined by the geographical location of Shenyang (see Table 2.1).

## 2.2 Livelihood and Financial Balance

Food, clothing, shelter and transportation are the basics of a social person. Only the basics of material were met, a person can make the pursuit of development. According to Maslow's hierarchy, physiological needs are not only the most primitive and basic needs of human beings, but also powerful driving forces for human behaviour, such as food, clothing, housing, medical treatment, etc., which successively above which are security needs, social needs, respect needs and self-actualization needs.[1] These demands are needed on a material basis, and their consumption is supported by income.

On the analysis of the financial balance of labour immigrant, factors including monthly disposable income, monthly expenditure (general, expenditures list, top three expenses, etc.), food expense (the largest expense for labour immigrant), CPI (Consumer Price Index) satisfaction (indicators including satisfaction on accommodation cost, daily necessities cost, average medical cost, leisure cost, transportation cost, socializing cost, educational and training cost) (Table 2.2).

In Table 2.2, "monthly disposable income" refers to the current monthly average income of respondents. According to the survey data, the monthly disposable income of labour immigrant is ranged between RMB1,001–1,500, and the average is RMB1,370 (the standard deviation is RMB735, and the sample size is 1,320). The average value of Hangzhou is slightly higher than that of other regions, but its standard deviation is also higher. The income values of Guangzhou and Dongguan are relatively close and higher than the overall average. Shenyang, Chengdu, Zhengzhou and other places are lower than the overall average level, especially, Zhengzhou has the lowest average.

Labour immigrant plays an important role in the development of cities, they overwhelmed as labourers and staffs, yet their monthly disposable income is lower than that of urban residents. According to the survey data and government statistics, there is still a large gap between the income of urban residents and labour immigrants. Taking Guangzhou as an example, the average monthly disposable income of labour immigrant in Guangzhou is RMB1,465, accounting for only one-fifth of urban residents' income (RMB7,080.78).

---

[1] Maslow, Abraham H., Xu Jinsheng(trans), *Motivation and Personality*, Beijing: Huaxia Publishing House, 1987.

**Table 2.1** Basic characteristics of labour immigrants (Unit:%)

| Items | | Guangzhou | Dongguan | Shenyang | Chengdu | Hangzhou | Zhengzhou | Total |
|---|---|---|---|---|---|---|---|---|
| Gender | Female | 46.2 | 62.4 | 52.5 | 49.3 | 43.9 | 50.0 | 51.4 |
| | Male | 53.8 | 37.6 | 47.5 | 50.7 | 56.1 | 50.0 | 48.6 |
| | Total | 100 | 100 | 100 | 100 | 100 | 100 | 100 |
| Age | 18 | 1.8 | 0.4 | 5.4 | 1.4 | 0.0 | 5.0 | 2.4 |
| | 19 | 4.7 | 4.6 | 8.8 | 6.5 | 1.6 | 7.6 | 5.8 |
| | 20 | 8.7 | 9.5 | 9.3 | 7.4 | 8.1 | 14.0 | 9.6 |
| | 21 | 7.2 | 6.5 | 5.9 | 6.0 | 1.6 | 18.2 | 8.1 |
| | 22 | 9.0 | 9.1 | 8.3 | 7.8 | 9.8 | 16.1 | 10.1 |
| | 23 | 6.1 | 9.5 | 7.4 | 4.6 | 8.9 | 9.7 | 7.7 |
| | 24 | 6.1 | 6.8 | 9.3 | 6.0 | 6.5 | 5.1 | 6.6 |
| | 25 | 5.4 | 4.2 | 5.9 | 3.2 | 4.9 | 4.2 | 4.6 |
| | 26 | 6.1 | 5.7 | 4.4 | 5.1 | 5.7 | 3.4 | 5.1 |
| | 27 | 4.7 | 5.7 | 4.9 | 6.9 | 4.1 | 3.0 | 4.9 |
| | 28 | 4.7 | 6.8 | 4.4 | 2.8 | 6.5 | 3.0 | 4.6 |
| | 29 | 3.6 | 3.4 | 2.0 | 0.9 | 0.8 | 3.0 | 2.5 |
| | 30 | 5.4 | 4.6 | 4.4 | 0.5 | 4.1 | 1.7 | 3.5 |
| | 31 | 1.4 | 1.5 | 1.5 | 2.3 | 3.3 | 0.0 | 1.5 |
| | 32 | 1.4 | 2.3 | 2.5 | 1.8 | 2.4 | 0.8 | 1.8 |
| | 33 | 0.7 | 1.1 | 1.5 | 0.9 | 0.8 | 0.4 | 0.9 |
| | 34 | 1.8 | 1.5 | 1.5 | 0.5 | 0.0 | 0.0 | 1.0 |

(continued)

**Table 2.1** (continued)

| Items | | Guangzhou | Dongguan | Shenyang | Chengdu | Hangzhou | Zhengzhou | Total |
|---|---|---|---|---|---|---|---|---|
| | 35 | 2.5 | 3.8 | 1.5 | 3.2 | 0.8 | 1.3 | 2.3 |
| | 36 | 1.8 | 0.8 | 1.5 | 0.5 | 2.4 | 0.4 | 1.1 |
| | 37 | 1.4 | 0.8 | 0.5 | 2.3 | 4.9 | 0.0 | 1.4 |
| | 38 | 1.1 | 4.2 | 0.5 | 0.9 | 0.8 | 0.8 | 1.5 |
| | 39 | 1.4 | 1.5 | 0.5 | 0.5 | 0.0 | 0.0 | 0.8 |
| | 40 | 3.6 | 2.3 | 1.5 | 1.4 | 1.6 | 0.4 | 1.9 |
| | 41 | 1.1 | 1.1 | 0.5 | 2.3 | 0.8 | 0.0 | 1.0 |
| | 42 | 1.1 | 0.4 | 1.0 | 0.9 | 0.8 | 0.4 | 0.8 |
| | 43 | 1.1 | 1.1 | 1.0 | 3.7 | 3.3 | 0.4 | 1.6 |
| | 44 | 0.4 | 0.0 | 0.5 | 0.9 | 0.8 | 0.8 | 0.5 |
| | 45 | 2.2 | 0.0 | 1.0 | 1.8 | 4.9 | 0.0 | 1.4 |
| | 46 | 0.4 | 0.0 | 0.5 | 1.8 | 3.3 | 0.0 | 0.8 |
| | 47 | 0.7 | 0.4 | 0.5 | 1.8 | 1.6 | 0.0 | 0.8 |
| | 48 | 0.7 | 0.0 | 0.5 | 0.9 | 3.3 | 0.0 | 0.7 |
| | 49 | 0.0 | 0.0 | 0.0 | 0.0 | 0.0 | 0.0 | 0.0 |
| | 50 or above | 1.4 | 0.4 | 1.5 | 12.4 | 1.6 | 0.0 | 2.8 |
| | Total | 100 | 100 | 100 | 100 | 100 | 100 | 100 |
| | Averages | 27.9 | 27.0 | 26.2 | 31.6 | 30.4 | 23.2 | 27.5 |
| | Standard deviations | 8.0 | 6.5 | 7.6 | 11.8 | 9.4 | 4.7 | 8.5 |

(continued)

**Table 2.1** (continued)

| Items | | Guangzhou | Dongguan | Shenyang | Chengdu | Hangzhou | Zhengzhou | Total |
|---|---|---|---|---|---|---|---|---|
| Ethnic groups | Han | 97.8 | 95.1 | 78.9 | 94.9 | 96.7 | 98.3 | 93.9 |
| | Manchuian | 0.0 | 0.0 | 15.7 | 0.0 | 0.0 | 0.0 | 2.4 |
| | Hui | 0.0 | 0.0 | 0.5 | 2.3 | 0.0 | 1.3 | 0.7 |
| | Zhuang | 0.7 | 2.3 | 0.0 | 0.0 | 0.0 | 0.0 | 0.6 |
| | Dong | 0.4 | 0.0 | 0.0 | 0.0 | 0.8 | 0.0 | 0.2 |
| | Mongolian | 0.0 | 0.0 | 3.4 | 0.0 | 0.0 | 0.0 | 0.5 |
| | Miao | 0.7 | 0.8 | 0.0 | 0.0 | 0.0 | 0.0 | 0.3 |
| | Tujia | 0.4 | 1.1 | 0.0 | 0.5 | 0.0 | 0.0 | 0.4 |
| | Yi | 0.0 | 0.4 | 0.0 | 0.9 | 0.8 | 0.4 | 0.4 |
| | Others | 0.0 | 0.4 | 1.5 | 1.4 | 1.6 | 0.0 | 0.7 |
| | Total | 100 | 100 | 100 | 100 | 100 | 100 | 100 |
| Education | Elementary or below | 6.5 | 3.4 | 5.9 | 23.0 | 5.7 | 3.4 | 7.9 |
| | Middle school | 52.7 | 41.1 | 53.4 | 34.6 | 28.5 | 24.2 | 40.2 |
| | High school | 40.4 | 54.8 | 35.8 | 41.5 | 55.3 | 71.2 | 49.6 |
| | Junior college or above | 0.4 | 0.8 | 4.9 | 1.0 | 10.5 | 1.3 | 2.3 |
| | Total | 100 | 100 | 100 | 100 | 100 | 100 | 100 |
| Marital status | Unmarried | 59.2 | 49.8 | 70.1 | 46.1 | 51.2 | 83.1 | 60.4 |
| | Married | 40.1 | 49.4 | 28.4 | 53.5 | 48.0 | 16.9 | 38.9 |
| | Divorced or widowed | 0.7 | 0.8 | 1.5 | 0.5 | 0.8 | 0.0 | 0.7 |
| | Total | 100 | 100 | 100 | 100 | 100 | 100 | 100 |
| Samples | | 277 | 263 | 204 | 217 | 123 | 236 | 1,320 |

**Table 2.2** Monthly disposable income of labour immigrants (Unit:%)

| Items | | Guangzhou | Dongguan | Shenyang | Chengdu | Hangzhou | Zhengzhou | Total |
|---|---|---|---|---|---|---|---|---|
| Monthly disposable income (RMB) | 1,000 and below | 13.7 | 11.4 | 28.4 | 41.5 | 11.4 | 36.0 | 23.9 |
| | 1,001–1,500 | 45.1 | 47.9 | 45.6 | 33.2 | 42.3 | 43.6 | 43.3 |
| | 1,501–2,000 | 26.0 | 23.2 | 13.7 | 13.4 | 18.7 | 13.6 | 18.6 |
| | 2,001–2,500 | 9.4 | 11.0 | 5.4 | 6.0 | 15.4 | 3.8 | 8.1 |
| | 2,501–3,000 | 4.3 | 4.6 | 3.9 | 2.3 | 4.1 | 2.1 | 3.6 |
| | 3,001–3,500 | 1.4 | 0.4 | 2.5 | 1.4 | 1.6 | 0.8 | 1.3 |
| | 3,501–4,000 | 0.0 | 0.8 | 0.0 | 0.0 | 1.6 | 0.0 | 0.3 |
| | 4,001–4,500 | 0.0 | 0.0 | 0.0 | 0.5 | 1.6 | 0.0 | 0.2 |
| | 4,501–5,000 | 0.0 | 0.0 | 0.5 | 0.5 | 1.6 | 0.0 | 0.3 |
| | 5,001 and above | 0.0 | 0.8 | 0.0 | 1.4 | 1.6 | 0.0 | 0.5 |
| | Total | 100 | 100 | 100 | 100 | 100 | 100 | 100 |
| | Averages | 1,465 | 1,516 | 1,284 | 1,213 | 1,717 | 1,135 | 1,370 |
| | Standard deviations | 589 | 671 | 704 | 869 | 958 | 587 | 735 |
| Samples | | 277 | 263 | 204 | 217 | 123 | 236 | 1,320 |

*Note* According to Guangzhou Statistics Bureau, the per capita disposable income of urban residents in Guangzhou was RMB 7,080.78 in June 2010

According to Karl Marx's theory, production and reproduction is required to maintain a continuous labour force, which means, means of livelihood should not only be enough to enable individual labourer to maintain a normal living conditions, but also should include the means for family reproduce the next generation of labour force.

Along with the rapid development of China's socio-political economy, labour migration is increasing day by day. People who work outside their homeland should not only bread-winning for their own lives, but also provide for their families and relatives. The vast majority of labour immigrant have relatives who live in their homeland, and more than half of the population have elderly or children who depend on them. As a result, a significant proportion of their monthly income, in addition to their expenses in cities, will be sent back to their homeland to support the family, education for their children, medical expenses for the elderly, etc., the survey found about half of the total expenses for labour immigrant is remittance (see Table 2.5).

According to the survey results of the research group (see Table 2.3 and Table 2.5), one labour immigrant needs to provide 1.4 people (standard deviation is 1.5) on average, and send home RMB370.2 per month. Among them, the number of dependents who are supported by labour immigrant in Hangzhou was higher than the overall average, reaching 1.8 (standard deviation is 1.6), while the labour immigrants in Chengdu had the lowest number of dependents, with an average of 1.1. Compare the monthly average income and remittance, the monthly average income of the labour immigrant in Zhengzhou was the lowest (mean is RMB1,135), and their remittance expenses accounted for the lowest income ratio (10.34%). The average monthly income of Dongguan labour immigrants was the second lowest (mean is RMB1,516), and the proportion of remittance was the highest (40.54%).

Consumption is necessary for human survival and reproduction, and it is also the foundation of social and economic framework. From the perspective of the progress, immigrant's consumption, in addition to ensuring basic survival, has gradually expanded to social intercourse, recreation, communication and other aspects. Therefore, labour immigrant's urban life consumption is also the emphasis of this section. In this research, we have made a detailed understanding of the household expenditure of labour immigrant, including the cost of food, accommodation, communications, transportation, recreation, social intercourse (entertainment, gifts, etc.), and the amount of remittance.

In the monthly expenditure composition of labour immigrant, food and accommodation is the basic subsistence consumption, accounting for more than 45% of total expenditure, and nearly 60% of whom in Hangzhou. The consumption of communication, recreation, social intercourse and social and spiritual relief accounts for about 30% of the total consumption, among which the proportion of recreation and social intercourse expenses accounts for more than 10%.

Table 2.4 and Table 2.5 shows the monthly expenditures composition of labour immigrant in all areas (sampled cities). Hangzhou has a higher income, but the expenditure on basic living expenses such as food expenses and housing expenses is relatively high as well, which is significantly higher than that in Shenyang and Zhengzhou. In addition, the expenditure of food in Chengdu is also relatively high,

**Table 2.3** Dependents support by labour immigrants

| Items | | Guangzhou | Dongguan | Shenyang | Chengdu | Hangzhou | Zhengzhou | Total |
|---|---|---|---|---|---|---|---|---|
| Number of dependents (%) | 0(no dependent) | 44.0 | 44.5 | 36.8 | 56.2 | 26.0 | 46.6 | 43.8 |
| | 1 | 11.2 | 10.3 | 3.4 | 10.6 | 15.4 | 6.4 | 9.2 |
| | 2 | 21.7 | 20.9 | 34.8 | 17.1 | 32.5 | 29.7 | 25.2 |
| | 3 | 10.8 | 12.2 | 15.7 | 6.9 | 13.0 | 11.4 | 11.5 |
| | 4 | 9.0 | 8.4 | 7.4 | 6.5 | 7.3 | 3.0 | 7.0 |
| | 5 | 2.2 | 3.8 | 1.5 | 2.3 | 3.3 | 1.7 | 2.4 |
| | 6 | 1.1 | 0.0 | 0.0 | 0.0 | 0.8 | 1.3 | 0.5 |
| | 7 | 0.0 | 0.0 | 0.0 | 0.0 | 0.8 | 0.0 | 0.1 |
| | 8 | 0.0 | 0.0 | 0.5 | 0.5 | 0.8 | 0.0 | 0.2 |
| | Total | 100 | 100 | 100 | 100 | 100 | 100 | 100 |
| | Average | 1.4 | 1.4 | 1.6 | 1.1 | 1.8 | 1.3 | 1.4 |
| | Standard deviations | 1.5 | 1.5 | 1.5 | 1.5 | 1.6 | 1.4 | 1.5 |
| Samples | | 277 | 263 | 204 | 217 | 123 | 236 | 1,320 |

ranking first among sampling cities. Guangzhou has the highest communication expense. As for social intercourse expense, Chengdu is significantly higher than Guangzhou, Hangzhou and Zhengzhou.

As for dinning occasions, labour immigrant mainly eats in canteen at their workplace or cook their own meals (47.3% and 43.3% respectively). In term of regional deviations, Dongguan and Guangzhou have a significantly higher proportion of dinning at canteen in their workplace, 60.5% and 56.3% respectively. Hangzhou and Shenyang have a higher rate of 56.1% and 52.0% respectively of cook and eat for oneself. In addition, the proportion of cooking with roommates in Shenyang was significantly higher than that in other regions, reaching 31.4%. The proportion of eat in cafeterias in Zhengzhou was significantly higher, at 55.9% (See Table 2.6).

The comparatively comprehensive price of all kinds of living substance, including accommodation and utility price, necessities price, average medical care price, recreation price, transportation price, social intercourse price and education price, shows labour immigrant generally have low satisfaction degree on accommodation price and medical care price, with the average satisfaction degree of 2.3 and 2.4 respectively.[2] Satisfaction with transportation price and social intercourse price was relatively high, with an average value of 3.3 and 3.2 respectively (see Table 2.7).

In term of regional deviations, the satisfaction with accommodation and utility price in Hangzhou was the significant lowest, with an average of only 2.0 and the second lowest was 4.9. The satisfaction with social intercourse price was slightly lower in Hangzhou than that in other regions as well, with an average of 2.9. In addition, Dongguan's satisfaction with transportation price was slightly lower, with an average of 3.2.

Of the three biggest daily expenses for labour immigrant, more than 60% chose food, while more than 40% said clothing and accommodation were also among the top three expenses (see Table 2.8).

The circumstance of income and expenses for labour immigrant is generally balance, 21.7% of the respondents' income covers their expense with a stable margin. 52% of respondents said they earn just enough to make ends meet and somewhat make a little savings. 53.5% of respondents believed that their income and expenses could be balanced at the wage level of RMB1,001–2,000, 26% of respondents thought it was necessary to make balance with the wage level of RMB2,001–3,000, with an average of RMB2,366, RMB996 higher than the overall actual average income.

As for regional deviations, Chengdu have the highest percentage of the respondents whose income covers their expense with a stable margin, 28.6%, while Zhengzhou have the lowest percentage of whom (see Table 2.9).

In general, the overall income level of labour migration is not high, with the average of about RMB1,000, especially in Zhengzhou, where the average income is only RMB1,135, significantly lower than that of other regions. More than half of the labour immigrant needs to support others, among which the proportion in Hangzhou,

[2] In the survey, the degree of satisfaction of the labour immigrant to all aspects of prices was divided into 5 grades, 5–very satisfied, 4–somewhat satisfied, 3–neither satisfied nor dissatisfied, 2–somewhat dissatisfied, 1–very dissatisfied.

**Table 2.4** Expenditures composition of labour immigrants (Unit:%)

| Expenditures | Amount (RMB) | Guangzhou | Dongguan | Shenyang | Chengdu | Hangzhou | Zhengzhou | Total |
|---|---|---|---|---|---|---|---|---|
| Food | ≤ 150 | 30.0 | 27.4 | 25.5 | 15.2 | 9.8 | 11.4 | 21.1 |
| | 151–300 | 14.1 | 17.5 | 25.5 | 17.5 | 12.2 | 37.3 | 21.1 |
| | 301–450 | 9.0 | 10.6 | 6.4 | 11.1 | 12.2 | 19.5 | 11.4 |
| | 451–600 | 20.9 | 15.6 | 22.5 | 20.7 | 31.7 | 19.9 | 20.9 |
| | 601–750 | 4.0 | 3.0 | 2.9 | 7.8 | 4.1 | 3.8 | 4.2 |
| | 751–900 | 6.5 | 8.0 | 6.9 | 6.5 | 13.0 | 3.4 | 6.9 |
| | ≥ 901 | 15.5 | 17.9 | 10.3 | 21.2 | 17.1 | 4.7 | 14.3 |
| | Total | 100 | 100 | 100 | 100 | 100 | 100 | 100 |
| Accommodation | ≤ 200 | 54.9 | 64.6 | 62.3 | 54.4 | 36.6 | 67.4 | 58.4 |
| | 201–400 | 24.5 | 24.3 | 15.2 | 25.3 | 25.2 | 24.2 | 23.2 |
| | 401–600 | 10.5 | 5.7 | 10.8 | 8.8 | 22.8 | 4.2 | 9.3 |
| | 601–800 | 4.7 | 0.4 | 4.4 | 3.7 | 8.1 | 1.3 | 3.3 |
| | 801–1,000 | 1.8 | 1.5 | 2.5 | 1.8 | 1.6 | 0.8 | 1.7 |
| | ≥ 1,001 | 3.6 | 3.4 | 4.9 | 6.0 | 5.7 | 2.1 | 4.1 |
| | Total | 100 | 100 | 100 | 100 | 100 | 100 | 100 |
| Communication | ≤ 50 | 25.6 | 29.7 | 42.2 | 37.8 | 34.1 | 35.2 | 33.5 |
| | 51–100 | 40.4 | 41.4 | 33.3 | 33.6 | 35.0 | 44.9 | 38.7 |
| | 101–150 | 9.0 | 7.6 | 8.3 | 7.4 | 13.8 | 8.5 | 8.7 |
| | 151–200 | 11.9 | 14.8 | 10.3 | 9.7 | 11.4 | 6.4 | 10.8 |
| | 201–300 | 7.9 | 3.4 | 3.9 | 8.3 | 3.3 | 3.8 | 5.3 |
| | ≥ 301 | 5.1 | 3.0 | 2.0 | 3.2 | 2.4 | 1.3 | 3.0 |

(continued)

**Table 2.4** (continued)

| Expenditures | Amount (RMB) | Guangzhou | Dongguan | Shenyang | Chengdu | Hangzhou | Zhengzhou | Total |
|---|---|---|---|---|---|---|---|---|
| | Total | 100 | 100 | 100 | 100 | 100 | 100 | 100 |
| Transportation | ≤ 50 | 63.9 | 62.7 | 58.3 | 69.6 | 74.0 | 69.5 | 65.7 |
| | 51–100 | 21.3 | 18.6 | 29.4 | 16.1 | 20.3 | 23.7 | 21.5 |
| | 101–200 | 11.2 | 9.1 | 7.8 | 7.4 | 3.3 | 4.2 | 7.7 |
| | 201–300 | 2.2 | 2.3 | 1.5 | 1.8 | 1.6 | 1.3 | 1.8 |
| | 301–600 | 0.4 | 4.6 | 2.0 | 3.2 | 0.8 | 1.3 | 2.1 |
| | ≥ 601 | 1.1 | 2.7 | 1.0 | 1.8 | 0 | 0 | 1.2 |
| | Total | 100 | 100 | 100 | 100 | 100 | 100 | 100 |
| Recreation | ≤ 100 | 64.3 | 57.8 | 67.6 | 62.2 | 58.5 | 62.7 | 62.3 |
| | 101–200 | 12.6 | 16.7 | 16.7 | 13.8 | 19.5 | 19.5 | 16.1 |
| | 201–300 | 9.0 | 8.0 | 6.4 | 7.8 | 8.9 | 9.3 | 8.3 |
| | 301–500 | 8.3 | 9.5 | 4.4 | 7.8 | 9.8 | 5.9 | 7.6 |
| | 501–1,000 | 4.0 | 5.7 | 4.4 | 5.1 | 2.4 | 2.5 | 4.2 |
| | ≥ 1,001 | 1.8 | 2.3 | 0.5 | 3.2 | 0.8 | 0 | 1.5 |
| | Total | 100 | 100 | 100 | 100 | 100 | 100 | 100 |
| Social intercourse | ≤ 100 | 65.7 | 52.5 | 55.4 | 50.7 | 63.4 | 58.1 | 57.4 |
| | 101–300 | 22.4 | 34.6 | 27.9 | 25.8 | 22.0 | 30.5 | 27.7 |
| | 301–500 | 8.3 | 8.0 | 9.8 | 8.3 | 8.1 | 9.3 | 8.6 |
| | 501–1,000 | 3.2 | 2.7 | 4.4 | 11.1 | 6.5 | 1.7 | 4.6 |
| | ≥ 1,001 | 0.4 | 2.3 | 2.5 | 4.1 | 0 | 0.4 | 1.7 |
| | Total | 100 | 100 | 100 | 100 | 100 | 100 | 100 |

(continued)

**Table 2.4** (continued)

| Expenditures | Amount (RMB) | Guangzhou | Dongguan | Shenyang | Chengdu | Hangzhou | Zhengzhou | Total |
|---|---|---|---|---|---|---|---|---|
| Remittance | ≤ 100 | 52.3 | 47.1 | 59.3 | 63.6 | 62.6 | 80.1 | 60.2 |
| | 101–500 | 20.9 | 19.4 | 22.5 | 14.7 | 21.1 | 12.7 | 18.4 |
| | 501–1,000 | 20.6 | 19.0 | 13.2 | 14.7 | 13.8 | 6.4 | 15.0 |
| | ≥ 1,001 | 6.1 | 14.4 | 4.9 | 6.9 | 2.4 | 0.8 | 6.4 |
| | Total | 100 | 100 | 100 | 100 | 100 | 100 | 100 |

**Table 2.5** Expenditures comparation of labour immigrants (Unit:RMB)

| Regions | Indicator | Food | Accommodation (including utility) | Communication | Transportation | Recreation | Social intercourse | Remittance |
|---|---|---|---|---|---|---|---|---|
| Guangzhou | Average | 498.5 | 264.7 | 137.9 | 72.4 | 172.5 | 132.1 | 454.4 |
| | Standard deviation | 588.1 | 372.2 | 136.7 | 151.9 | 343.8 | 212.5 | 827.7 |
| Dongguan | Average | 507.8 | 242.2 | 116.6 | 106.3 | 216.0 | 204.6 | 614.6 |
| | Standard deviation | 499.5 | 475.8 | 87.6 | 212.3 | 380.2 | 396.5 | 937.9 |
| Shenyang | Average | 431.2 | 323.9 | 101.7 | 88.0 | 137.9 | 232.4 | 340.7 |
| | Standard deviation | 441.8 | 620.7 | 85.7 | 231.7 | 225.8 | 470.6 | 677.0 |
| Chengdu | Average | 610.3 | 308.4 | 112.4 | 81.2 | 209.3 | 294.1 | 344.6 |
| | Standard deviation | 531.1 | 481.1 | 99.0 | 162.3 | 410.7 | 534.1 | 716.8 |
| Hangzhou | Average | 598.2 | 419.5 | 108.3 | 47.7 | 170.1 | 153.5 | 237.0 |
| | Standard deviation | 391.9 | 566.9 | 82.1 | 63.8 | 316.9 | 223.8 | 380.3 |
| Zhengzhou | Average | 394.5 | 206.3 | 100.6 | 58.4 | 138.3 | 152.3 | 117.4 |
| | Standard deviation | 224.9 | 310.0 | 114.3 | 71.1 | 169.0 | 183.1 | 284.5 |
| Total | Average | 499.0 | 280.5 | 114.4 | 78.2 | 175.5 | 194.3 | 370.2 |
| | Standard deviation | 476.5 | 470.7 | 106.3 | 167.2 | 322.9 | 367.8 | 726.1 |

**Table 2.6** Dinning occasions of labour immigrants (Unit:%)

| Categories | Guangzhou | Dongguan | Shenyang | Chengdu | Hangzhou | Zhengzhou | Total |
|---|---|---|---|---|---|---|---|
| Canteen (including boarding) | 56.3 | 60.5 | 45.6 | 44.2 | 36.6 | 32.2 | 47.3 |
| Cook and eat for oneself | 38.3 | 39.9 | 52.0 | 38.7 | 56.1 | 42.8 | 43.3 |
| Cafeteria | 10.1 | 22.8 | 30.4 | 27.2 | 16.3 | 55.9 | 27.3 |
| Cook and eat with family | 22.7 | 32.7 | 23.0 | 34.6 | 28.5 | 17.8 | 26.4 |
| Take-out | 23.8 | 22.4 | 22.1 | 17.1 | 27.6 | 28.0 | 23.3 |
| Cook and eat with roommate | 11.9 | 14.8 | 31.4 | 18.9 | 10.6 | 16.9 | 17.4 |
| Canteen outside | 1.8 | 0.4 | 2.9 | 0.5 | 5.7 | 2.1 | 1.9 |
| Others | 0.4 | 0.0 | 0.0 | 0.0 | 0.0 | 0.4 | 0.2 |
| Non-duplicate | 100 | 100 | 100 | 100 | 100 | 100 | 100 |
| Total | 165 | 194 | 207 | 181 | 181 | 196 | 187 |
| Samples | 277 | 263 | 204 | 217 | 123 | 236 | 1,320 |

Shenyang is higher, reaching 74% and 63.2% respectively. As for balance of income and expenses, about half of the labour immigrant thought they make ends meet and somewhat make a little savings. A quarter of respondent in short or debt situation, they thought that their income should reach RMB2,366 (the overall average) to make balance, which was nearly one thousand yuan higher than the current average income of labour immigrant. The income expectancy and reality make a strong contra, and the expenditures on living substance also indicate that the living condition for labour immigrant is relatively poor.

## 2.3 Housing Conditions

In all kinds of resources supporting human survival and development, house ownership is crucial in today's China due to the land policy in this nation. With the acceleration of social urbanization, the population of large and medium-sized cities is increasing rapidly, and the housing problem is becoming more and more prominent. For labour immigrants, it is very difficult for them to own their own houses and settle down in cities.

The analysis of the housing conditions of labour immigrants includes the research on the housing types and the reasons for renting, the people living together and the housing area, the types of renting, the equipment in the house and the surrounding

**Table 2.7** Labour immigrants' satisfaction on prices

| Regions | Indicators | Accommodation | Necessities | Medical care | Recreation | Transportation | Social intercourse | Education and training |
|---|---|---|---|---|---|---|---|---|
| Guangzhou | Average | 2.3 | 3.2 | 2.3 | 2.9 | 3.4 | 3.2 | 2.7 |
| | TOP2(%) | 5.1 | 28.2 | 7.2 | 15.5 | 41.5 | 27.1 | 14.4 |
| | Standard deviation | 0.2 | 0.2 | 0.2 | 0.3 | 0.2 | 0.3 | 0.2 |
| Dongguan | Average | 2.5 | 3.0 | 2.3 | 2.9 | 3.2 | 3.2 | 2.8 |
| | TOP2(%) | 9.1 | 20.5 | 9.5 | 10.6 | 35.0 | 23.2 | 16.7 |
| | Standard deviation | 0.2 | 0.2 | 0.2 | 0.3 | 0.2 | 0.3 | 0.2 |
| Shenyang | Average | 2.4 | 3.2 | 2.5 | 2.9 | 3.5 | 3.3 | 2.8 |
| | TOP2(%) | 8.3 | 32.4 | 12.7 | 21.6 | 51.0 | 34.8 | 21.1 |
| | Standard deviation | 0.2 | 0.2 | 0.2 | 0.2 | 0.2 | 0.2 | 0.2 |
| Chengdu | Average | 2.4 | 3.0 | 2.3 | 2.9 | 3.5 | 3.2 | 2.9 |
| | TOP2(%) | 6.0 | 23.5 | 7.4 | 21.2 | 50.2 | 34.6 | 23.5 |
| | Standard deviation | 0.2 | 0.2 | 0.2 | 0.2 | 0.2 | 0.2 | 0.2 |
| Hangzhou | Average | 2.0 | 3.0 | 2.5 | 2.8 | 3.3 | 2.9 | 2.7 |
| | TOP2(%) | 4.9 | 15.4 | 12.2 | 14.6 | 31.7 | 13.0 | 13.0 |
| | Standard deviation | 0.2 | 0.3 | 0.2 | 0.2 | 0.2 | 0.3 | 0.2 |

(continued)

**Table 2.7** (continued)

| Regions | Indicators | Accommodation | Necessities | Medical care | Recreation | Transportation | Social intercourse | Education and training |
|---|---|---|---|---|---|---|---|---|
| Zhengzhou | Average | 2.2 | 2.9 | 2.3 | 2.8 | 3.3 | 3.1 | 2.7 |
| | TOP2(%) | 3.8 | 13.6 | 5.9 | 11.4 | 35.2 | 28.0 | 13.1 |
| | Standard deviation | 0.2 | 0.3 | 0.2 | 0.2 | 0.2 | 0.2 | 0.2 |
| Total | Average | 2.3 | 3.0 | 2.4 | 2.9 | 3.3 | 3.2 | 2.8 |
| | TOP2(%) | 6.3 | 22.7 | 8.8 | 15.6 | 41.1 | 27.6 | 17.0 |
| | Standard deviation | 0.2 | 0.2 | 0.2 | 0.2 | 0.2 | 0.2 | 0.2 |

*Note* A score of 1–5 is assigned to the satisfaction level of 1–5, and the average score range is 1–5. The higher the score, the higher the average satisfaction degree. TOP2 indicates the sum of the proportions of "very satisfied" and "somewhat satisfied"

**Table 2.8** Selection on three biggest daily expenses for labour immigrants (Unit:%)

| Categories | Guangzhou | Dongguan | Shenyang | Chengdu | Hangzhou | Zhengzhou | Overall |
|---|---|---|---|---|---|---|---|
| Food | 62.1 | 54.3 | 57.4 | 63.1 | 70.7 | 66.5 | 61.6 |
| Clothing | 37.2 | 42.2 | 53.4 | 33.1 | 42.3 | 59.0 | 44.4 |
| Accommodation | 39.3 | 38.0 | 42.7 | 50.7 | 46.3 | 43.7 | 42.8 |
| Communication | 33.9 | 28.5 | 33.3 | 24.0 | 26.0 | 28.4 | 29.4 |
| Child Support | 32.2 | 32.4 | 15.2 | 25.8 | 26.0 | 11.5 | 24.2 |
| Social intercourse | 17.7 | 16.0 | 25.0 | 24.9 | 17.1 | 20.4 | 20.0 |
| Recreation | 18.8 | 15.2 | 20.1 | 19.9 | 19.5 | 22.5 | 19.2 |
| Family support | 23.9 | 24.3 | 15.7 | 17.5 | 12.2 | 9.7 | 18.1 |
| Transportation | 7.6 | 10.6 | 10.8 | 4.6 | 8.9 | 8.9 | 8.6 |
| Education and training | 6.8 | 11.0 | 6.4 | 5.1 | 3.2 | 9.3 | 7.4 |
| Medical care | 10.9 | 7.3 | 4.9 | 11.5 | 4.9 | 3.8 | 7.4 |
| Cosmetics | 2.9 | 11.5 | 5.9 | 4.6 | 2.4 | 8.9 | 6.3 |
| Mortgage | 3.6 | 4.9 | 6.9 | 2.8 | 11.5 | 5.9 | 5.4 |
| Insurance | 1.8 | 2.6 | 2.5 | 1.9 | 5.7 | 1.7 | 2.4 |
| Others | 1.4 | 1.2 | 0.0 | 10.5 | 3.2 | 0.0 | 2.6 |
| Total | 300.0 | 300.0 | 300.0 | 300.0 | 300.0 | 300.0 | 300.0 |

traffic conditions, the views on buying houses (including the housing price, the willingness to buy houses, the factors affecting buying houses) and other indicators.

According to our survey data (see Table 2.10), the living mode of labour immigrants is renting or workplace providing (boarding dormitory). Among them, renting a house is the most important way of housing. (sample cities) the proportion of this group is about 60%. Only 6.1% of the surveyed households bought their own homes. In Dongguan, the proportion of people living in dormitories was significantly higher. The proportion of rented houses in Hangzhou and Zhengzhou is higher than that in other regions. In Shenyang, the proportion of labour immigrants who own their own houses is slightly higher, reaching more than 10%. The reason why the group of labour immigrants choose to rent a house is that the accommodation is close to the workplace, they can live with their families or friends, and the price of renting a house is relatively cheap (see Table 2.10).

30% to 40% of labour migrants live with friends/colleagues or family members, with the majority sharing accommodation with two (29.2%); besides, more than 16% of the group live alone. In the case of shared housing, the number of people living together will vary according to different housing types. The average number of people living together in dormitories provided by the workplace is 5. In the case of renting or free stay, the number of people living together is about 3 (see Table 2.11).

**Table 2.9** Balance of income and expenses for labour immigrants (Unit:%)

| Itmes | | Guangzhou | Dongguan | Shenyang | Chengdu | Hangzhou | Zhengzhou | Total |
|---|---|---|---|---|---|---|---|---|
| Income/expense balance | Margined | 19.1 | 26.2 | 23.5 | 28.6 | 22.8 | 11.4 | 21.7 |
| | Balanced | 56.3 | 53.2 | 49.5 | 54.8 | 47.2 | 47.9 | 52.0 |
| | Slightly short | 18.8 | 18.6 | 23.0 | 13.4 | 23.6 | 32.6 | 21.4 |
| | In debt | 4.3 | 1.5 | 3.4 | 2.3 | 5.7 | 5.9 | 3.7 |
| | Not sure | 1.4 | 0.4 | 0.5 | 0.9 | 0.8 | 2.1 | 1.1 |
| | Total | 100 | 100 | 100 | 100 | 100 | 100 | 100 |
| Samples | | 277 | 263 | 204 | 217 | 123 | 236 | 1,320 |
| The anticipation of wage level necessary to make balance | RMB0-1,000 | 10.3 | 5.6 | 14.5 | 0.0 | 2.7 | 5.2 | 6.9 |
| | RMB1,001–2,000 | 51.5 | 50.0 | 58.2 | 63.9 | 40.5 | 55.2 | 53.5 |
| | RMB2,001–3,000 | 29.4 | 31.5 | 20.0 | 13.9 | 40.5 | 22.9 | 26.0 |
| | RMB3,001–4,000 | 1.5 | 7.4 | 1.8 | 13.9 | 2.7 | 7.3 | 5.5 |
| | RMB4,001–5,000 | 7.4 | 3.7 | 3.6 | 5.6 | 8.1 | 5.2 | 5.5 |
| | RMB5,001 and above | 0.0 | 1.9 | 1.8 | 2.8 | 5.4 | 4.2 | 2.6 |
| | Average | 2,181.0 | 2,330.0 | 2,036.0 | 2,494.0 | 2,785.0 | 2,498.0 | 2,366.0 |
| | Total percentage | 100 | 100 | 100 | 100 | 100 | 100 | 100 |
| Samples | | 68 | 54 | 55 | 36 | 37 | 96 | 346 |

**Table 2.10** Housing conditions and reasons for labour immigrants (Unit:%)

| Items | | Guangzhou | Dongguan | Shenyang | Chengdu | Hangzhou | Zhengzhou | Total |
|---|---|---|---|---|---|---|---|---|
| Housing categories | Rent | 55.6 | 50.2 | 55.4 | 59.4 | 62.6 | 65.3 | 57.5 |
| | Dormitory | 34.3 | 46.0 | 30.4 | 25.3 | 25.2 | 22.5 | 31.6 |
| | Own house | 3.6 | 2.7 | 12.3 | 5.5 | 5.7 | 8.1 | 6.1 |
| | Stay for free | 6.1 | 1.1 | 1.5 | 1.8 | 3.3 | 3.8 | 3.0 |
| | Others | 0.4 | 0.0 | 0.5 | 7.8 | 3.3 | 0.4 | 1.8 |
| | Total percentage | 100 | 100 | 100 | 100 | 100 | 100 | 100 |
| Samples | | 277 | 263 | 204 | 217 | 123 | 236 | 1,320 |
| Reasons for renting | Close to workplace | 44.8 | 35.6 | 51.3 | 46.5 | 31.2 | 39.0 | 41.9 |
| | Convenient living | 32.5 | 48.5 | 38.1 | 39.5 | 28.6 | 39.6 | 38.3 |
| | Stay with friends or families | 31.2 | 57.6 | 31.9 | 36.4 | 23.4 | 23.4 | 34.4 |
| | Less renting price | 27.3 | 15.2 | 39.8 | 45.0 | 29.9 | 30.5 | 31.0 |
| | Workplace doesn't provide dormitory | 24.7 | 18.9 | 31.9 | 34.9 | 37.7 | 37.7 | 30.4 |
| | Poor dormitory living condition | 5.8 | 4.5 | 3.5 | 0.0 | 2.6 | 1.9 | 3.2 |
| | Less restriction | 0.6 | 1.5 | 0.0 | 0.0 | 0.0 | 0.6 | 0.5 |
| | Others | 0.0 | 0.0 | 0.9 | 1.6 | 1.3 | 0.0 | 0.5 |
| | Total percentage | 100 | 100 | 100 | 100 | 100 | 100 | 100 |
| Samples | | 154 | 132 | 113 | 129 | 77 | 154 | 759 |

**Table 2.11** Housing companions for labour immigrants (Unit:%)

| Items | | Guangzhou | Dongguan | Shenyang | Chengdu | Hangzhou | Zhengzhou | Total |
|---|---|---|---|---|---|---|---|---|
| Companion types | Friends/colleagues | 39.7 | 38.8 | 51.5 | 34.6 | 24.4 | 35.6 | 38.3 |
| | Families | 37.5 | 42.2 | 25.0 | 47.9 | 45.5 | 24.6 | 36.7 |
| | Alone | 16.6 | 12.5 | 12.3 | 12.4 | 17.9 | 26.3 | 16.3 |
| | Cohabitation | 3.2 | 7.2 | 12.7 | 6.0 | 13.8 | 7.6 | 7.7 |
| | Kinsfolk | 7.9 | 6.5 | 3.9 | 2.3 | 2.4 | 5.5 | 5.2 |
| | Housemate | 0.4 | 0.0 | 1.0 | 0.9 | 0.0 | 0.0 | 0.4 |
| | Schoolmate | 0.0 | 0.0 | 1.0 | 0.5 | 0.0 | 0.4 | 0.3 |
| | Others | 0.4 | 0.0 | 0.5 | 0.9 | 1.6 | 0.0 | 0.5 |
| | Total percentage | 105.8 | 107.2 | 107.8 | 105.5 | 105.7 | 100.0 | 105.3 |
| Total number of people living together | 1 | 15.9 | 13.3 | 10.8 | 12.9 | 21.1 | 26.3 | 16.4 |
| | 2 | 19.9 | 33.1 | 24.5 | 31.8 | 35.0 | 34.7 | 29.2 |
| | 3 | 22.0 | 13.3 | 22.1 | 14.7 | 21.1 | 16.5 | 18.0 |
| | 4 | 14.1 | 14.1 | 14.7 | 16.6 | 13.8 | 10.2 | 13.9 |
| | 5 | 8.3 | 7.2 | 4.4 | 7.8 | 4.9 | 4.7 | 6.4 |
| | 6 or above | 19.9 | 19.0 | 23.5 | 16.1 | 4.1 | 7.6 | 16.0 |
| Samples | | 277 | 263 | 204 | 217 | 123 | 236 | 1,320 |

**Table 2.12** Housing types for labour immigrants (Unit:%)

| Categories | Guangzhou | Dongguan | Shenyang | Chengdu | Hangzhou | Zhengzhou | Total |
|---|---|---|---|---|---|---|---|
| Urban village | 81.8 | 78.8 | 34.5 | 38.0 | 66.2 | 81.8 | 65.2 |
| Residence commercial | 11.7 | 8.3 | 45.1 | 50.4 | 19.5 | 12.3 | 23.6 |
| Immigrant Apartment | 1.9 | 12.1 | 8.8 | 2.3 | 5.2 | 1.3 | 5.0 |
| Workplace dormitory | 2.6 | 0.8 | 2.7 | 4.7 | 6.5 | 2.6 | 3.0 |
| Public rental housing | 0.6 | 0.0 | 1.8 | 2.3 | 1.3 | 1.9 | 1.3 |
| Workplace | 0.6 | 0.0 | 0.9 | 0.8 | 1.3 | 0.0 | 0.5 |
| Others | 0.6 | 0.0 | 6.2 | 1.6 | 0.0 | 0.0 | 1.3 |
| Non-repeat number | 100 | 100 | 100 | 100 | 100 | 100 | 100 |
| Total | 100 | 100 | 100 | 100 | 100 | 100 | 100 |
| Samples | 154 | 132 | 113 | 129 | 77 | 154 | 759 |

For labour immigrants, most of their houses are rented from villages or urban villages, accounting for 65.2% in total, and more than 80% in Guangzhou and Zhengzhou. Nonetheless, commercial residence is in the majority. The commercial residence in Shenyang, Chengdu and other places is more than 10% higher than that in the urban village/village rental housing, ranking the first in each city. In contrast, the overall proportion of rural migrants' apartments and government low-rent housing was 5.0% and 1.3% respectively, which had little impact (see Table 2.12).

According to the survey, the basic water and electricity facilities are nearly 100% in the housing of rural migrant. In terms of entertainment facilities, the TV ownership rate was 72.6%, while the proportion of DVD players, computers, stereos and broadband internet access was 34.8%, 30.8%, 26.4% and 25.0%, respectively. Other living facilities were scarce, with only 51.5%, 30.9%, 24.6% and 14.6% of households owning water heaters, washing machines, refrigerators and microwaves, respectively (see Table 2.13).

Labour immigrants housing is closely related to their work, they generally prefer to rental housing or workplace dormitory, they give high priority to the distance to the workplace, the living convenience, and convenient public transportation (see Table 2.11). 97.7% of the housing location have bus stops, 87.3% location has taxi stops, other transportation means such as motorcycle or bicycle are 35.5% and 24.2%, respectively (see Table 2.14).

Given the consideration of labour immigrants to purchase houses, this survey collected data on their housing price sensation, their willingness to purchase houses and the factors affecting the purchase of houses. According to the statistical results,

**Table 2.13** Household facilities for labour immigrants (Unit:%)

| Categories | | Guangzhou | Dongguan | Shenyang | Chengdu | Hangzhou | Zhengzhou | Total |
|---|---|---|---|---|---|---|---|---|
| Household facilities ownership | Electricity | 98.9 | 99.2 | 99.0 | 99.1 | 100.0 | 99.6 | 99.2 |
| | Tap-water | 98.9 | 97.3 | 97.5 | 96.8 | 94.3 | 97.9 | 97.4 |
| | Sewer | 84.8 | 89.0 | 85.3 | 82.9 | 68.3 | 80.5 | 83.1 |
| | Television | 74.7 | 74.5 | 75.5 | 68.7 | 74.8 | 67.8 | 72.6 |
| | Flush-toilet | 50.2 | 55.1 | 72.1 | 43.8 | 62.6 | 71.6 | 58.5 |
| | Gas | 66.4 | 47.9 | 64.2 | 72.4 | 57.7 | 35.6 | 57.0 |
| | Water heater | 63.9 | 50.6 | 43.1 | 60.8 | 39.0 | 43.2 | 51.5 |
| | Bicycle | 33.9 | 31.2 | 34.8 | 41.9 | 49.6 | 41.5 | 37.7 |
| | DVD player | 38.3 | 46.4 | 29.4 | 36.4 | 24.4 | 26.7 | 34.8 |
| | Telephone | 32.9 | 33.5 | 35.3 | 33.2 | 22.8 | 26.7 | 31.4 |
| | Washing machine | 28.9 | 15.6 | 33.8 | 44.7 | 35.8 | 32.6 | 30.9 |
| | Computer | 31.0 | 34.2 | 21.1 | 29.0 | 40.7 | 31.4 | 30.8 |
| | Stereo | 32.1 | 31.2 | 20.1 | 18.4 | 23.6 | 28.8 | 26.4 |
| | Broadband internet | 21.3 | 28.5 | 13.7 | 28.6 | 39.8 | 24.2 | 25.0 |
| | Refrigerator | 27.8 | 18.3 | 26.5 | 31.8 | 33.3 | 15.3 | 24.6 |
| | Camera | 14.4 | 23.2 | 14.2 | 18.4 | 33.3 | 16.1 | 18.9 |
| | Microwave | 14.1 | 10.3 | 16.7 | 17.5 | 17.1 | 14.4 | 14.6 |
| | Electrical vehicle | 4.0 | 1.1 | 12.3 | 23.5 | 39.8 | 21.2 | 14.3 |
| | Motorcycle | 5.1 | 1.9 | 4.9 | 6.5 | 0.8 | 1.7 | 3.6 |
| | Personal Car | 2.2 | 4.2 | 1.5 | 2.3 | 0.8 | 2.1 | 2.3 |
| | Cargo vehicle | 0.7 | 1.5 | 2.5 | 1.4 | 0.8 | 0.8 | 1.3 |

(continued)

**Table 2.13** (continued)

| Categories | | Guangzhou | Dongguan | Shenyang | Chengdu | Hangzhou | Zhengzhou | Total |
|---|---|---|---|---|---|---|---|---|
| | Non-repeat number | 100 | 100 | 100 | 100 | 100 | 100 | 100 |
| | Total | 916 | 888 | 888 | 939 | 860 | 869 | 896 |
| Samples | | 277 | 263 | 204 | 217 | 123 | 236 | 1,320 |

**Table 2.14** Transportation means for labour immigrants (Unit:%)

| Categories | Guangzhou | Dongguan | Shenyang | Chengdu | Hangzhou | Zhengzhou | Total |
|---|---|---|---|---|---|---|---|
| Public bus | 98.2 | 97.3 | 97.5 | 96.8 | 95.9 | 99.6 | 97.7 |
| Taxi | 85.6 | 87.8 | 91.2 | 95.4 | 48.0 | 98.3 | 87.3 |
| Motorcycle | 33.2 | 62.0 | 24.5 | 46.5 | 8.1 | 22.0 | 35.5 |
| Bicycle | 18.1 | 25.9 | 18.6 | 62.7 | 5.7 | 8.5 | 24.2 |
| Community bus | 12.6 | 29.3 | 24.5 | 3.2 | 6.5 | 6.8 | 14.6 |
| Subway | 33.2 | 0.0 | 1.0 | 0.0 | 0.8 | 0.0 | 7.2 |
| Electrical vehicle | 0.4 | 0.0 | 0.0 | 0.0 | 0.0 | 0.0 | 0.1 |
| Others | 0.7 | 0.0 | 1.5 | 0.9 | 0.8 | 0.4 | 0.7 |
| None | 0.0 | 0.0 | 0.5 | 0.0 | 0.8 | 0.0 | 0.2 |
| Non-repeat number | 100 | 100 | 100 | 100 | 100 | 100 | 100 |
| Total | 282 | 302 | 259 | 306 | 167 | 236 | 267 |
| Samples | 277 | 263 | 204 | 217 | 123 | 236 | 1,320 |

42.4% of the labour immigrants could not accept the current housing price, contrast 28.0% of the respondents said they could. More than half of the respondents said they would be willing to buy a house in a city if the policy allowed. Among the factors that affect their home purchase and settlement in cities, respondents think that excessively high housing price and unstable job play an important role (65.3% and 51.4% respectively) in their decision making. Secondly, the reasons such as the unclear intention of settling down and the restriction of household registration system also have some influence (see Table 2.15).

All the aforementioned statistics show that labour immigrants have the highest proportion of renters, between 50 and 60%, in all surveyed locations. Secondly, the proportion of living in the dormitory is higher; comparatively speaking, the proportion of labour immigrants who buy their own houses is generally low, only 6.1% overall (Shenyang is slightly higher, 12.3%). However, among the group who choose to rent a house, most of them choose to rent a house for the reasons of being close to the workplace, convenient living, being able to live with family or friends, and a relatively lesser expense. More than 80% of rural migrants live with others (26.3% in Zhengzhou and 21.1% in Hangzhou), mainly friends, colleagues and family. The housing of labour immigrants was mostly rented in urban villages or villages (accounting for 65.2%), followed by commercial residence. As most of the groups rent houses, most of which are cheaper to rent and have convenient transportation, they are welcomed by labour immigrants. Corresponding to the income and expenditure status of labour immigrants in the above section, as well as the current situation of living mainly in rented houses, the ownership rate of basic water and electricity facilities in the housing of labour immigrants is close to 100%. However, the proportion of other entertainment facilities (except TV) is low, and other living

**Table 2.15** Opinions on house purchasing for labour immigrants (Unit:%)

| Opinions | | Guangzhou | Dongguan | Shenyang | Chengdu | Hangzhou | Zhengzhou | Total |
|---|---|---|---|---|---|---|---|---|
| On housing price | Acceptable | 25.6 | 31.6 | 34.8 | 32.7 | 15.4 | 22.9 | 28.0 |
| | No idea | 28.9 | 39.5 | 34.8 | 15.7 | 20.3 | 31.8 | 29.5 |
| | Unacceptable | 45.5 | 28.9 | 30.4 | 50.7 | 64.2 | 45.3 | 42.4 |
| | Others | 0.0 | 0.0 | 0.0 | 0.9 | 0.0 | 0.0 | 0.2 |
| | Total | 100 | 100 | 100 | 100 | 100 | 100 | 100 |
| If the policy allows, are you willing to purchase | Willing to | 46.9 | 47.5 | 67.2 | 53.9 | 38.2 | 70.8 | 54.8 |
| | Willing to, but couldn't afford | 26.4 | 22.1 | 22.1 | 21.2 | 46.3 | 10.6 | 23.0 |
| | Unwilling | 14.4 | 16.7 | 4.9 | 16.1 | 8.9 | 10.2 | 12.4 |
| | Not necessary | 11.9 | 13.7 | 5.9 | 8.3 | 6.5 | 8.5 | 9.6 |
| | Others | 0.4 | 0.0 | 0.0 | 0.5 | 0.0 | 0.0 | 0.2 |
| | Total | 100 | 100 | 100 | 100 | 100 | 100 | 100 |
| Factors affect house purchase decision making | High housing price | 62.1 | 58.6 | 63.7 | 73.3 | 69.9 | 68.2 | 65.3 |
| | Unstable job | 48.0 | 58.9 | 51.5 | 48.8 | 47.2 | 51.7 | 51.4 |
| | Settlement intentions are not yet clear | 22.4 | 32.7 | 27.9 | 31.8 | 22.0 | 22.0 | 26.7 |
| | Household registration restriction | 19.9 | 31.2 | 18.6 | 20.7 | 14.6 | 15.7 | 20.8 |

(continued)

**Table 2.15** (continued)

| Opinions | | Guangzhou | Dongguan | Shenyang | Chengdu | Hangzhou | Zhengzhou | Total |
|---|---|---|---|---|---|---|---|---|
| | Unsatisfied living conditions | 0.7 | 0.0 | 0.5 | 0.5 | 0.0 | 0.0 | 0.3 |
| | Others | 1.4 | 2.3 | 1.5 | 2.8 | 1.6 | 0.4 | 1.7 |
| | No idea | 1.1 | 0.4 | 0.0 | 0.5 | 0.0 | 0.0 | 0.4 |
| | Non-repeat number | 100 | 100 | 100 | 100 | 100 | 100 | 100 |
| | Total | 156 | 184 | 164 | 178 | 155 | 158 | 167 |
| Samples | | 277 | 263 | 204 | 217 | 123 | 236 | 1,320 |

facilities (washing machines, refrigerators, microwaves) are scarce. Although under the circumstances of the policy, the willingness of labour immigrants to purchase houses is relatively high, but more than 40% of the respondents think the current housing price is unacceptable; Accordingly, high housing prices and job instability remain the biggest obstacles to labour immigrant's house purchasing decision.

## 2.4 Working Conditions

In the process of urban–rural dual economic development, one of the prominent features is the surplus of rural labour force provides a cheap labour supply for industrialization, and the wage growth is slow, so the employment relationship is not conducive to labour immigrants. At the same time, due to the asymmetric information and status of both sides of the market and the instability of labour relations, labour immigrants have great uncertainty in employment.

Analysis on the working condition of labour immigrants, including the aspects of stability (change of working city, change of employer in the same city, length of each job), working time (fixed hours or not, working hours and vacation), what are the more important factors in job hunting, job satisfaction in all aspects, unequal pay for equal work due to different identity, and influencing factors of career. According to the survey conducted by the research group, 59% of respondents have worked in different cities, and 68.3% have changed jobs within the same city. The average job length is only around 2 years (see Table 2.16).

Among the labour immigrant with regular jobs,[3] 45.4% said income was the most important factor in choosing a job, followed by job prospects, personal interests and working environment, with only a small proportion of people take skill counterparts and social status into account as the important factors in job choosing. Considering the educational background and age background of labour immigrants, most of them are under the condition of low age and low education. At the same time, they belong to the urban immigrant group and need to bear greater living pressure. Therefore, in the process of choosing a job, more of them will consider the practical income factor (see Table 2.17).

The weekly working hours and daily working hours of labour immigrants were fixed (79.0% and 84.5% of respondents worked fixed weekly and daily hours respectively, as shown in Table 2.18). However, the working intensity was relatively high, with 79.8% of respondents working 6–7 days a week, 6 days a week on average. 51.4% of respondents work 7–8 h per day, and 46.3% work more than 9 h per day, with an overall average of 9.2 h per day. More than half said they do not have a clear

[3] According to the survey, the proportion of labour migrants with regular jobs reached 83.9%, 82.7% in Guangzhou, 93.2% in Dongguan, 57.4% in Shenyang, 88% in Chengdu, 90.2% in Hangzhou and 90.7% in Zhengzhou. The analysis and data presented below in this section are for those interviewed who have regular jobs.

**Table 2.16** Job moving situation of labour immigrants (Unit:%)

| Items | | Guangzhou | Dongguan | Shenyang | Chengdu | Hangzhou | Zhengzhou | Total |
|---|---|---|---|---|---|---|---|---|
| How many cities have you been working in? | 1 | 42.2 | 43.3 | 39.2 | 47.0 | 26.8 | 40.3 | 41.0 |
| | 2 | 27.8 | 29.7 | 26.5 | 18.9 | 28.5 | 20.8 | 25.3 |
| | 3 | 17.0 | 13.7 | 21.6 | 10.6 | 24.4 | 20.8 | 17.3 |
| | 4 | 6.5 | 7.6 | 7.4 | 9.7 | 12.2 | 10.2 | 8.6 |
| | 5 | 0.7 | 1.9 | 3.4 | 4.1 | 4.9 | 3.4 | 2.8 |
| | 6 or above | 5.8 | 3.8 | 2.0 | 9.7 | 3.3 | 4.7 | 5.0 |
| | Average | 2 | 2 | 2 | 2 | 2 | 2 | 2 |
| | Total | 100 | 100 | 100 | 100 | 100 | 100 | 100 |
| How many times of job change in the present city? | 0 | 38.6 | 36.1 | 28.4 | 39.6 | 26.0 | 16.9 | 31.7 |
| | 1 | 17.0 | 22.1 | 12.7 | 11.1 | 17.1 | 13.1 | 15.7 |
| | 2 | 20.9 | 24.7 | 21.1 | 21.7 | 24.4 | 31.8 | 24.1 |
| | 3 | 11.2 | 11.0 | 21.1 | 15.7 | 11.4 | 20.8 | 15.2 |
| | 4 | 3.6 | 3.4 | 7.4 | 4.1 | 5.7 | 6.8 | 5.0 |
| | 5 or above | 8.7 | 2.7 | 9.3 | 7.8 | 15.4 | 10.6 | 8.4 |
| | Average | 1 | 1 | 2 | 1 | 2 | 2 | 2 |
| | Total | 100 | 100 | 100 | 100 | 100 | 100 | 100 |
| Length(month) of the job current occupied | 0–3 | 11.2 | 5.7 | 9.3 | 13.4 | 7.3 | 19.1 | 11.2 |
| | 4–6 | 5.8 | 7.2 | 12.3 | 9.7 | 10.6 | 18.2 | 10.4 |
| | 7–12 | 7.9 | 12.2 | 15.7 | 9.2 | 14.6 | 14.0 | 11.9 |

(continued)

Table 2.16 (continued)

| Items | | Guangzhou | Dongguan | Shenyang | Chengdu | Hangzhou | Zhengzhou | Total |
|---|---|---|---|---|---|---|---|---|
| | 13–24 | 24.9 | 25.9 | 34.3 | 31.8 | 22.8 | 30.9 | 28.6 |
| | 25–36 | 26.0 | 27.8 | 15.2 | 19.8 | 18.7 | 11.9 | 20.5 |
| | 37–48 | 10.8 | 12.9 | 5.9 | 6.9 | 11.4 | 3.8 | 8.6 |
| | 49 months or above | 13.4 | 8.4 | 7.4 | 9.2 | 14.6 | 2.1 | 8.9 |
| | Average (month) | 28.0 | 27.4 | 22.4 | 23.5 | 25.4 | 16.1 | 23.9 |
| | Average (year) | 2.3 | 2.3 | 1.9 | 2.0 | 2.1 | 1.3 | 2.0 |
| | Total | 100 | 100 | 100 | 100 | 100 | 100 | 100 |
| Samples | | 277 | 263 | 204 | 217 | 123 | 236 | 1,320 |

Table 2.17 Job choosing factors for labour immigrants (Unit:%)

| Items | | Guangzhou | Dongguan | Shenyang | Chengdu | Hangzhou | Zhengzhou | Total |
|---|---|---|---|---|---|---|---|---|
| Most important factor in job choosing | Income | 43.3 | 42.6 | 51.1 | 47.1 | 48.7 | 43.2 | 45.4 |
| | Job prospects | 19.0 | 22.3 | 18.3 | 15.5 | 10.9 | 13.5 | 17.2 |
| | Personal interests | 20.2 | 16.0 | 14.0 | 16.0 | 12.6 | 16.2 | 16.3 |
| | Working environment | 11.4 | 13.3 | 9.7 | 14.1 | 10.1 | 16.2 | 12.7 |
| | Skill counterparts | 2.7 | 3.5 | 4.3 | 0.5 | 5.0 | 4.8 | 3.3 |
| | Social status | 0.4 | 0.8 | 0.0 | 0.0 | 0.8 | 0.4 | 0.4 |
| | Others | 3.0 | 1.6 | 2.7 | 6.8 | 11.8 | 5.7 | 4.6 |
| | Total | 100 | 100 | 100 | 100 | 100 | 100 | 100 |
| Samples | | 263 | 256 | 186 | 206 | 119 | 229 | 1,259 |

break (either a fixed break or a shift), while 13.8% do not have any vacation and have to work all the time.

In terms of employment stability, income, working hours and intensity, there is still a certain gap between the working conditions of labour immigrants in manufacturing factories or other enterprises and the overall level of cities and towns. Their overall satisfaction level is 3.4.[4] In terms of job satisfaction, we take into account factors such as income, working environment, social status, promotion opportunities, work pressure, interpersonal relationships and career prospects. Labour immigrants have the lowest negative satisfaction with the promotion opportunity, with an average score of 2.5, followed by the social status of the work (an average score of 2.8, with a low overall sense of social identity and enterprise belongingness). Positive Satisfaction was highest in workplace relationships, averaging 3.9, followed by work environment (averaging 3.3) (see Table 2.19).

In terms of equality, 48.9% of labour immigrants regard the income is equal compare to local colleagues, 23.0% regard there is some difference in social welfare and public insurance, 20.7% think they are in the situation of inequal paid for equal work. Among the above, immigrants in Hangzhou are significantly high regard themselves in the situation of inequality, 30.3% for inequality in social welfare and public insurance, and 36.1% for inequal paid for equal work(see Table 2.20).

As for the factors affecting career prospects, 63.8% of the respondents chose education. Considering the background of labour immigrants, most of them have relatively low education, and the number of workers with tertiary degree or above in the labour market is increasing year by year, which also causes corresponding competition and psychological pressure. This proportion is more obvious in the Dongguan area. Second, 57.6% of the respondents chose skill and training, with respondent in Dongguan, Shenyang and Hangzhou, agreeing with this factor slightly more important. Besides, respondents' approval of opportunities and interpersonal relationships was also high, with the overall choice rate both exceeding 30% (see Table 2.21).

Generally, labour-based migrants showed low job stability, with nearly 60% of respondents have worked in different cities and nearly 70% have changed jobs in the city, where the average length of employment is now just two years. In combination with the above-mentioned conditions such as labour immigrants' education level and income and expenditure pressure, it can be seen that labour immigrants are burdened with great living pressure. When they choose a job, they first focus on their income. Therefore, nearly 50% of the respondents believe that income is the most important factor in choosing a job. No more than 20% of respondents put their job prospects, personal interests or work environment at the top of their list. On the other hand, labour immigrants are still the weak party in the labour market, and the labour immigrants with fixed jobs have relatively heavy working intensity. Nearly 80% of

[4] The survey divided labour immigrants' satisfaction with all aspects of their jobs into five levels, with 5 being "very satisfied," 4 being "satisfied," 3 being "average," 2 being "dissatisfied," and 1 being "very dissatisfied." In Table 2.19, TOP2 represents the sum of the ratios of "very satisfied" and "satisfied".

**Table 2.18** Working hours for labour immigrants (Unit:%)

| Items | | Guangzhou | Dongguan | Shenyang | Chengdu | Hangzhou | Zhengzhou | Total |
|---|---|---|---|---|---|---|---|---|
| Weekly fixed working days | Fixed | 71.1 | 83.2 | 82.8 | 67.0 | 89.1 | 86.0 | 79.0 |
| | Flexible | 28.9 | 16.8 | 17.2 | 33.0 | 10.9 | 14.0 | 21.0 |
| | Total | 100 | 100 | 100 | 100 | 100 | 100 | 100 |
| Samples | | 263 | 256 | 186 | 206 | 119 | 229 | 1,259 |
| Weekly working days | 4 or below | 1.6 | 0.0 | 2.6 | 2.2 | 4.7 | 2.5 | 2.0 |
| | 5 | 16.0 | 30.0 | 14.9 | 11.6 | 22.6 | 12.2 | 18.2 |
| | 6 | 40.1 | 53.5 | 38.3 | 42.0 | 54.7 | 44.7 | 45.4 |
| | 7 | 42.2 | 16.4 | 44.2 | 44.2 | 17.9 | 40.6 | 34.4 |
| | Average | 6 | 6 | 6 | 6 | 6 | 6 | 6 |
| | Total | 100 | 100 | 100 | 100 | 100 | 100 | 100 |
| Samples | | 187 | 213 | 154 | 138 | 106 | 197 | 995 |
| Daily fixed working hours | Fixed | 81.4 | 89.8 | 82.8 | 77.7 | 92.4 | 85.6 | 84.5 |
| | Flexible | 18.6 | 10.2 | 17.2 | 22.3 | 7.6 | 14.4 | 15.5 |
| | Total | 100 | 100 | 100 | 100 | 100 | 100 | 100 |
| Samples | | 263 | 256 | 186 | 206 | 119 | 229 | 1,259 |
| Daily working hours | 6 or below | 2.8 | 1.3 | 3.2 | 1.3 | 0.9 | 3.6 | 2.3 |
| | 7–8 | 40.2 | 69.6 | 35.7 | 43.1 | 53.6 | 60.2 | 51.4 |
| | 9–10 | 31.3 | 16.1 | 35.1 | 25.6 | 29.1 | 24.0 | 26.1 |
| | 11–13 | 23.4 | 12.6 | 14.9 | 21.9 | 11.8 | 11.2 | 16.2 |
| | 13 or above | 2.3 | 0.4 | 11.0 | 8.1 | 4.5 | 1.0 | 4.0 |
| | Average | 9.4 | 8.6 | 9.7 | 9.7 | 9.1 | 8.7 | 9.2 |

(continued)

**Table 2.18** (continued)

| Items | | Guangzhou | Dongguan | Shenyang | Chengdu | Hangzhou | Zhengzhou | Total |
|---|---|---|---|---|---|---|---|---|
| | Total | 100 | 100 | 100 | 100 | 100 | 100 | 100 |
| Samples | | 214 | 230 | 154 | 160 | 110 | 196 | 1,064 |
| Leaves | No scheduled leave | 36.9 | 17.2 | 28.0 | 36.9 | 13.4 | 33.2 | 28.7 |
| | Shift leave | 24.3 | 21.1 | 21.5 | 20.9 | 37.8 | 16.6 | 22.6 |
| | Leave on Sunday | 9.1 | 27.7 | 14.5 | 8.3 | 10.9 | 15.7 | 14.9 |
| | No leave | 12.5 | 4.3 | 17.2 | 21.8 | 10.1 | 17.9 | 13.8 |
| | Leave on Saturday and Sunday | 8.4 | 20.7 | 10.2 | 7.3 | 20.2 | 6.1 | 11.7 |
| | Scheduled leave days weekly or monthly | 8.7 | 9.0 | 8.6 | 4.9 | 7.6 | 10.5 | 8.3 |
| | Non-repeat number | 100 | 100 | 100 | 100 | 100 | 100 | 100 |
| | Total | 100 | 100 | 100 | 100 | 100 | 100 | 100 |
| Samples | | 263 | 256 | 186 | 206 | 119 | 229 | 1,259 |

**Table 2.19** General satisfaction for labour immigrants

| Regions | Index | Income | Working environment | Social status | Promotion opportunities | Work pressure | Interpersonal relationships | Career prospects | Overall |
|---|---|---|---|---|---|---|---|---|---|
| Guangzhou | Average | 3.0 | 3.4 | 2.8 | 2.4 | 3.1 | 3.9 | 2.9 | 3.3 |
| | TOP2 (%) | 22.1 | 41.1 | 20.2 | 12.9 | 35.0 | 67.3 | 27.4 | 34.6 |
| | Standard deviation | 0.2 | 0.2 | 0.2 | 0.1 | 0.1 | 0.2 | 0.1 | 0.2 |
| Dongguan | Average | 3.0 | 3.5 | 2.9 | 2.6 | 3.1 | 4.0 | 3.0 | 3.5 |
| | TOP2 (%) | 24.2 | 49.2 | 25.8 | 18.8 | 35.5 | 78.1 | 32.8 | 49.2 |
| | Standard deviation | 0.2 | 0.1 | 0.2 | 0.1 | 0.1 | 0.2 | 0.1 | 0.2 |
| Shenyang | Average | 3.0 | 3.2 | 2.7 | 2.5 | 3.2 | 4.1 | 2.8 | 3.6 |
| | TOP2 (%) | 21.5 | 39.8 | 23.1 | 23.7 | 47.3 | 75.3 | 29.0 | 48.9 |
| | Standard deviation | 0.2 | 0.1 | 0.1 | 0.1 | 0.0 | 0.2 | 0.1 | 0.2 |
| Chengdu | Average | 3.0 | 3.3 | 2.7 | 2.3 | 2.8 | 3.8 | 2.6 | 3.3 |
| | TOP2 (%) | 26.2 | 40.8 | 18.4 | 14.6 | 30.6 | 67.0 | 18.0 | 36.4 |
| | Standard deviation | 0.2 | 0.2 | 0.2 | 0.1 | 0.1 | 0.2 | 0.1 | 0.2 |
| Hangzhou | Average | 3.0 | 3.3 | 2.8 | 2.6 | 3.1 | 3.5 | 2.8 | 3.2 |
| | TOP2 (%) | 25.2 | 42.0 | 21.8 | 11.8 | 24.4 | 52.9 | 17.6 | 32.8 |

(continued)

**Table 2.19** (continued)

| Regions | Index | Income | Working environment | Social status | Promotion opportunities | Work pressure | Interpersonal relationships | Career prospects | Overall |
|---|---|---|---|---|---|---|---|---|---|
| | Standard deviation | 0.2 | 0.2 | 0.2 | 0.2 | 0.2 | 0.2 | 0.2 | 0.2 |
| Zhengzhou | Average | 2.6 | 3.1 | 2.6 | 2.5 | 3.0 | 4.0 | 2.9 | 3.2 |
| | TOP2 (%) | 14.8 | 28.8 | 15.3 | 15.3 | 34.5 | 71.6 | 26.6 | 31.4 |
| | Standard deviation | 0.1 | 0.2 | 0.2 | 0.1 | 0.1 | 0.2 | 0.1 | 0.2 |
| Overall | Average | 2.9 | 3.3 | 2.8 | 2.5 | 3.1 | 3.9 | 2.9 | 3.4 |
| | TOP2 (%) | 22.1 | 40.3 | 20.7 | 16.3 | 35.1 | 70.1 | 26.1 | 39.2 |
| | Standard deviation | 0.2 | 0.1 | 0.1 | 0.1 | 0.1 | 0.2 | 0.1 | 0.2 |

**Table 2.20** Inequality in work for labour immigrants (Unit:%)

| Regions | Guangzhou | Dongguan | Shenyang | Chengdu | Hangzhou | Zhengzhou | Total |
|---|---|---|---|---|---|---|---|
| Equal | 50.6 | 50.4 | 41.9 | 52.9 | 37.8 | 53.3 | 48.9 |
| Inequality in social welfare & public insurance | 23.2 | 23.4 | 25.3 | 19.4 | 30.3 | 19.7 | 23.0 |
| Inequal pay | 18.3 | 15.6 | 22.6 | 18 | 36.1 | 21.8 | 20.7 |
| Inequal promotion | 11.8 | 17.6 | 21.0 | 11.7 | 14.3 | 20.1 | 16.0 |
| Inequal training | 9.9 | 13.7 | 23.1 | 10.7 | 16 | 16.6 | 14.5 |
| No local colleague | 8.4 | 2.0 | 0.5 | 4.4 | 1.7 | 0.0 | 3.1 |
| No idea | 0.4 | 1.6 | 0.0 | 2.4 | 0.8 | 0.0 | 0.9 |
| Others | 0.0 | 3.5 | 0.5 | 3.9 | 0.8 | 2.2 | 1.9 |
| Non-repeat number | 100 | 100 | 100 | 100 | 100 | 100 | 100 |
| Total | 122.4 | 127.7 | 134.9 | 123.3 | 137.8 | 133.6 | 129.0 |
| Samples | 263 | 256 | 186 | 206 | 119 | 229 | 1,259 |

respondents worked six to seven days a week, on average six days a week; nearly half of the respondents worked more than nine hours a day, with an overall average of 9.2 h. More than half said they do not have a scheduled leave (neither a fixed one nor a shift one), and more than one in 10 do not have any vacation and need to work all the time. On the other hand, about 50% of the respondents think that there are differences in the treatment of immigrant workers and their local colleagues (including differences in social welfare and public insurance, unequal pay for equal work, fewer promotion opportunities, less training opportunities, etc.). Among them, the proportion that thinks social welfare and public insurance has difference is slightly higher than other aspects. From the perspective of labour immigrants' career prospects, respondents are more pragmatic. About 60% believe that education, skills and training are one of the factors affecting career prospects, while only 40% and 30% believe that opportunities and interpersonal relationships are important to career prospects.

## 2.5 Social Welfare and Public Insurance

China's social welfare and public insurance system is designed to ensure its residents' income security and basic living support, prevent predictable risks they may encounter in work and daily life, and providing certain support for their future. Before

**Table 2.21** Factors affecting career for labour immigrants (Unit:%)

| Items | | Guangzhou | Dongguan | Shenyang | Chengdu | Hangzhou | Zhengzhou | Total |
|---|---|---|---|---|---|---|---|---|
| Factors affecting career for labour immigrants | Education | 60.1 | 71.5 | 66.7 | 62.1 | 63.9 | 58.5 | 63.8 |
| | Skill and training | 57.4 | 60.5 | 59.7 | 56.8 | 61.3 | 51.5 | 57.6 |
| | Opportunity | 36.9 | 38.7 | 40.9 | 44.2 | 29.4 | 45.4 | 39.9 |
| | Interpersonal relationship | 26.6 | 28.1 | 33.9 | 32.0 | 26.1 | 38.0 | 30.9 |
| | Personality | 16.7 | 21.9 | 22.6 | 29.1 | 23.5 | 26.6 | 23.1 |
| | Family background | 16.3 | 16.4 | 21.5 | 23.8 | 23.5 | 15.7 | 18.9 |
| | Supervisor | 11.8 | 19.1 | 16.7 | 20.4 | 17.6 | 18.3 | 17.2 |
| | Age | 14.1 | 15.2 | 23.7 | 19.9 | 5.0 | 9.6 | 15.0 |
| | Household registration | 12.2 | 11.7 | 19.9 | 10.7 | 22.7 | 7.4 | 13.1 |
| | Gender | 1.5 | 2.7 | 9.1 | 3.4 | 0.8 | 3.1 | 3.4 |
| | Marriage status | 0.8 | 3.1 | 6.5 | 4.9 | 0.0 | 2.2 | 2.9 |
| | Maternity status | 0.0 | 2.3 | 2.7 | 3.4 | 0.0 | 0.9 | 1.6 |
| | No specific factor | 2.3 | 1.6 | 0.5 | 2.4 | 0.0 | 0.0 | 1.3 |
| | No idea | 0.0 | 0.0 | 0.0 | 1.9 | 0.0 | 0.0 | 0.3 |
| | Other | 0.8 | 0.8 | 0.5 | 1.5 | 1.7 | 1.7 | 1.1 |
| | Non-repeat number | 100 | 100 | 100 | 100 | 100 | 100 | 100 |
| | Total | 257 | 294 | 325 | 317 | 276 | 279 | 290 |
| Samples | | 263 | 256 | 186 | 206 | 119 | 229 | 1,259 |

the 1990s, China's social security system only covers employees of state-owned and collective enterprises. With the in-depth development of China, the labour immigration has increased, and labour immigrants have become an important force in urban construction. The social welfare and public insurance system have also experienced intense discussions in various places and a variety of models have been developed.

The social welfare and public insurance analysis for labour migration, was conducted in the following aspects: household registration status (urban or rural, the types of household registration in the city's registration category); weight of different types of social welfare in respondent's opinion; coverage of public insurance, other type of welfare asides from bonus and insurance, the inequality on social welfare (compare to local colleagues, in the fields of education, health, recreation, housing), and the reasons for inequality.

In our survey, the respondents are largely non-local household registration, only 3.3% possess local household registration, and among this minority, only 34.9% possess local household registration as individual, the rest are in the situation of collective household registration or non-kinship affiliated household registration (see Table 2.22).

There are many types of social welfare, in our survey, housing, pension, education, health, recreation, social security (such as unemployment etc.), were taken into account. Among 1,320 respondents, 42.7% take housing as their priority, greatly exceeding other types of welfare, with education follows, which were given more attention in Dongguan (33.8%) than Shenyang (26.0%), Chengdu (25.8%), and Zhengzhou (21.2%) (see Table 2.23).

The above table shows the respondents are lesser concern with health insurance, maybe because the health insurance coverage is the highest among all sectors of social welfare, around 50%. Besides, respondents' workplace injury insurance and pension coverage are around 30%, with respondents' spouse coverage 26.5% and 18.9% respectively, respondents' parents were lesser cover by pension, only 22.3%. Generally speaking, labour immigrants has a low coverage of social welfare and public insurance, with 37.1% of who has no coverage at all, and the types of coverage were relatively fewer than their urban counterpart. For regions, Dongguan and Hangzhou significantly high coverage than other cities (see Table 2.24).

Aside from bonus and insurance, 47% of the respondents were given free lunch from the workplace, moreover, 39.4% has paid leave and sick leave, 25% of the respondents have annual physical, maternity leave, and skill training. As for regions, Guangzhou and Shenyang provides more free lunch, and Dongguan is generally higher in every aspect (see Table 2.25).

Only 8.6% of the respondents regards they are equal with local counterparts in educational welfare, 19.8% for health, 17.4% for recreation, and 10.2% for housing, thus the labour immigrant thinks it is highly inequal in nearly all aspects of welfare. Among which, education and housing is the most prominent two aspects, respectively 54.3% and 65.4% of the respondents regards as "no coverage" or "highly inequality",

**Table 2.22** Household registration status for labour immigrants (Unit:%)

| Items | | Guangzhou | Dongguan | Shenyang | Chengdu | Hangzhou | Zhengzhou | Total |
|---|---|---|---|---|---|---|---|---|
| household registration status | Non-local | 98.6 | 98.9 | 93.1 | 96.8 | 94.3 | 96.6 | 96.7 |
| | Local | 1.4 | 1.1 | 6.9 | 3.2 | 5.7 | 3.4 | 3.3 |
| | Total | 100 | 100 | 100 | 100 | 100 | 100 | 100 |
| Sample | | 277 | 263 | 204 | 217 | 123 | 236 | 1,320 |
| Local household registration status | Collective | 0.0 | 0.0 | 0.0 | 14.3 | 28.6 | 50.0 | 16.3 |
| | Individual | 0.0 | 33.3 | 35.7 | 71.4 | 28.6 | 25.0 | 34.9 |
| | Affiliated | 100.0 | 33.3 | 7.1 | 0.0 | 14.3 | 25.0 | 20.9 |
| | Other | 0.0 | 33.3 | 57.1 | 14.3 | 28.6 | 0.0 | 27.9 |
| | Total | 100 | 100 | 100 | 100 | 100 | 100 | 100 |
| Samples | | 4 | 3 | 14 | 7 | 7 | 8 | 43 |

**Table 2.23** Social welfare weighting by labour immigrants (Unit:%)

| Items | Guangzhou | Dongguan | Shenyang | Chengdu | Hangzhou | Zhengzhou | Total |
|---|---|---|---|---|---|---|---|
| Housing | 37.2 | 31.2 | 55.9 | 37.8 | 44.7 | 54.2 | 42.7 |
| Education | 32.1 | 33.8 | 26.0 | 25.8 | 30.9 | 21.2 | 28.4 |
| Heath | 16.6 | 17.5 | 11.3 | 17.5 | 11.4 | 11.4 | 14.7 |
| Recreation | 8.3 | 13.3 | 4.4 | 7.4 | 5.7 | 9.7 | 8.6 |
| Social security | 0.0 | 0.8 | 0.5 | 0.9 | 0.0 | 0.0 | 0.4 |
| Pension | 0.0 | 0.0 | 0.0 | 0.0 | 0.0 | 0.4 | 0.1 |
| No idea | 1.4 | 0.8 | 0.5 | 1.4 | 0.0 | 0.4 | 0.8 |
| None | 2.9 | 0.8 | 0.5 | 7.4 | 3.3 | 1.3 | 2.6 |
| Other | 1.4 | 1.9 | 1.0 | 1.8 | 4.1 | 1.3 | 1.7 |
| Non-repeat number | 100 | 100 | 100 | 100 | 100 | 100 | 100 |
| Total | 100 | 100 | 100 | 100 | 100 | 100 | 100 |
| Samples | 277 | 263 | 204 | 217 | 123 | 236 | 1,320 |

As for regions, Dongguan has the highest satisfaction for equality, the average[5] is 2.3.

In Dongguan, 31.2% of respondents regards "somehow equal" in education, with Shenyang and Zhengzhou felt "inequal" more than other regions, respectively 37.7% and 37.3%.

For health, 27.8% of the respondents in Dongguan felt "equal", and 35.4% felt "somehow equal". While 26.8% of the respondents in Hangzhou felt "somehow inequal", and 43.6% of the respondents in Zhengzhou felt "inequal".

For recreation, 29.3% of the respondents in Dongguan felt "equal", while majority of respondents in Shenyang and Zhengzhou felt "inequal", respectively 41.2% and 42.8%.

For housing, 23.6% of the respondents in Dongguan felt "somehow equal", while 32.3% of respondents in Chengdu felt "somehow inequal", and majority of respondents in Shenyang and Zhengzhou felt "inequal", respectively 50.5% and 54.7% (see Table 2.26).

In terms of the cognition of the reasons for the inequality in social welfare, nearly 70% of the respondents think that the reason is the household registration. Secondly, the proportion of the respondents think that the difference in welfare sharing caused by the uneven distribution of public welfare resources and the lack of public welfare resources also reaches 28.5% and 24.8% respectively (see Table 2.27).

[5] The average of satisfaction represents the equality in welfares for labour immigrants and locals. The value is between 1 and 4. The higher the value is, the more equality labour immigrants were felt. Score 1 indicates that the welfare is most inequal, while a score of 4 indicates that the welfares are most equal accordance to the local people.

**Table 2.24** Public insurance coverage for labour immigrants (Unit:%)

| Items | | Guangzhou | Dongguan | Shenyang | Chengdu | Hangzhou | Zhengzhou | Total |
|---|---|---|---|---|---|---|---|---|
| Personal insurance coverage | Health | 44.4 | 67.3 | 43.6 | 56.7 | 65.0 | 42.4 | 52.4 |
| | Workplace injury | 28.5 | 52.5 | 25.5 | 22.1 | 36.6 | 14.4 | 30.0 |
| | Pension | 19.1 | 42.6 | 27.5 | 24.0 | 64.2 | 13.6 | 29.1 |
| | Unemployment | 10.1 | 18.6 | 17.6 | 12.4 | 27.6 | 8.1 | 14.6 |
| | Maternity | 7.2 | 8.0 | 7.4 | 9.2 | 16.3 | 0.4 | 7.3 |
| | Commercial insurance | 6.5 | 3.4 | 4.4 | 2.8 | 8.1 | 4.7 | 4.8 |
| | Social security | 1.1 | 2.3 | 0.5 | 0.5 | 0.0 | 0.0 | 0.8 |
| | Compensation | 0.0 | 0.0 | 0.0 | 0.5 | 0.8 | 0.0 | 0.2 |
| | Public housing fund | 0.4 | 0.0 | 0.0 | 0.0 | 0.0 | 0.0 | 0.1 |
| | Others | 0.0 | 0.0 | 0.0 | 1.4 | 0.0 | 1.3 | 0.5 |
| | None of above | 44.0 | 21.7 | 39.2 | 35.9 | 22.8 | 53.0 | 37.1 |
| | Non-repeat number | 100 | 100 | 100 | 100 | 100 | 100 | 100 |
| | Total | 161.4 | 216.3 | 165.7 | 165.4 | 241.5 | 137.7 | 176.9 |
| Samples | | 277 | 263 | 204 | 217 | 123 | 236 | 1,320 |
| Spouse insurance coverage | Health | 38.7 | 56.9 | 43.1 | 44.8 | 55.9 | 55.0 | 48.4 |
| | Workplace injury | 13.5 | 31.5 | 15.5 | 9.5 | 27.1 | 12.5 | 18.9 |
| | Pension | 13.5 | 33.8 | 34.5 | 14.7 | 47.5 | 30.0 | 26.5 |

(continued)

**Table 2.24** (continued)

| Items | | Guangzhou | Dongguan | Shenyang | Chengdu | Hangzhou | Zhengzhou | Total |
|---|---|---|---|---|---|---|---|---|
| | Unemployment | 4.5 | 12.3 | 10.3 | 3.4 | 22.0 | 15.0 | 9.7 |
| | Maternity | 5.4 | 5.4 | 10.3 | 4.3 | 23.7 | 10.0 | 8.2 |
| | Commercial insurance | 2.7 | 1.5 | 3.4 | 3.4 | 3.4 | 5.0 | 2.9 |
| | Social security | 0.0 | 1.5 | 0.0 | 0.0 | 0.0 | 0.0 | 0.4 |
| | Compensation | 0.0 | 0.0 | 0.0 | 0.9 | 0.0 | 0.0 | 0.2 |
| | Public housing fund | 0.9 | 3.8 | 0.0 | 1.7 | 0.0 | 0.0 | 1.6 |
| | Others | 0.0 | 0.8 | 0.0 | 0.0 | 0.0 | 0.0 | 0.2 |
| | None of above | 53.2 | 28.5 | 43.1 | 46.6 | 37.3 | 37.5 | 41.2 |
| | Non-repeat number | 100 | 100 | 100 | 100 | 100 | 100 | 100 |
| | Total | 132 | 176 | 160 | 129 | 217 | 165 | 158 |
| Samples | | 111 | 130 | 58 | 116 | 59 | 40 | 514 |
| Filial insurance coverage | Health | 39.0 | 45.8 | 40.4 | 44.8 | 20.8 | 33.3 | 39.8 |
| | Workplace injury | 5.7 | 1.9 | 4.3 | 10.5 | 2.1 | 3.3 | 5.2 |
| | Pension | 3.8 | 3.7 | 10.6 | 14.3 | 2.1 | 3.3 | 6.8 |
| | Unemployment | 2.9 | 0.9 | 4.3 | 8.6 | 0.0 | 0.0 | 3.4 |
| | Maternity | 2.9 | 0.0 | 2.1 | 7.6 | 0.0 | 0.0 | 2.7 |
| | Commercial insurance | 2.9 | 5.6 | 10.6 | 3.8 | 2.1 | 13.3 | 5.2 |
| | Social security | 1.0 | 0.9 | 0.0 | 0.0 | 0.0 | 0.0 | 0.5 |
| | Compensation | 0.0 | 0.0 | 0.0 | 1.9 | 0.0 | 0.0 | 0.5 |
| | Educational welfare | 1.0 | 1.9 | 0.0 | 1.9 | 2.1 | 3.3 | 1.6 |
| | No idea | 1.0 | 2.8 | 0.0 | 6.7 | 0.0 | 0.0 | 2.5 |

(continued)

**Table 2.24** (continued)

| Items | | Guangzhou | Dongguan | Shenyang | Chengdu | Hangzhou | Zhengzhou | Total |
|---|---|---|---|---|---|---|---|---|
| | Others | 1.0 | 2.8 | 2.1 | 5.7 | 0.0 | 0.0 | 2.5 |
| | None of above | 56.2 | 41.1 | 46.8 | 35.2 | 75.0 | 53.3 | 48.4 |
| | Non-repeat number | 100 | 100 | 100 | 100 | 100 | 100 | 100 |
| | Total | 117 | 108 | 121 | 141 | 104 | 110 | 119 |
| Samples | | 105 | 107 | 47 | 105 | 48 | 30 | 442 |
| Parents insurance coverage | Health | 42.6 | 49.8 | 49.0 | 47.9 | 22.0 | 41.9 | 43.9 |
| | Workplace injury | 4.7 | 6.1 | 5.4 | 6.9 | 4.9 | 2.1 | 5.0 |
| | Pension | 19.1 | 27.4 | 40.2 | 23.0 | 16.3 | 7.6 | 22.3 |
| | Unemployment | 2.2 | 2.3 | 6.4 | 5.1 | 1.6 | 1.7 | 3.2 |
| | Maternity | 2.9 | 1.1 | 3.4 | 3.2 | 0.8 | 0.4 | 2.0 |
| | Commercial insurance | 1.1 | 1.9 | 1.0 | 2.8 | 4.1 | 0.4 | 1.7 |
| | Social security | 0.4 | 1.1 | 0.0 | 0.0 | 0.0 | 0.0 | 0.3 |
| | Compensation | 0.0 | 0.0 | 0.0 | 0.0 | 0.0 | 0.4 | 0.1 |
| | No idea | 2.2 | 1.5 | 0.0 | 2.8 | 0.8 | 0.0 | 1.3 |
| | Others | 0.0 | 0.4 | 0.0 | 0.0 | 0.0 | 0.0 | 0.1 |
| | None of above | 48.0 | 33.1 | 32.4 | 31.3 | 59.3 | 53.8 | 42.0 |
| | Missed | 1.1 | 3.0 | 8.8 | 10.6 | 9.8 | 1.3 | 5.1 |
| | Non-repeat number | 100 | 100 | 100 | 100 | 100 | 100 | 100 |
| | Total | 124 | 128 | 147 | 134 | 120 | 110 | 127 |
| Samples | | 277 | 263 | 204 | 217 | 123 | 236 | 1,320 |

**Table 2.25** Other welfares for labour immigrants (Unit:%)

| Items | | Guangzhou | Dongguan | Shenyang | Chengdu | Hangzhou | Zhengzhou | Total |
|---|---|---|---|---|---|---|---|---|
| Other welfares | Free lunch | 57.0 | 39.2 | 58.8 | 44.7 | 41.5 | 38.6 | 47.0 |
| | Sick leave and other paid leave | 32.1 | 62.4 | 40.2 | 38.2 | 37.4 | 23.7 | 39.4 |
| | Annual physical | 25.3 | 39.5 | 26.5 | 23.0 | 30.9 | 14.4 | 26.5 |
| | Maternity leave | 14.4 | 61.2 | 17.6 | 16.1 | 15.4 | 15.7 | 24.8 |
| | Skill training | 15.9 | 45.2 | 20.6 | 25.8 | 17.1 | 16.1 | 24.2 |
| | Migratory expense reimbursement | 20.2 | 23.2 | 13.2 | 6.9 | 11.4 | 6.4 | 14.2 |
| | Kindergarten | 0.4 | 3.0 | 1.0 | 1.8 | 0.0 | 0.0 | 1.1 |
| | No workplace | 0.0 | 0.0 | 0.5 | 0.9 | 0.0 | 0.4 | 0.3 |
| | Holiday gift (cash or in kind) | 0.0 | 0.0 | 0.5 | 0.0 | 0.0 | 0.0 | 0.1 |
| | Phone bill reimbursement | 0.0 | 0.0 | 0.0 | 0.0 | 0.0 | 0.0 | 0.0 |
| | Not appliable | 0.0 | 0.0 | 0.0 | 0.9 | 0.8 | 2.1 | 0.6 |
| | Others | 1.8 | 0.8 | 0.5 | 1.8 | 2.4 | 0.8 | 1.3 |
| | None | 24.5 | 10.3 | 21.6 | 30.9 | 22.0 | 34.3 | 23.8 |
| | Non-repeat number | 100 | 100 | 100 | 100 | 100 | 100 | 100 |
| | Total | 192 | 285 | 201 | 191 | 179 | 153 | 203 |
| Samples | | 277 | 263 | 204 | 217 | 123 | 236 | 1,320 |

**Table 2.26** Inequality in social welfare for labour immigrants (Unit:%)

| Items | | Guangzhou | Dongguan | Shenyang | Chengdu | Hangzhou | Zhengzhou | Total |
|---|---|---|---|---|---|---|---|---|
| Education equality | Equal | 9.0 | 8.0 | 10.8 | 9.7 | 4.9 | 7.6 | 8.6 |
| | Somehow equal | 26.7 | 31.2 | 27.5 | 22.1 | 19.5 | 23.7 | 25.8 |
| | Somehow inequal | 22.4 | 25.1 | 18.6 | 23.5 | 22.8 | 20.3 | 22.2 |
| | Inequal | 33.2 | 26.2 | 37.7 | 27.6 | 30.9 | 37.3 | 32.1 |
| | No idea | 8.7 | 9.5 | 5.4 | 17.1 | 22.0 | 11.0 | 11.4 |
| | Total | 100 | 100 | 100 | 100 | 100 | 100 | 100 |
| Average | | 1.9 | 2.0 | 2.1 | 1.7 | 1.6 | 1.8 | 1.8 |
| Health equality | Equal | 19.9 | 27.8 | 22.5 | 19.4 | 11.4 | 13.1 | 19.8 |
| | Somehow equal | 27.8 | 35.4 | 24.5 | 29.5 | 22.8 | 15.3 | 26.4 |
| | Somehow inequal | 18.1 | 15.6 | 15.7 | 18.4 | 26.8 | 17.4 | 18.0 |
| | Inequal | 29.2 | 14.8 | 33.8 | 21.2 | 26.0 | 43.6 | 28.0 |
| | No idea | 5.1 | 6.5 | 3.4 | 11.5 | 13.0 | 10.6 | 7.9 |
| | Total | 100 | 100 | 100 | 100 | 100 | 100 | 100 |
| Average | | 2.3 | 2.6 | 2.3 | 2.2 | 1.9 | 1.8 | 2.2 |
| Recreation equality | Equal | 15.9 | 29.3 | 16.2 | 15.2 | 10.6 | 12.7 | 17.4 |
| | Somehow equal | 22.0 | 28.1 | 25.0 | 27.6 | 20.3 | 15.3 | 23.3 |
| | Somehow inequal | 22.7 | 11.8 | 14.2 | 23.0 | 21.1 | 17.4 | 18.2 |
| | Inequal | 33.6 | 20.9 | 41.2 | 24.9 | 30.1 | 42.8 | 32.1 |
| | No idea | 5.8 | 9.9 | 3.4 | 9.2 | 17.9 | 11.9 | 9.0 |
| | Total | 100 | 100 | 100 | 100 | 100 | 100 | 100 |
| Average | | 2.1 | 2.5 | 2.1 | 2.1 | 1.8 | 1.7 | 2.1 |

(continued)

**Table 2.26** (continued)

| Items | | Guangzhou | Dongguan | Shenyang | Chengdu | Hangzhou | Zhengzhou | Total |
|---|---|---|---|---|---|---|---|---|
| Housing equality | Equal | 11.9 | 12.9 | 14.2 | 8.8 | 4.1 | 5.9 | 10.2 |
| | Somehow equal | 14.8 | 23.6 | 17.6 | 11.5 | 14.6 | 10.2 | 15.6 |
| | Somehow inequal | 24.2 | 22.4 | 15.2 | 32.3 | 22.0 | 16.5 | 22.2 |
| | Inequal | 44.4 | 30.8 | 50.5 | 36.9 | 43.9 | 54.7 | 43.2 |
| | No idea | 4.7 | 10.3 | 2.5 | 10.6 | 15.4 | 12.7 | 8.9 |
| | Total | 100 | 100 | 100 | 100 | 100 | 100 | 100 |
| Average | | 1.8 | 2.0 | 1.9 | 1.7 | 1.5 | 1.4 | 1.8 |
| Overall | Equal | 3.6 | 5.3 | 7.4 | 1.8 | 3.3 | 3.8 | 4.2 |
| | Somehow equal | 31.4 | 40.3 | 34.8 | 34.6 | 22.8 | 16.5 | 30.8 |
| | Somehow inequal | 34.7 | 36.9 | 30.9 | 39.2 | 36.6 | 36.4 | 35.8 |
| | Inequal | 24.9 | 12.9 | 25.5 | 17.5 | 21.1 | 36.0 | 23.0 |
| | No idea | 5.4 | 4.6 | 1.5 | 6.9 | 16.3 | 7.2 | 6.2 |
| | Average | 100 | 100 | 100 | 100 | 100 | 100 | 100 |
| Average | | 2.0 | 2.3 | 2.2 | 2.1 | 1.8 | 1.7 | 2.0 |
| Samples | | 277 | 263 | 204 | 217 | 123 | 236 | 1,320 |

**Table 2.27** Labour immigrants' cognition of the reasons for the inequality in social welfare (Unit:%)

| Items | Guangzhou | Dongguan | Shenyang | Chengdu | Hangzhou | Zhengzhou | Total |
|---|---|---|---|---|---|---|---|
| Non-local household registration | 72.9 | 74.1 | 69.6 | 64.5 | 65.0 | 66.9 | 69.5 |
| Uneven distribution of public welfare resources | 21.7 | 28.1 | 31.9 | 33.6 | 30.9 | 28.0 | 28.5 |
| Lack of public welfare resources | 17.7 | 21.3 | 29.4 | 18.9 | 34.1 | 33.5 | 24.8 |
| Local household registration but with some restrictions | 6.5 | 12.9 | 13.7 | 1.8 | 13.8 | 10.2 | 9.5 |
| No inequality | 2.2 | 1.1 | 1.0 | 0.5 | 0.8 | 3.0 | 1.5 |
| No idea | 2.9 | 3.0 | 0.0 | 6.0 | 4.1 | 0.4 | 2.7 |
| Others | 2.2 | 0.4 | 2.9 | 4.6 | 0.8 | 1.7 | 2.1 |
| Non-repeat number | 100 | 100 | 100 | 100 | 100 | 100 | 100 |
| Total | 126 | 141 | 149 | 130 | 150 | 144 | 139 |
| Samples | 277 | 263 | 204 | 217 | 123 | 236 | 1,320 |

Among the samples randomly selected in this survey, the household registration of most of the respondents is not local (nearly 95% or more in every region). Even for those with permanent household registration in the city, only a few (0% in Guangzhou) have independent personal household registration. Since the public welfare is still closely related to the household registration system to a considerable extent, the labour immigrants who cannot obtain the local household registration cannot obtain the same social welfare as the local citizens to a considerable extent. In fact, according to our survey results, nearly 70% of respondents believe that the main reason for the difference between their welfare and that of locals is that they do not have local household registration. Among the social and public welfare that labour immigrants care about, housing welfare takes the first place (Dongguan area is slightly different, and the attention to housing welfare and education welfare is the same), followed by education welfare. Health benefits were the least of the concerns, at 0%. However, from the perspective of the social welfare coverage for labour immigrants and their families, the ratio is the opposite. Labour immigrants and their families have a relatively high proportion of actual access to health benefits, while the actual access to housing and education benefits of concern is very low. On the one hand, labour immigrants may focus on benefits they already have (such as health benefits). On the other hand, due to the desire of labour immigrants to integrate into the city, the welfare of housing and education needed for integration into the city is naturally highly concerned. In addition to bonuses and insurance, nearly 50% of the respondents received free lunch from the workplace. Besides, nearly 40% of the respondents are entitled to sick leave and paid leave. Also, the proportion of annual physical, maternity leave and annual vocational training is about a quarter. From the perspective of welfare differences, most labour immigrants believe that the various welfare (education welfare, health welfare, cultural and recreational welfare, housing welfare) they have been different from those of local people. Among them, education welfare and housing welfare are particularly prominent, with 54.3% and 65.4% of respondents believing that they are not covered at all and there is a great inequality.

## 2.6 Filial Education

A great concern for the labour immigrant is education for their children. At the early stage of rural–urban migration, most individual choose to leave one's child at their rural home, went to cities alone or with one's spouse or relatives. After the period of establishment, some immigrants will bring their children along and take education in cities, yet the majority still left their child in their household registered region for education. Labour immigrants are not only bother by the filial relation maintenance, but also facing the future development of their children.

The analysis on labour immigrants' filial education, constructed by the factors such as whether one have child, how many children one has, how many children are taking education, where the children are taking education, reasons for whether taking education in cities, what improvement is needed for the school, and extra fee for education (sponsor fee is usually require for non-local student).

Among the 1,320 labour immigrants in this survey, 33.5% of them have child, 63.1% have one child, and 29.2% have 2 children, Chengdu is significantly high on 2 children (see Table 2.28).

Among the labour immigrants who have child, 62.9% (278 cases) of the respondents' children are taking education, mostly 1 or 2 children (see Table 2.29).

Among those who have child taking education, 63.3% (176 cases) of the respondents have their child taking education in their household registered location, only 32% attend schools in their living cities. 77.2% of the respondents in Dongguan have their children taking education in their household registered location, higher than those of Guangzhou, Chengdu and Zhengzhou (see Table 2.30).

Among the labour immigrant's family with children and taking education in the living city, 68.3% (69 cases) of those regard the educational expense as high, follow by low quality of school and hard to get a place in the school, each taking 20%. As for the evaluation of school that these children attending, around 26% of the respondents

**Table 2.28** Number of children for labour immigrants (Unit:%)

| Items | | Guangzhou | Dongguan | Shenyang | Chengdu | Hangzhou | Zhengzhou | Total |
|---|---|---|---|---|---|---|---|---|
| Whether one have child | No | 62.1 | 59.3 | 77.0 | 51.6 | 61.0 | 87.3 | 66.5 |
| | Yes | 37.9 | 40.7 | 23.0 | 48.4 | 39.0 | 12.7 | 33.5 |
| | Total | 100 | 100 | 100 | 100 | 100 | 100 | 100 |
| Sample | | 277 | 263 | 204 | 217 | 123 | 236 | 1,320 |
| How many children one has | 1 | 64.8 | 66.4 | 78.7 | 47.6 | 64.6 | 73.3 | 63.1 |
| | 2 | 24.8 | 27.1 | 17.0 | 42.9 | 27.1 | 26.7 | 29.2 |
| | 3 | 9.5 | 5.6 | 4.3 | 9.5 | 8.3 | 0.0 | 7.2 |
| | 4 | 1.0 | 0.9 | 0.0 | 0.0 | 0.0 | 0.0 | 0.5 |
| | 5 or above | 0.0 | 0.0 | 0.0 | 0.0 | 0.0 | 0.0 | 0.0 |
| | Average | 1 | 1 | 1 | 2 | 1 | 1 | 1 |
| | Total | 100 | 100 | 100 | 100 | 100 | 100 | 100 |
| Samples | | 105 | 107 | 47 | 105 | 48 | 30 | 442 |

**Table 2.29** Filial education for labour immigrants (Unit:%)

| Items | | Guangzhou | Dongguan | Shenyang | Chengdu | Hangzhou | Zhengzhou | Total |
|---|---|---|---|---|---|---|---|---|
| Whether one's child is taking education | No | 34.3 | 26.2 | 38.3 | 52.4 | 27.1 | 46.7 | 37.1 |
| | Yes | 65.7 | 73.8 | 61.7 | 47.6 | 72.9 | 53.3 | 62.9 |
| | Total | 100 | 100 | 100 | 100 | 100 | 100 | 100 |
| Samples | | 105 | 107 | 47 | 105 | 48 | 30 | 442 |
| How many children are taking education | 1 | 76.8 | 78.5 | 96.6 | 74.0 | 62.9 | 75.0 | 77.0 |
| | 2 | 18.8 | 17.7 | 3.4 | 16.0 | 31.4 | 25.0 | 18.3 |
| | 3 | 4.3 | 2.5 | 0.0 | 10.0 | 5.7 | 0.0 | 4.3 |
| | 4 | 0.0 | 1.3 | 0.0 | 0.0 | 0.0 | 0.0 | 0.4 |
| | 5 or above | 0.0 | 0.0 | 0.0 | 0.0 | 0.0 | 0.0 | 0.0 |
| | Average | 1 | 1 | 1 | 1 | 1 | 1 | 1 |
| | Total | 100 | 100 | 100 | 100 | 100 | 100 | 100 |
| Samples | | 69 | 79 | 29 | 50 | 35 | 16 | 278 |

believe that there is still room for improvement in the quality of teaching, teaching facilities, level of schooling and management practices (see Table 2.31).

Among the labour immigrant's family with children and taking education in the living city, 57.4% (58 cases) of those didn't paid extra expense for education, yet the annual expense on filial education is large fall in the category of RMB2,001–6,000, generally, the average of annual expense on filial education is RMB6,184 (see Table 2.32).

**Table 2.30** Location of filial education for labour immigrants (Unit:%)

| Items | | Guangzhou | Dongguan | Shenyang | Chengdu | Hangzhou | Zhengzhou | Total |
|---|---|---|---|---|---|---|---|---|
| Where one child is taking education | household registration registered location | 60.9 | 77.2 | 58.6 | 56.0 | 57.1 | 50.0 | 63.3 |
| | Public school in the living cities | 15.9 | 8.9 | 17.2 | 26.0 | 11.4 | 37.5 | 16.5 |
| | Private school in the living cities | 20.3 | 5.1 | 6.9 | 8.0 | 5.7 | 12.5 | 10.1 |
| | Special funded immigrants' school | 0.0 | 6.3 | 6.9 | 0.0 | 22.9 | 0.0 | 5.4 |
| | Others | 1.4 | 2.5 | 10.3 | 10.0 | 2.9 | 0.0 | 4.3 |
| | Missed | 1.4 | 0.0 | 0.0 | 0.0 | 0.0 | 0.0 | 0.4 |
| | Non-repeat number | 100 | 100 | 100 | 100 | 100 | 100 | 100 |
| | Total | 100 | 100 | 100 | 100 | 100 | 100 | 100 |
| Samples | | 69 | 79 | 29 | 50 | 35 | 16 | 278 |

**Table 2.31** Factors on taking education in the cities for labour immigrants (Unit:%)

| Items | | Guangzhou | Dongguan | Shenyang | Chengdu | Hangzhou | Zhengzhou | Total |
|---|---|---|---|---|---|---|---|---|
| Factors affecting children taking education in the cities | High expense | 76.9 | 72.2 | 33.3 | 72.7 | 80.0 | 50.0 | 68.3 |
| | Low teaching quality | 23.1 | 38.9 | 25.0 | 9.1 | 6.7 | 50.0 | 22.8 |
| | Hard to get in | 23.1 | 33.3 | 33.3 | 0.0 | 6.7 | 37.5 | 19.8 |
| | Inadaptation for child | 11.5 | 0.0 | 25.0 | 4.5 | 20.0 | 12.5 | 10.9 |
| | No affecting factor | 3.8 | 0.0 | 0.0 | 13.6 | 0.0 | 12.5 | 5.0 |
| | Other | 0.0 | 0.0 | 16.7 | 13.6 | 0.0 | 0.0 | 5.0 |
| | Non-repeat number | 100 | 100 | 100 | 100 | 100 | 100 | 100 |
| | Total | 139 | 144 | 133 | 114 | 113 | 163 | 132 |
| Improvements needed for the school | Teaching quality | 34.6 | 33.3 | 41.7 | 18.2 | 20.0 | 25.0 | 28.7 |
| | Teaching facility | 15.4 | 33.3 | 41.7 | 31.8 | 13.3 | 62.5 | 28.7 |
| | Schooling | 23.1 | 27.8 | 25.0 | 18.2 | 46.7 | 25.0 | 26.7 |
| | Administration | 42.3 | 33.3 | 8.3 | 9.1 | 13.3 | 37.5 | 24.8 |
| | Prestige | 11.5 | 5.6 | 25.0 | 9.1 | 13.3 | 0.0 | 10.9 |
| | No improvement needed | 0.0 | 5.6 | 8.3 | 13.6 | 0.0 | 0.0 | 5.0 |
| | Expense | 0.0 | 5.6 | 0.0 | 4.5 | 0.0 | 0.0 | 2.0 |
| | Other | 0.0 | 0.0 | 0.0 | 0.0 | 0.0 | 0.0 | 0.0 |
| | No idea | 38.5 | 22.2 | 8.3 | 40.9 | 46.7 | 25.0 | 32.7 |
| | Non-repeat number | 100 | 100 | 100 | 100 | 100 | 100 | 100 |
| | Total | 165 | 167 | 158 | 146 | 153 | 175 | 159 |
| Samples | | 26 | 18 | 12 | 22 | 15 | 8 | 101 |

**Table 2.32** Expense on filial education for labour immigrants (Unit:%)

| Items | | Guangzhou | Dongguan | Shenyang | Chengdu | Hangzhou | Zhengzhou | Total |
|---|---|---|---|---|---|---|---|---|
| Extra expense | Yes | 30.8 | 55.6 | 8.3 | 45.5 | 53.3 | 62.5 | 41.6 |
| | No | 65.4 | 44.4 | 91.7 | 54.5 | 46.7 | 37.5 | 57.4 |
| | No idea | 3.8 | 0.0 | 0.0 | 0.0 | 0.0 | 0.0 | 1.0 |
| | Total | 100 | 100 | 100 | 100 | 100 | 100 | 100 |
| Samples | | 26 | 18 | 12 | 22 | 15 | 8 | 101 |
| Annual expense on filial education | RMB0-2,000 | 0.0 | 11.1 | 8.3 | 31.8 | 0.0 | 12.5 | 10.9 |
| | RMB2,001–4,000 | 26.9 | 50.0 | 16.7 | 27.3 | 26.7 | 0.0 | 27.7 |
| | RMB4,001–6,000 | 38.5 | 33.3 | 16.7 | 13.6 | 33.3 | 25.0 | 27.7 |
| | RMB6,001–8,000 | 11.5 | 0.0 | 0.0 | 9.1 | 0.0 | 12.5 | 5.9 |
| | RMB8,001–10,000 | 7.7 | 5.6 | 41.7 | 4.5 | 13.3 | 50.0 | 14.9 |
| | RMB10,001–15,000 | 3.8 | 0.0 | 8.3 | 9.1 | 6.7 | 0.0 | 5.0 |
| | RMB15,001–20,000 | 3.8 | 0.0 | 8.3 | 4.5 | 13.3 | 0.0 | 5.0 |
| | RMB20,001 or above | 7.7 | 0.0 | 0.0 | 0.0 | 6.7 | 0.0 | 3.0 |
| | Average(RMB) | 6,228 | 4,101 | 8,850 | 4,875 | 7,807 | 7,500 | 6,184 |
| | Total | 100 | 100 | 100 | 100 | 100 | 100 | 100 |
| Samples | | 26 | 18 | 12 | 22 | 15 | 8 | 101 |

In our labour immigrants' samples, one-third of the respondents have children, in which two-thirds of them are taking education. Among the respondents with school-going children, only about 30% of their children went to school in the living city. About 26% of the respondents believe that the quality of teachers, teaching facilities, schooling and management standards of their children's schools need to be improved. As can be seen from the data, the high cost of education remains an important obstacle to the schooling of the children of labour immigrants in cities. Nearly 70% of the respondents (whose children go to school in cities) believe that the high cost of education is one of the factors affecting their children's schooling in the city, which takes the first place among all factors. At the same time, due to household registration restriction and other reasons, a considerable part of the high tuition fees faced by labour immigrants are sponsorship fees/quota borrowing fees required for non-municipal household registration. Generally, the annual education expenses of labour immigrants for their children in the cities reach RMB6,184, and even close to RMB9,000 in Shenyang, which is far higher than the average annual income of labour immigrants.

## 2.7 Recreation

Recreation is very necessary for people, it release one from daily weary, and giving spiritual comfort and relax. The recreational life for labour immigrants is conducted in the following aspects: recreational occasion, companionship, three best friends, recourse helper.

In Sect. 2.4 (see Table 2.18), nearly 90% of labour immigrants more or less have some time for themselves, 77.3% of the respondents said they choose to watch TV or video for leisure, follow by 64.5% listen to music or radio, 63.3% go shopping, 62.4% visit friends or chatting, 61.4% take their time online with PC, and 54.2% spend their leisure time on smartphones, and the tick-box of "drinking tea or eating, sleeping or just being idle" also have 50.6% chooser. As for region difference, respondents in Dongguan apparently have more choices, all the leisure activities are more likely to be selected than other cities, and respondents in Zhengzhou tend to do online leisure than others (see Table 2.33).

From Table 2.34, we learnt that labour immigrants tend to choose recreations with less expenses, such as watching television or listening to music, more than 80% of the respondents stay in their home or their friends' home for vacations, besides, around 50% of the respondents go to parks or go to some gathering place (restaurants) for leisure. As for companionship, more than 70% of the respondents take their vacation with friends or colleagues, while around 50% take the time with family, relatives or fellow countryman.

In the survey, labour immigrants were asked about their 3 best friends, besides Guangzhou and Hangzhou, the gender ratio in other cities are slightly falls to female, with the age around 26, Hangzhou and Chengdu are higher in age, around 30. The location of 3 best friends are more than 60% in the city where the respondent lives in,

**Table 2.33** Recreational activities for labour immigrants (Unit:%)

| Items | Guangzhou | Dongguan | Shenyang | Chengdu | Hangzhou | Zhengzhou | Total |
|---|---|---|---|---|---|---|---|
| Watch TV or video | 83.4 | 89.7 | 82.8 | 68.2 | 74.0 | 61.9 | 77.3 |
| Listen to music or radio | 63.9 | 79.1 | 68.6 | 56.2 | 48.8 | 61.4 | 64.5 |
| Shopping | 62.8 | 81.7 | 69.6 | 57.1 | 39.0 | 56.4 | 63.3 |
| Visit friends and chatting | 62.5 | 86.3 | 64.7 | 49.3 | 47.2 | 53.8 | 62.4 |
| Online with PC | 54.5 | 63.1 | 63.2 | 52.1 | 62.6 | 74.2 | 61.4 |
| Smartphone | 52.3 | 74.5 | 58.3 | 51.2 | 29.3 | 45.8 | 54.2 |
| Drinking tea or eating, sleeping or just being idle | 46.9 | 73.0 | 51.0 | 53.5 | 39.0 | 33.1 | 50.6 |
| Chore | 43.0 | 69.2 | 56.9 | 59.4 | 39.0 | 25.0 | 49.5 |
| Reading, photography or calligraphy | 27.1 | 52.1 | 33.3 | 26.7 | 39.8 | 27.1 | 34.2 |
| Play chess, cards or mah-jong | 35.0 | 35.4 | 31.4 | 36.9 | 22.0 | 22.5 | 31.4 |
| Sports, exercise | 34.3 | 45.2 | 23.0 | 23.5 | 22.8 | 25.0 | 30.2 |
| Karaoke, night club | 20.9 | 24.3 | 30.9 | 19.8 | 8.9 | 19.9 | 21.7 |
| Go to barber shop or beauty salon | 17.3 | 29.3 | 21.6 | 19.8 | 8.9 | 10.2 | 18.7 |
| Billiard | 15.5 | 21.3 | 21.1 | 12.0 | 13.0 | 18.2 | 17.2 |
| Lottery | 12.6 | 16.0 | 20.1 | 24.0 | 10.6 | 8.1 | 15.3 |
| Gardening, flower or pet | 10.5 | 24.0 | 16.7 | 9.2 | 6.5 | 6.4 | 12.8 |
| Fishing | 0.0 | 0.0 | 0.5 | 0.5 | 0.0 | 0.0 | 0.2 |
| Go to park | 0.0 | 0.0 | 0.5 | 0.0 | 0.0 | 0.0 | 0.1 |
| Travels | 0.0 | 0.4 | 0.0 | 0.0 | 0.0 | 0.0 | 0.1 |
| Others | 0.4 | 0.0 | 0.0 | 0.5 | 0.0 | 0.0 | 0.2 |
| None of above | 0.0 | 0.0 | 0.0 | 0.9 | 0.0 | 0.0 | 0.2 |

(continued)

**Table 2.33** (continued)

| Items | Guangzhou | Dongguan | Shenyang | Chengdu | Hangzhou | Zhengzhou | Total |
|---|---|---|---|---|---|---|---|
| Non-repeat number | 100 | 100 | 100 | 100 | 100 | 100 | 100 |
| Total | 643 | 865 | 714 | 621 | 511 | 549 | 665 |
| Samples | 277 | 263 | 204 | 217 | 123 | 236 | 1,320 |

with Zhengzhou reaches 71.0%. As for the relationship to 3 best friends, colleague takes 35.7%, schoolmate takes 27.4%, and countryman takes 25.8% (see Table 2.35).

When labour immigrants encounter difficulties, they turn to different people for help, depending on the nature of the situation. When they need to borrow money, see a doctor, being unemployed (or temporarily have no income), or being scammed, they seek help from their family and friends more often, followed by relatives and colleagues. When they have no place to live and need help in dealing with disputes with others, they will turn to their friends for help more often. In the case of traffic accidents, more people will turn to relevant departments of the municipal government and their families for help. When encounter industrial injury, more than 60% of the respondents seek help in the workplace. Besides, when the occurrence of scam, disputes and industrial injury, nearly 30% of the respondents will choose to seek help from the relevant departments of the municipal government.

Among all the difficulties and disputes, trouble for filial education is less likely to happen and hard to find help. In general, about 20% of the respondents seek help from families and relatives (see Table 2.36).

The above data show that leisure activities of labour immigrants are mainly low-cost indoor activities (in their own or friends' homes) where acquaintances gather. More than half of them are watching TV/video (77.3%), listening to music/radio (64.5%), going shopping (63.3%), visiting friends/chatting (62.4%), online leisure with PC (61.4%), smartphone (54.2%), drinking tea/eating, sleeping, be idle (50.6%), etc. In addition to their own or a friend's home, about 50% would choose a park or restaurant. Including friends, colleagues, family relatives, fellow townsman and so on. Among the three best friends of working migrants, the ratio of men to women was about the same, with an average age of about 26; their best friends are located in the city, more than 60%; Looking from the friend type, the colleague, the classmate, the fellow countryman occupies the first three steadily. In the labour immigrants encounter different difficulties, will turn to different people for help; among them, when borrowing money, the target range of recourse is wider, while when children go to school, the target range is narrower.

**Table 2.34** Occasions and companions for labour immigrants' leisure (Unit:%)

| Items | | Guangzhou | Dongguan | Shenyang | Chengdu | Hangzhou | Zhengzhou | Total |
|---|---|---|---|---|---|---|---|---|
| Occasions | Home or friends' home | 76.9 | 87.5 | 79.4 | 87.1 | 79.7 | 70.8 | 80.2 |
| | Park | 52.0 | 77.9 | 48.5 | 55.3 | 43.9 | 53.4 | 56.7 |
| | Restaurant | 39.4 | 55.5 | 66.2 | 37.8 | 43.9 | 46.2 | 48.1 |
| | Internet cafe | 24.2 | 26.6 | 47.1 | 31.8 | 19.5 | 35.6 | 31.1 |
| | Playground | 19.1 | 37.6 | 23.5 | 32.3 | 12.2 | 26.3 | 26.3 |
| | Pub or teahouse | 15.5 | 15.2 | 10.3 | 24.9 | 24.4 | 10.2 | 16.1 |
| | Ballroom | 7.9 | 14.4 | 19.1 | 12.0 | 6.5 | 13.1 | 12.4 |
| | Gymnasium | 9.7 | 12.9 | 7.4 | 9.2 | 4.9 | 9.7 | 9.5 |
| | Mah-jong house | 6.1 | 7.2 | 6.9 | 14.7 | 6.5 | 3.0 | 7.3 |
| | Workplace | 1.1 | 0.8 | 0.0 | 0.5 | 0.0 | 0.4 | 0.5 |
| | Shopping mall | 0.0 | 0.4 | 1.5 | 0.5 | 0.0 | 0.0 | 0.4 |
| | Street | 0.7 | 0.0 | 0.0 | 0.5 | 0.0 | 0.0 | 0.2 |
| | Karaoke | 0.0 | 0.0 | 0.5 | 0.0 | 0.0 | 0.0 | 0.1 |
| | Others | 0.4 | 1.1 | 0.5 | 0.5 | 0.8 | 0.0 | 0.5 |
| | None of above | 2.9 | 0.8 | 0.0 | 0.5 | 0.8 | 0.8 | 1.1 |
| | Non-repeat number | 100 | 100 | 100 | 100 | 100 | 100 | 100 |
| | Total | 256 | 338 | 311 | 307 | 243 | 270 | 291 |
| Samples | | 277 | 263 | 204 | 217 | 123 | 236 | 1,320 |
| Companions | Friends | 72.5 | 83.5 | 83.3 | 66.7 | 71.3 | 79.5 | 76.6 |
| | Colleagues | 71.7 | 83.9 | 80.9 | 67.1 | 66.4 | 73.1 | 74.6 |
| | Family or relatives | 53.5 | 66.3 | 45.1 | 64.8 | 51.6 | 37.6 | 53.6 |

(continued)

**Table 2.34** (continued)

| Items | | Guangzhou | Dongguan | Shenyang | Chengdu | Hangzhou | Zhengzhou | Total |
|---|---|---|---|---|---|---|---|---|
| | Countryman | 57.2 | 70.5 | 45.6 | 43.1 | 49.2 | 30.8 | 50.2 |
| | Schoolmate | 29.0 | 39.1 | 43.1 | 31.5 | 34.4 | 54.7 | 38.7 |
| | Neighbour | 14.9 | 30.3 | 20.1 | 24.1 | 15.6 | 5.1 | 18.6 |
| | Business partner | 9.7 | 11.5 | 13.2 | 11.6 | 6.6 | 7.3 | 10.2 |
| | Others | 0.0 | 0.0 | 0.0 | 0.0 | 0.8 | 0.0 | 0.1 |
| | None of above | 0.7 | 0.0 | 0.5 | 1.4 | 0.0 | 0.4 | 0.5 |
| | Missed | 0.4 | 0.0 | 0.0 | 0.0 | 0.8 | 0.4 | 0.2 |
| | Non-repeat number | 100 | 100 | 100 | 100 | 100 | 100 | 100 |
| | Total | 310 | 385 | 332 | 310 | 297 | 289 | 323 |
| Samples | | 269 | 261 | 204 | 216 | 122 | 234 | 1,306 |

**Table 2.35** Information on the "3 best friends" for labour immigrants (Unit:%)

| Regions | | Guangzhou | Dongguan | Shenyang | Chengdu | Hangzhou | Zhengzhou | Total |
|---|---|---|---|---|---|---|---|---|
| Gender | Female | 44.2 | 53.9 | 50.0 | 50.8 | 44.7 | 52.8 | 49.7 |
| | Male | 48.7 | 41.8 | 48.5 | 46.2 | 50.7 | 46.3 | 46.7 |
| | Missed | 7.1 | 4.3 | 1.5 | 2.9 | 4.6 | 0.8 | 3.6 |
| | Total | 100 | 100 | 100 | 100 | 100 | 100 | 100 |
| Age | Average | 25.4 | 25.2 | 26.3 | 29.6 | 29.8 | 23.2 | 26.2 |
| | Total | 100 | 100 | 100 | 100 | 100 | 100 | 100 |
| Location | In the city | 55.6 | 58.3 | 63.4 | 68.2 | 64.2 | 71.0 | 63.0 |
| | In the hometown | 22.4 | 17.4 | 21.2 | 16.0 | 13.5 | 15.5 | 18.1 |
| | Another place | 15.7 | 19.5 | 14.4 | 13.7 | 7.8 | 12.1 | 14.6 |
| | Missed | 6.3 | 4.8 | 1.0 | 2.1 | 14.4 | 1.3 | 4.3 |
| | Total | 100 | 100 | 100 | 100 | 100 | 100 | 100 |
| Relationship | Colleague | 33.4 | 40.9 | 35.1 | 36.6 | 35.5 | 32.4 | 35.7 |
| | Schoolmate | 23.3 | 23.8 | 31.4 | 21.7 | 25.5 | 39 | 27.4 |
| | Countryman | 33 | 31.2 | 20.8 | 22.1 | 32 | 15.8 | 25.8 |
| | Friend | 14.6 | 10 | 14.1 | 15.1 | 1.6 | 12.7 | 12.1 |
| | Neighbour | 6.6 | 4.8 | 6.2 | 5.7 | 5.2 | 3.8 | 5.4 |
| | Local | 2.1 | 2.5 | 5.4 | 2 | 7.6 | 2.7 | 3.3 |
| | Supervisor | 1.1 | 3 | 2.6 | 0.9 | 4.4 | 0.8 | 1.9 |
| | Business partner | 0.7 | 0.9 | 0.8 | 0.9 | 1.9 | 0.7 | 0.9 |
| Samples | | 277 | 263 | 204 | 217 | 123 | 236 | 1,320 |

**Table 2.36** Help seeking target for labour immigrants (Unit:%)

| Targets | Borrow money | Sick | Scam | Unemployment | Shelter | Filial education | Dispute | Traffic accident | Industrial injury |
|---|---|---|---|---|---|---|---|---|---|
| Family | 52.3 | 54.4 | 30.4 | 38.0 | 25.0 | 20.3 | 20.5 | 32.0 | 13.9 |
| Friend | 42.4 | 24.9 | 24.5 | 39.3 | 40.9 | 13.1 | 33.9 | 22.3 | 8.9 |
| Relative | 35.8 | 26.2 | 12.4 | 22.0 | 25.2 | 18.4 | 13.7 | 14.5 | 5.9 |
| Colleague | 34.9 | 21.2 | 17.3 | 19.0 | 21.0 | 5.6 | 19.9 | 8.9 | 7.3 |
| Schoolmate | 23.9 | 10.8 | 11.4 | 22.4 | 19.8 | 5.2 | 12.7 | 6.0 | 2.3 |
| Countryman | 22.1 | 11.8 | 12.0 | 20.6 | 21.4 | 4.6 | 12.1 | 7.7 | 3.9 |
| Employer | 10.1 | 9.3 | 6.1 | 1.5 | 2.6 | 3.0 | 9.8 | 12.2 | 63.5 |
| Others | 0.4 | 0.0 | 0.0 | 0.0 | 0.0 | 0.0 | 0.0 | 0.0 | 0.0 |
| Insurance | 0.0 | 4.5 | 0.0 | 0.0 | 0.0 | 0.0 | 0.0 | 13.4 | 17.2 |
| Municipal government | 0.0 | 5.8 | 28.4 | 3.3 | 3.1 | 15.7 | 25.1 | 41.0 | 17.5 |
| Local government | 0.0 | 1.2 | 2.0 | 0.0 | 0.0 | 6.9 | 2.3 | 2.0 | 1.1 |
| Social relief | 0.0 | 1.9 | 1.7 | 1.0 | 0.0 | 2.5 | 0.0 | 1.9 | 2.0 |
| No help | 0.2 | 2.4 | 5.2 | 2.2 | 1.7 | 1.9 | 2.0 | 0.0 | 0.0 |
| No idea | 2.3 | 2.8 | 7.4 | 3.5 | 4.8 | 38.7 | 5.4 | 5.6 | 5.4 |

## 2.8 Mental Conditions

People encounter all kinds of difficulties, setbacks, failures and other situations, and may also produce a variety of negative emotions due to these circumstances. Labour immigrants, in particular, who live in cities where they were not native to, are likely to suffer more. The analysis of the mental state of labour immigrants includes three indicators of well-being: their most distressing events, their mental/psychological problems and the degree of these problems.

According to our survey, the biggest mental obstacles of more than half of the respondents were low income and insufficient knowledge (56.7 and 52.1% respectively). Follow by job pressure and drab lifestyle, both accounted for 30% above. From a regional perspective, the proportion of respondents in Hangzhou suffering from low income is significantly higher (66.7%). Dongguan and Shenyang had a higher proportion of those suffering from inadequate knowledge, 62.7 and 57.8%, respectively (see Table 2.37).

**Table 2.37** The most stressful things for labour immigrants (Unit:%)

| Items | Guangzhou | Dongguan | Shenyang | Chengdu | Hangzhou | Zhengzhou | Total |
|---|---|---|---|---|---|---|---|
| Low income | 49.8 | 56.3 | 55.4 | 56.2 | 66.7 | 61.4 | 56.7 |
| Insufficient knowledge | 45.1 | 62.7 | 57.8 | 45.2 | 47.2 | 52.5 | 52.1 |
| High pressure | 33.9 | 38.4 | 39.2 | 41.9 | 26.8 | 34.7 | 36.4 |
| Drab lifestyle | 26.4 | 31.6 | 36.8 | 29.5 | 29.3 | 35.6 | 31.4 |
| Instability | 26.4 | 20.2 | 38.2 | 33.2 | 36.6 | 25.4 | 28.9 |
| Lack of professional skill | 30.7 | 27.8 | 36.3 | 30.0 | 30.9 | 15.3 | 28.1 |
| Lack of goal | 13.7 | 20.9 | 25.0 | 18.4 | 17.1 | 15.3 | 18.3 |
| Lack of accomplishment | 17.0 | 9.1 | 14.2 | 14.7 | 14.6 | 8.9 | 13.0 |
| Loneliness | 6.1 | 6.1 | 13.7 | 7.8 | 11.4 | 7.2 | 8.3 |
| Lovesickness | 4.0 | 5.7 | 7.4 | 8.8 | 4.1 | 10.6 | 6.8 |
| Trouble by sick | 3.6 | 9.1 | 6.9 | 7.8 | 5.7 | 2.1 | 5.8 |
| Trouble by family dissension | 2.2 | 2.7 | 4.4 | 2.3 | 4.9 | 2.1 | 2.9 |
| Others | 0.4 | 0.4 | 0.5 | 2.8 | 0.0 | 0.4 | 0.8 |
| None of above | 4.7 | 2.7 | 2.5 | 4.6 | 3.3 | 3.4 | 3.6 |
| Non-repeat number | 100 | 100 | 100 | 100 | 100 | 100 | 100 |
| Total | 264 | 294 | 338 | 303 | 298 | 275 | 293 |
| Samples | 277 | 263 | 204 | 217 | 123 | 236 | 1,320 |

We evaluated labour immigrants' mental condition in 9 aspects, they are "insomnia", "feeling exhausted physically and mentally", "irritable", "crying easily or prone to cry", "vanish prospect", "feeling lonely", "feeling useless", "feeling that life is hard" and "feeling that life is boring". There are four levels of severity: none, sometimes, often, and severe (categorise situations where the respondent is unable to articulate as unclear). The proportion of labour immigrants who have experienced the above situations (no matter the severity) is 53.3%, 71.0%, 64.5%, 32.0%, 59.3%, 45.0%, 41.0% and 60.3% respectively, which reflects that labour immigrants bear great pressure in their daily work and life.

The percentage of "feeling exhausted physically and mentally" was higher in Dongguan and Chengdu. The proportion of labour immigrants in Zhengzhou who often felt "uncertain" was significantly higher (17.8%). Besides, the proportion of sometimes "feeling lonely" was significantly higher in Chengdu (41.0%) and 12.2% of respondents in Hangzhou feel this emotion "often" (see Table 2.38).

In the survey of three indicators of satisfactory, respondents ranked "filial piety of children", "family happiness" and "being respected by others" as the top three, with 52.2%, 45.8% and 20.1% respectively. Regionally, the proportion of Chengdu and Hangzhou choosing "content" is significantly higher, reaching 29.5 and 20.3% respectively (see Table 2.39).

Thus, the results of a survey of labour immigrants show that their biggest mental pressure source is low income and lack of knowledge, followed by high work pressure and monotonous lifestyle. According to the analysis results in the previous part, only about 20% of the labour immigrants have margined income over expense. Besides, 97.7% of the respondents have an education level of high school or below, and their inherent background characteristics cause their troubles in life. Type at the same time, most of the labour migration state of mind in a state of sub-health, in "insomnia", "feeling exhausted physically and mentally", "irritable", "crying easily or prone to cry", "vanish prospect", "feeling lonely", "feeling useless", "feeling that life is hard", "feeling that life is boring" were 53.3%, 71.0%, 64.5%, 32.0%, 59.3%, 45.0%, 41.0% and 60.3% more or less problems. "Filial piety of children", "family happiness" and "being respected by others" took the top three places in the survey of three indicators of happiness rated by the labourers.

**Table 2.38** Mental problems and severity for labour immigrants (Unit:%)

| Items | | Guangzhou | Dongguan | Shenyang | Chengdu | Hangzhou | Zhengzhou | Total |
|---|---|---|---|---|---|---|---|---|
| Insomnia | None | 56.0 | 41.4 | 44.6 | 39.6 | 43.9 | 48.7 | 46.2 |
| | Sometimes | 36.5 | 49.4 | 47.1 | 47.9 | 42.3 | 43.6 | 44.4 |
| | Often | 6.9 | 6.5 | 7.4 | 11.1 | 10.6 | 5.9 | 7.7 |
| | Severe | 0.7 | 2.3 | 0.5 | 0.9 | 1.6 | 1.3 | 1.2 |
| | Unclear | 0.0 | 0.4 | 0.5 | 0.5 | 1.6 | 0.4 | 0.5 |
| | Total | 100 | 100 | 100 | 100 | 100 | 100 | 100 |
| Feeling exhausted physically and mentally | None | 33.2 | 29.3 | 25.5 | 18.9 | 35.8 | 28.8 | 28.3 |
| | Sometimes | 54.9 | 60.8 | 52.5 | 60.8 | 43.9 | 49.2 | 54.6 |
| | Often | 10.8 | 9.1 | 19.1 | 18.4 | 17.1 | 18.6 | 15.0 |
| | Severe | 1.1 | 0.4 | 2.9 | 1.4 | 0.8 | 1.7 | 1.4 |
| | Unclear | 0.0 | 0.4 | 0.0 | 0.5 | 2.4 | 1.7 | 0.7 |
| | Total | 100 | 100 | 100 | 100 | 100 | 100 | 100 |
| Irritable | None | 38.3 | 31.9 | 25.5 | 33.6 | 49.6 | 33.1 | 34.4 |
| | Sometimes | 51.3 | 55.9 | 55.4 | 51.6 | 38.2 | 50.4 | 51.5 |
| | Often | 8.3 | 10.6 | 16.2 | 13.4 | 9.8 | 13.6 | 11.9 |
| | Severe | 1.4 | 0.4 | 2.5 | 0.5 | 0.0 | 1.7 | 1.1 |
| | Unclear | 0.7 | 1.1 | 0.5 | 0.9 | 2.4 | 1.3 | 1.1 |
| | Total | 100 | 100 | 100 | 100 | 100 | 100 | 100 |
| Crying easily or prone to cry | None | 74.4 | 65.4 | 58.3 | 67.3 | 69.9 | 64.4 | 66.7 |
| | Sometimes | 23.8 | 28.1 | 29.9 | 26.7 | 22.0 | 29.2 | 26.9 |
| | Often | 1.1 | 5.3 | 9.3 | 5.1 | 2.4 | 4.2 | 4.5 |

(continued)

**Table 2.38** (continued)

| Items | | Guangzhou | Dongguan | Shenyang | Chengdu | Hangzhou | Zhengzhou | Total |
|---|---|---|---|---|---|---|---|---|
| | Severe | 0.0 | 0.8 | 2.0 | 0.0 | 0.8 | 0.4 | 0.6 |
| | Unclear | 0.7 | 0.4 | 0.5 | 0.9 | 4.9 | 1.7 | 1.2 |
| | Total | 100 | 100 | 100 | 100 | 100 | 100 | 100 |
| Vanish prospect | None | 47.7 | 41.4 | 36.3 | 35.5 | 34.1 | 33.5 | 38.9 |
| | Sometimes | 39.0 | 43.0 | 38.7 | 46.1 | 41.5 | 41.1 | 41.5 |
| | Often | 9.4 | 9.9 | 15.7 | 12.0 | 15.4 | 17.8 | 13.0 |
| | Severe | 3.2 | 4.6 | 8.3 | 3.7 | 4.1 | 5.1 | 4.8 |
| | Unclear | 0.7 | 1.1 | 1.0 | 2.8 | 4.9 | 2.5 | 1.9 |
| | Total | 100 | 100 | 100 | 100 | 100 | 100 | 100 |
| Feeling lonely | None | 57.4 | 60.1 | 52.9 | 51.6 | 49.6 | 49.2 | 54.1 |
| | Sometimes | 36.8 | 31.9 | 32.8 | 41.0 | 31.7 | 38.1 | 35.7 |
| | Often | 5.4 | 6.8 | 9.8 | 6.0 | 12.2 | 9.3 | 7.8 |
| | Severe | 0.4 | 0.8 | 4.4 | 0.5 | 1.6 | 2.1 | 1.5 |
| | Unclear | 0.0 | 0.4 | 0.0 | 0.9 | 4.9 | 1.3 | 0.9 |
| | Total | 100 | 100 | 100 | 100 | 100 | 100 | 100 |
| Feeling useless | None | 57.8 | 60.1 | 57.4 | 50.7 | 48.8 | 66.5 | 57.7 |
| | Sometimes | 35.0 | 31.6 | 31.9 | 36.4 | 38.2 | 28.4 | 33.2 |
| | Often | 5.4 | 5.7 | 6.4 | 9.2 | 7.3 | 3.4 | 6.1 |
| | Severe | 1.1 | 1.9 | 3.9 | 1.8 | 1.6 | 0.4 | 1.7 |
| | Unclear | 0.7 | 0.8 | 0.5 | 1.8 | 4.1 | 1.3 | 1.3 |
| | Total | 100 | 100 | 100 | 100 | 100 | 100 | 100 |

(continued)

**Table 2.38** (continued)

| Items | | Guangzhou | Dongguan | Shenyang | Chengdu | Hangzhou | Zhengzhou | Total |
|---|---|---|---|---|---|---|---|---|
| Feeling that life is hard | None | 42.2 | 40.7 | 32.4 | 39.2 | 43.1 | 37.3 | 39.1 |
| | Sometimes | 41.2 | 48.3 | 47.1 | 41.0 | 33.3 | 44.9 | 43.4 |
| | Often | 13.0 | 8.7 | 14.7 | 14.7 | 12.2 | 12.7 | 12.6 |
| | Severe | 3.2 | 2.3 | 5.9 | 3.7 | 8.1 | 5.1 | 4.3 |
| | Unclear | 0.4 | 0.0 | 0.0 | 1.4 | 3.3 | 0.0 | 0.6 |
| | Total | 100 | 100 | 100 | 100 | 100 | 100 | 100 |
| Samples | | 277 | 263 | 204 | 217 | 123 | 236 | 1,320 |

**Table 2.39** Top three indicators for satisfactory selected by labour immigrants (Unit:%)

| Items | Guangzhou | Dongguan | Shenyang | Chengdu | Hangzhou | Zhengzhou | Total |
|---|---|---|---|---|---|---|---|
| Filial piety of children | 55.2 | 57.8 | 39.2 | 53.0 | 57.7 | 50.0 | 52.2 |
| Family happiness | 44.8 | 41.1 | 53.9 | 34.6 | 42.3 | 57.2 | 45.8 |
| Being respected by others | 22.0 | 18.6 | 21.1 | 16.6 | 18.7 | 22.5 | 20.1 |
| Accomplishment | 14.8 | 15.6 | 23.0 | 20.3 | 19.5 | 19.9 | 18.5 |
| Wealth life | 15.9 | 11.8 | 17.2 | 12.0 | 18.7 | 16.9 | 15.1 |
| Prospect fulfilled | 14.4 | 11.4 | 20.6 | 11.5 | 15.4 | 16.5 | 14.8 |
| Health | 15.2 | 13.3 | 13.7 | 12.9 | 13.8 | 17.4 | 14.5 |
| Contentment | 11.9 | 13.3 | 6.4 | 29.5 | 20.3 | 5.1 | 13.8 |
| Having good friends | 10.1 | 11.0 | 12.7 | 11.5 | 12.2 | 8.5 | 10.8 |
| Contribution to the society | 7.9 | 9.9 | 10.3 | 8.8 | 13.0 | 8.5 | 9.4 |
| Make a fortune | 6.1 | 9.1 | 5.4 | 4.6 | 10.6 | 6.4 | 6.8 |
| Good interpersonal relationship | 0.0 | 0.8 | 0.0 | 0.5 | 0.0 | 0.0 | 0.2 |
| Others | 52.0 | 56.3 | 58.3 | 60.8 | 56.1 | 69.5 | 58.8 |
| Non-repeat number | 100 | 100 | 100 | 100 | 100 | 100 | 100 |
| Total | 270 | 270 | 282 | 277 | 298 | 298 | 281 |
| Samples | 277 | 263 | 204 | 217 | 123 | 236 | 1,320 |

# Chapter 3
# Intellectual Immigrants

With the expansion of higher education and the reform of employment policies for college graduates, a new group of immigrants—intellectual immigrants—is forging. This chapter first makes a brief analysis of the reasons for the formation of intellectual immigrants, and then describes the living conditions of this immigrant group in combination with the actual investigation.

## 3.1 The Forming of Intellectual Immigrants

Intellectual immigrants refer to those who have received higher education (college degree or above, master's degree or below) and work in cities but do not have a permanent residence in the city. Under the background of the increasingly tense employment form of college graduates, intellectual immigrants have become an important new immigrant group, and their living conditions have attracted more and more attention. In brief, the formation of intelligent immigration is related to the expansion of college enrolment and the reform of employment policy for college graduates.

### *3.1.1 The Impact of College Enrolment Expansion*

In 1999, the Communist Party of China (CPC) Central Committee and the State Council held a national conference on education, which discussed and approved the decision to expand college enrolment. In that year, the enrolment of undergraduate and junior college students in ordinary institutions of higher learning reached 1,548,554, an increase of 42.9% over that of 1998. In 1997, the gross enrolment rate of China's higher education was only 5%. By 2002, the gross enrolment rate of China's higher education had reached 15%. China's higher education began to enter


D. Zhou, *Urban Migrants in China*,
https://doi.org/10.1007/978-981-99-3114-9_3

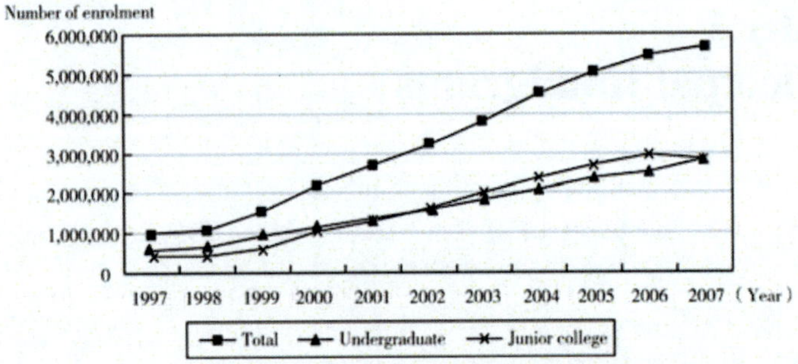


**Fig. 3.1** Changes in the enrolment of undergraduate and junior college students in Chinese universities from 1997 to 2007. *Source* Statistics from the website of the Ministry of Education, http://www.moe.edu.cn/

the development stage of popularization.[1] In 2008, the number of undergraduate and junior college students in China has reached 6,076,600, 5.6 times higher than that in 1998 before the enrolment expansion (see Fig. 3.1 for the changing trend of undergraduate and junior college students in China from 1997 to 2007).

With the enrolment expansion of colleges and universities, the enrolling proportion of college students from a rural area is increasing greatly. The research of some scholars shows that from 1989 to 2008, the proportion of rural enrolment in colleges and universities in China increased year by year. In 1989, the proportion of rural freshmen in colleges and universities in China was 43.4%. In 2003, the rural enrolment and urban enrolment were almost the same. In 2005, the proportion of rural students reached 53%.[2] *Guangming Daily* also reported that in 1998, 1.08 million students were enrolled, 400 thousand of them are from the rural area, accounting for 37%. In 2005, 4.47 million students were enrolled, among whom 2.3 million were from rural areas, accounting for 51%. The absolute number of rural children going to university has risen nearly six-fold, outpacing urban youth for the first time. The increase in rural college students is also reflected in some provinces. Take Hubei province as an example. In 2006, 530,000 students applied for the National College Entrance Examination (NCEE), rural students accounting for 51% of the total applicant (270,000). In 2007, 500,000 students applied for the NCEE, rural students accounting for 59% of the total applicant (290,000). In 2008, 520,000 students applied for the NCEE, of whom 330,000 were in rural areas, accounting for 63%.[3]

[1] Xia Fei, A Rational and Objective View on University Enrolment Expansion, *Guangming Daily*, December 10, 2008, retrieved at May 21, 2020, https://epaper.gmw.cn/gmrb/html/2008-12/10/nw.D110000gmrb_20081210_1-05.htm?div=-1.

[2] Yang Dongping, Monitoring Educational Equity and Conducting Background Checks on Students' Families, retrieved at March 19, 2009, http://blog.tianya.cn/blogger/post_show.asp?BlogID=296700&PostID=16802752.

[3] Xia Fei, A Rational and Objective View on University Enrolment Expansion, *Guangming Daily*, December 10, 2008, retrieved at May 21, 2020, https://epaper.gmw.cn/gmrb/html/2008-12/10/nw.D110000gmrb_20081210_1-05.htm?div=-1.

It should be pointed out that although the proportion of rural college students in the overall population has increased, the proportion of rural college students in national key universities has declined. Data from China Agricultural University show that from 1999 to 2001, the proportion of new students in rural areas was around 39%. After 2002, it began to decline, and in 2007 it dropped to a low of only 31.2%. According to the data of Nankai University in recent three years, the proportion of rural freshmen was 30% in 2006, 25% in 2007, and 24% in 2008, with an obvious downward trend. In addition, the proportion of rural freshmen in Peking University and Tsinghua University in previous years was less than 20%.[4]

While the proportion of rural students in national key universities has declined, the number of rural students in the second batch and third batch of universities has increased substantially. Some people once randomly selected four newly promoted undergraduate colleges in Anhui province and investigated 360 normal graduates of some specialized majors in 2008. According to the survey, only 14.7% of junior college students come from cities, 66% from rural areas and 19.3% from small towns. It can be seen that the proportion of college students from rural areas is much larger than that from urban areas. In fact, after the implementation of the expansion of China's institutions of higher learning, compared with the undergraduate enrolment, the number of specialized college enrolment grew relatively fast. As can be seen from Fig. 3.1, compared with 1998 before the enrolment expansion, in 2007, the annual enrolment of undergraduate students in universities increased by 4.3 times, while the annual enrolment of specialized subjects increased by 6.6 times. Even in the undergraduate enrolment, the proportion of enrolment expansion is very low. For example, in 2008, Henan province was designated by the state to recruit 503,000 students, including 210,000 undergraduates and 293,000 specialists. Later, colleges and universities in Henan province expanded their enrolment by more than 10,000 students, but only 1,240 students were enrolled.[5]

Generally, since the late 1990s, with the expansion of college enrolment, the proportion of rural students enrol in college has increased significantly. At the same time, the proportion of college students from rural areas in national key universities has declined significantly, and more and more college students from rural areas are concentrated in specialized colleges and undergraduate universities. The expansion of college enrolment and the gathering of rural students in colleges and universities are important background for the formation of intellectual immigration.

---

[4] Yang Dongping, Monitoring Educational Equity and Conducting Background Checks on Students' Families, retrieved at March 19, 2009, http://blog.tianya.cn/blogger/post_show.asp?BlogID=296700&PostID=16802752.

[5] Henan, 513,000 Students were Admitted to Ordinary High School in 2008, retrieved at May 21, 2020, http://gaokao.chsi.com.cn/gkxx/ss/200808/20080819/7954003.html.

### 3.1.2 The Impact of Graduate's Employment Policy Reform

The formation of intellectual immigration is also related to the reform of the employment policy for college graduates. After the founding of the People's Republic of China in 1949, the state implemented the policy of "deployment and allocation" for college graduates. Under this policy, according to the needs of social development, the government departments in charge of the distribution of college graduates according to the unified deployment. This policy, on the one hand, restricts the right of graduates to choose employment, and on the other hand, ensures that every college graduate can get a job. The policy of "deployment and allocation" continued to 1980s, in a document called "Measures for Management of Students in Full-time Higher Education Institutions" issued by the Ministry of Education in 1983, the graduated student must obey the state's deployment and allocation, and must report on time to the workplace one have been deployed to. For those who, in disregard of the needs of the state, insist on unreasonable demands of individuals and refuse to comply with the deployment after criticism and education, fail to report for duty within three months after the date of the announcement of the deployment ordered by the school, the school shall, upon the approval of the local deployment department, announce the cancellation of the deployment and order the student to leave the school within a time limit.

The "deployment and allocation" policy was gradually waiving along with the reform and opening up of China. In 1985, the CPC Central Committee issued "Decision of Reform on Educational System", the Decision clearly point out that reform the enrolment and deployment system of tertiary schools, given autonomy for schools in administration, for the majority of student sponsored by national fund, the guideline is under the guidance of the state plan, a system shall be implemented whereby the candidates shall apply for voluntary employment, be recommended by the school, and be selected on the basis of merit by the employer; for the minority of student who are self-sponsored, the guideline is after graduation, students can either be deployed by the school or can seek a job by oneself. In 1989, the State Council approved the Reform Plan for the Distribution System of College Graduates. Targeted enrolment will be gradually reduced. The plan states that in line with the long-term reform direction, graduates will choose their careers mainly through the talent (labour) market. In 1993, the CPC Central Committee and the State Council issued the Program for China's Education Reform and Development, the Program points out that the reform of graduates of institutions of higher learning "deployment and allocation" employment system, "besides some necessary deployment to pedagogical position, remote area and hardship industry, the majority of graduate will practice job seeking through labour market, independently choosing their employment under the guidance of national policies and regulations". The National Education Development Plan for the Ninth Five-Year Plan released in 1996 pointed out that "the autonomously of graduates and employer will be achieve in near future, except for those students who are under sponsorship contract for designated careers or positions, other students will basically autonomously choose their employers or be choose by the employers, under the guidance of national policies and regulations".

Since 1996, the employment policy for China's college graduates is shifting from "deployment and allocation" to "autonomously choose their employers or be choose by the employers". The changing of the employment policy has a profound influence on China's students, under the policy of "deployment and allocation", there is hardly any unemployment issue, nearly all graduates are guarantee a stable job. However, under the "autonomously choose their employers or be choose by the employers" policy, the employment is not a sure thing, especially in recent years, the pressure on employment is getting higher, the unemployment rate for graduates is growing fast, and the date released by the ministry of education shows the employment rate in 1996 is 93.7%, in 2009, the employment rate has dropped to 68%, Fig. 3.2 shows the change of employment rate for graduates from 1996 to 2009.

With the decline of employment rate, more and more college students have difficulty in finding suitable jobs. Among them, from the rural graduates from non-key universities account for a large proportion. In the traditional view, these children from the countryside go to college, which means they have an "iron rice bowl" in the city. However, faced with fierce competition in the job market, these college students from rural areas lack the necessary social network and find it difficult to find a stable job in the city. Even if some graduates do find jobs, they are mostly employed in the informal sector in cities, where they struggle to obtain household registration. In this sense, they are the same as traditional rural migrants—working in the city but

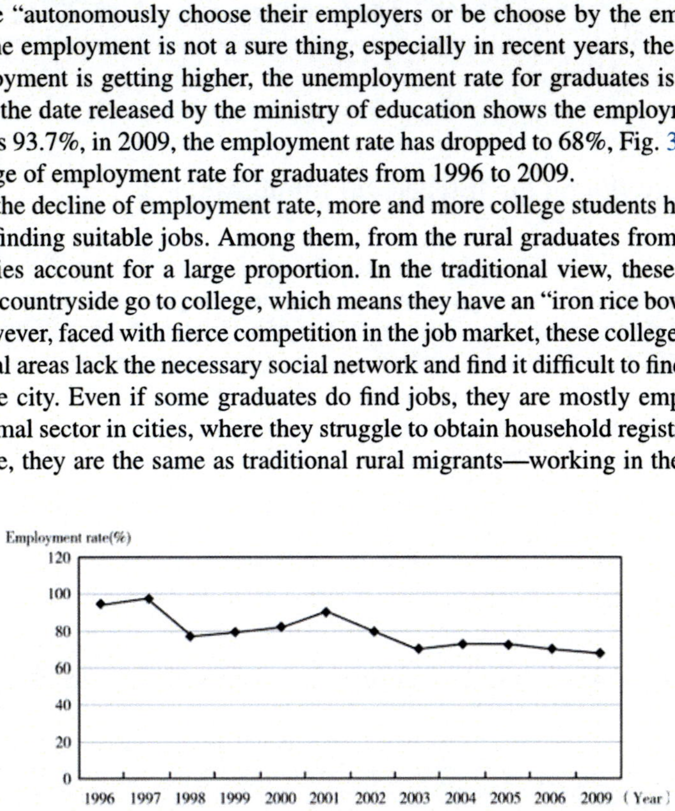


**Fig. 3.2** The change of employment rate for graduates from 1996 to 2009. *Source* Employment data from 1996 to 2002, "One million college students unemployed in 2003", *Economic Information Daily*, January 30, 2004; The employment rate from 2003 to 2004 was obtained from the following sources, 73% of the employment rate of college graduates this year, published in *People's Daily* (*overseas edition*) on September 30, 2004. Employment rate data in 2005 are from, "Zhou Ji Minister's Speech at the 2006 National College Graduates Employment Work Conference", contained in *Circular of the Ministry of Education*, December 8, 2005; The data of employment rate in 2006 are from, "Chen Zhili's Speech on the Teleconference on Employment of College Graduates in 2007", *China Education Daily*, April 26, 2007. The above data were all quoted from Lai Desheng, Government Responsibilities to Alleviate the Employment Dilemma of College Students, in *Chinese College Student Employment*, 8th issue, 2008. Employment in 2009 released by the Ministry of Education information, see the website, http://learning.sohu.com/20090812/n265900990.shtml

not with the city household registration, even though their status has shifted from agricultural to urban. But they don't want to go back to their places of origin, they don't want to go back to the countryside. Because at the beginning they chose to study is to change the fate, out of the countryside. In order to stay in the city, they have to work temporarily in informal sectors such as private enterprises. They make up the bulk of intellectual immigration.

Generally, the expansion of college enrolment, the increase of rural students, the reform of the employment system for college graduates, and the tension of the employment form of college graduates are the important reasons for the formation of intellectual immigration.

## 3.2 Livelihood for Intellectual Immigrants

In order to grasp the development trend of urban immigrants, from November 2008 to September 2009, we, the research team organized a survey of urban immigrants in Guangzhou, Dongguan, Shenyang, Chengdu, Hangzhou, Zhengzhou and other places. This survey regards intellectual immigrants as an important type of urban immigrants, and 984 valid questionnaires were collected. At the same time, the research group also conducted individual case interviews of intellectual immigrants in the above-mentioned cities. Based on these data, this chapter briefly describes the living conditions of intellectual immigrants.

### *3.2.1 Basics*

Our survey shows the intellectual immigrants are gender balanced, 49.19% of who are male, and 50.81% are female (see Fig. 3.3).

The age distribution of intellectual immigrants is concentrated between 21 and 30, which accounted for 91.87% of the intellectual immigrant, with only 0.81% age over 40 (see Fig. 3.4).

Most of our sampled intellectual immigrants are not yet married by the time we conducted the survey, mainly because they are newly graduated from colleges, and their jobs are still instable. 80.08% of the intellectual immigrant are not married (see Fig. 3.5).

Among them, 35.37% of the intellectual immigrants have lived in the city for 2 years, and 29.98% have lived in the city for 3 years. The proportion of the intellectual immigrants who have lived in the city for 4 years and 5 years are 16.57% and 18.09% respectively (see Fig. 3.6).

If we divide the reasons for immigration into three categories: residence, work and study, the survey shows that most of the intellectual immigrants move to cities for work, with the proportion reaching 69.41%, and 25.51% move to cities for study,

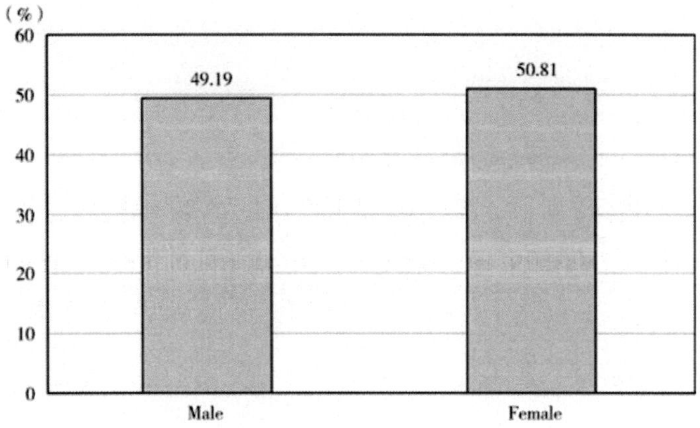


**Fig. 3.3** Gender ratio for intellectual immigrants

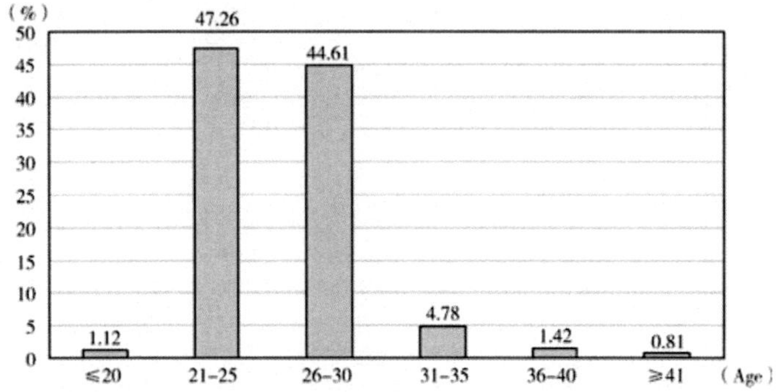


**Fig. 3.4** Age distribution for intellectual immigrants

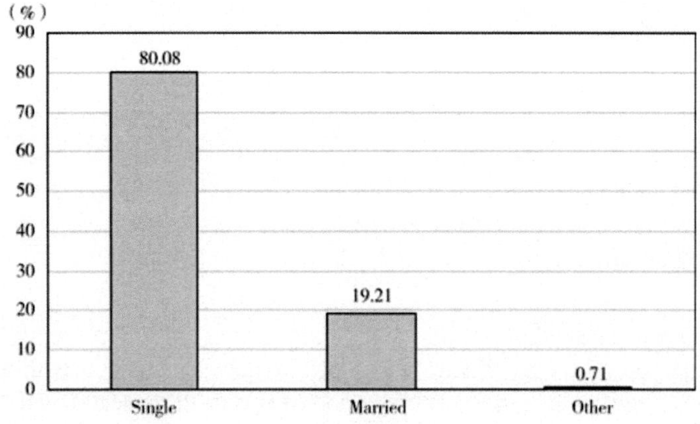


**Fig. 3.5** Marriage status for intellectual immigrants

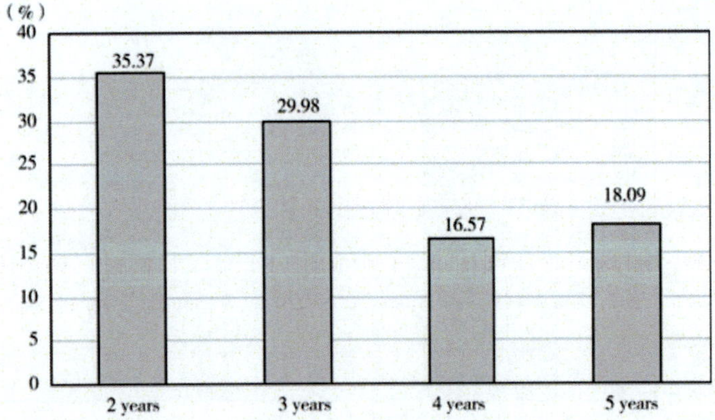


**Fig. 3.6** Immigration time for intellectual immigrants

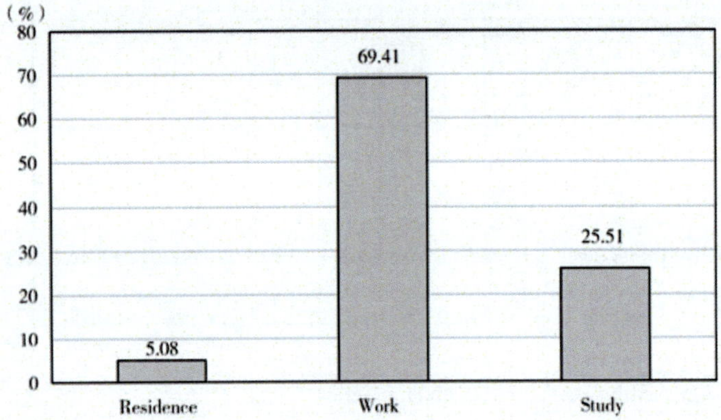


**Fig. 3.7** Initiative for migration

and stay in this city to work after graduation. Only 5.08% chose to settle directly. It can be seen that the main purpose of intellectual immigrants is to work (see Fig. 3.7).

### 3.2.2 *Livelihood*

In the survey, we divided the job stability into several types: "frequently changing" (with a period of six months or less), "occasionally changing" (with a period of one year), "rarely changing" (with a period of two or three years), and "stable and unchanged". According to the survey, only 29.88% of the intellectual immigrants have stable jobs, and most of them have varying degrees of job changes (see Fig. 3.8).

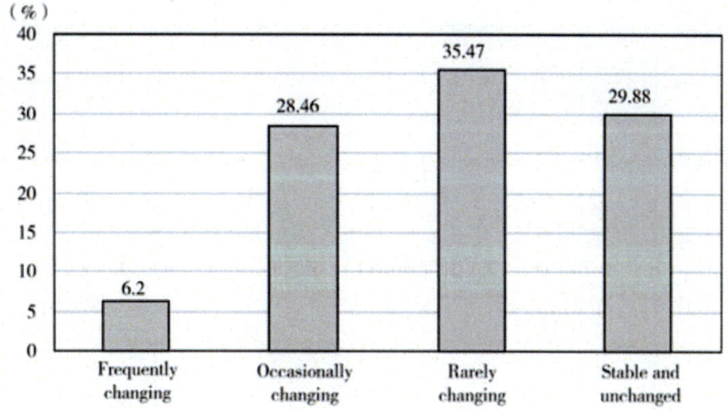


**Fig. 3.8** Job stability for intellectual immigrants

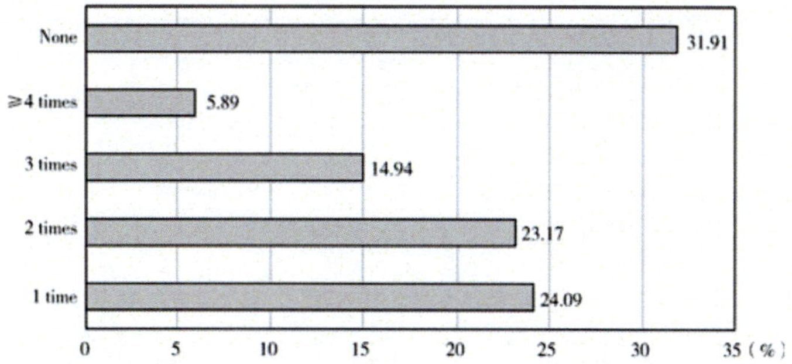


**Fig. 3.9** Job changing for intellectual immigrants

The number of job changes also proves this point. According to the survey, nearly 70% of the respondents in the cities where they currently work have changed jobs, and some have even changed jobs three or four times. Only 31.91% of the intellectual immigrants have not changed their job (see Fig. 3.9).

According to the survey, intellectual immigrants work an average of 5.5 days a week. More than half of all intellectual immigrants (52.42%) worked five or fewer days a week, 33.19% worked between 5.5 and 6.5 days a week, and 5.36% worked seven days a week (see Fig. 3.10).

On average, intellectual immigrants work 8.2 h per day, of which 72.58% work 8 h or less, and only 18.70% work more than 8 h per day. In general, the daily overtime of intellectual immigrants is not serious (see Fig. 3.11).

How do intellectual immigrants rate their job? According to the survey, the intellectual immigrants have a relatively high evaluation on the interpersonal relationship in the workplace, with 44.22% choosing to be relatively satisfied and 24.37% choosing to be very satisfied. Comparatively speaking, intellectual immigrants were dissatisfied with the opportunity for advancement in their jobs, with 13.55% choosing very dissatisfied, 23.32% choosing relatively dissatisfied, and only 4.52% choosing

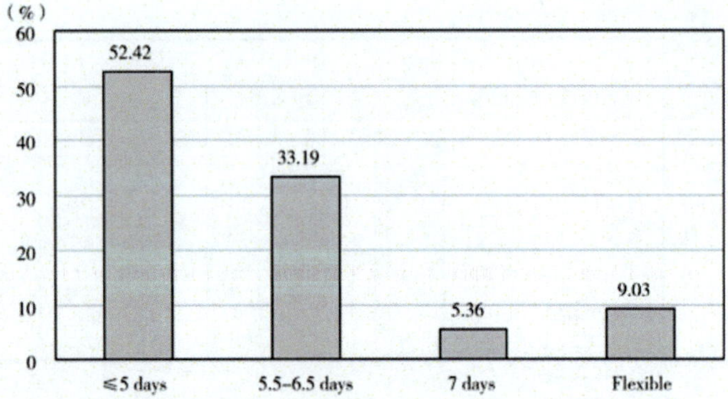


**Fig. 3.10** Working days for intellectual immigrants

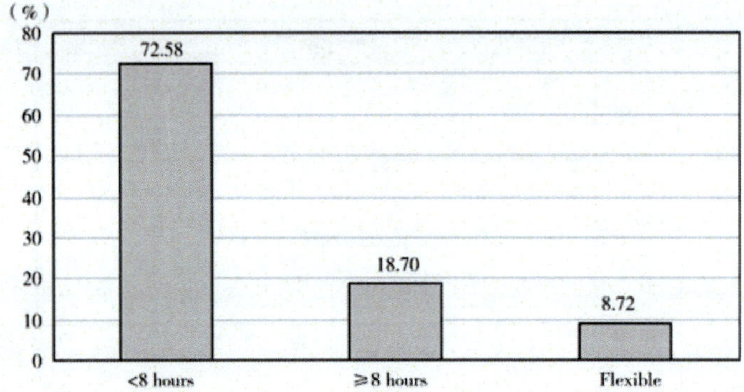


**Fig. 3.11** Working hours for intellectual immigrants

very satisfied. In terms of work pressure, the proportion of very dissatisfied was also relatively high, with 10.50% choosing very dissatisfied. In terms of income treatment, those who are very satisfied are the lowest, only 3.05%, and those who choose to be relatively satisfied are also very low, only 15.13%, and the sum of the two is 18.2% (see Table 3.1).

Overall, the percentage of intellectual immigrants who said they were very dissatisfied with their jobs was 1.26%, with more than half choosing average, 31.72% saying they were relatively satisfied and 8.8% saying they were very satisfied.

What factors influence the job choices of intellectual immigrants at the most? According to the survey, income is the most important factor for most intellectual immigrants when they find a job, with 41.70% believing that income is the most important factor in job finding. In addition, 27.63% of intellectual immigrants cited job prospects as the most important factor in choosing a job (see Fig. 3.12).

**Table 3.1** Job satisfaction evaluation for intellectual immigrants

| Satisfaction | Evaluation factors | | | | | | | |
|---|---|---|---|---|---|---|---|---|
| | Overall | Career prospect | Interpersonal relationship | Work pressure | Promotion opportunity | Social status | Working environment | Income |
| Very dissatisfied | 1.26 | 6.72 | 1.26 | 10.50 | 13.55 | 6.20 | 3.05 | 8.72 |
| Relatively dissatisfied | 8.09 | 15.02 | 4.73 | 14.81 | 23.32 | 19.01 | 12.29 | 19.64 |
| Average | 50.95 | 43.17 | 25.42 | 37.61 | 42.33 | 47.79 | 43.38 | 53.47 |
| Relatively satisfied | 31.72 | 25.42 | 44.22 | 23.42 | 16.28 | 22.90 | 31.83 | 15.13 |
| Very satisfied | 7.98 | 9.66 | 24.37 | 13.66 | 4.52 | 4.10 | 9.45 | 3.05 |

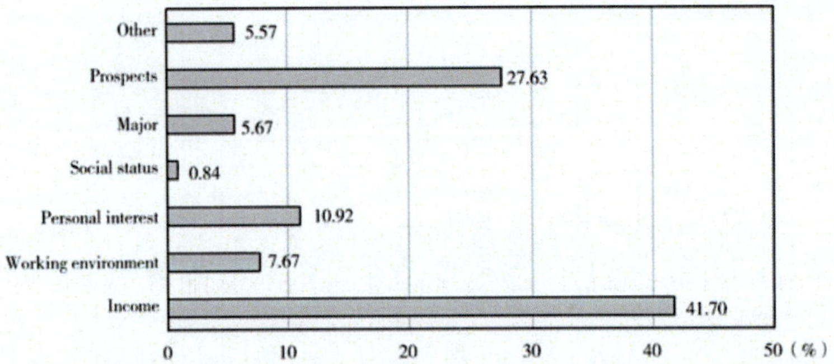


**Fig. 3.12** Factors on job finding

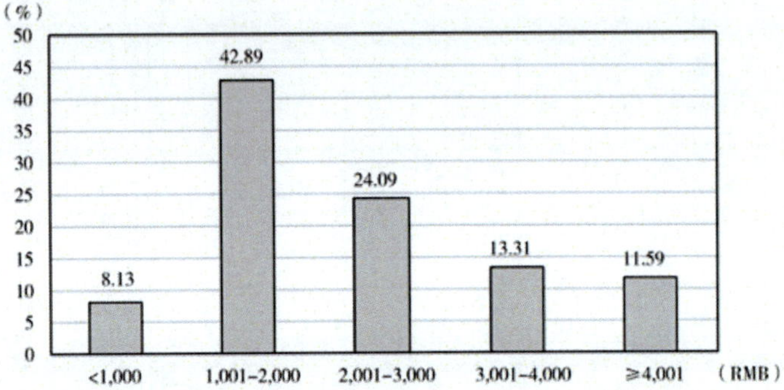


**Fig. 3.13** Monthly income for intellectual immigrants

### *3.2.3 Financial Balance*

Financial status is the most important factor to the living condition for intellectual immigrants. The survey shows more than half (51.0%) of the intellectual immigrant has the monthly income lower than RMB2,000, and 8.13% of them is lower than RMB1,000, only 11.59% monthly income exceeding RMB4,000. Overall, even for intellectual immigrants, the salary is relatively low (see Fig. 3.13).

In what sections do intellectual immigrants spend most of their money? Surveys show that food, housing and clothing are the main expenses of intellectual immigrants. In the first-place expenditure, 30.49% of intellectual immigrants chose housing rent, while 23.68% chose food. In the second-place, 27.64% of intellectual immigrants chose food, 15.35% chose housing rent and 14.53% chose clothing. In the third-place spending, 15.85% of intellectual immigrants chose food and 15.35% chose clothing. It can be seen that the expenditure of intellectual immigrants is mainly spent on "necessities of life" (see Table 3.2).

**Table 3.2** The expenditure percentage for intellectual immigrants (Unit: %)

| Expenditures | 1st place | 2nd place | 3rd place |
|---|---|---|---|
| Food | 23.68 | 27.64 | 15.85 |
| Clothing | 9.55 | 14.53 | 15.35 |
| Housing mortgage | 11.69 | 2.34 | 1.02 |
| Housing rent | 30.49 | 15.35 | 6.91 |
| Commute | 1.42 | 3.86 | 5.28 |
| Child support | 2.95 | 3.05 | 1.42 |
| Parent support | 1.83 | 3.15 | 4.78 |
| Interpersonal socialising | 8.33 | 9.55 | 12.60 |
| Learning/training | 2.85 | 3.96 | 5.39 |
| Cosmetics | 0.81 | 3.05 | 3.25 |
| Recreation | 3.76 | 7.32 | 11.79 |
| Health | 0.51 | 1.12 | 1.83 |
| Communications (Internet included) | 1.42 | 3.76 | 12.30 |
| Insurance | 0.20 | 0.91 | 1.73 |
| Others | 0.51 | 0.41 | 0.51 |

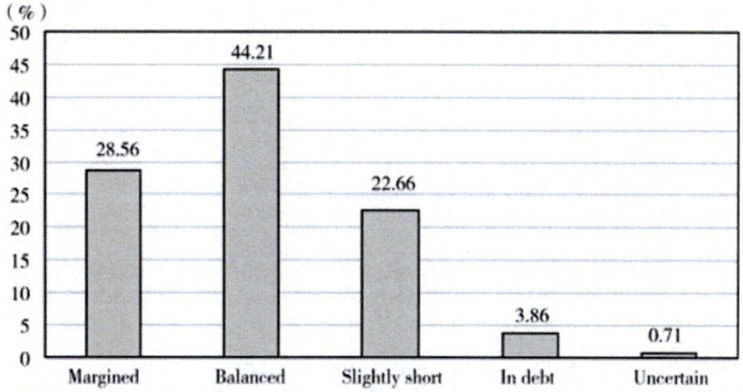


**Fig. 3.14** Financial balance for intellectual immigrants

In the balance of income/expense, the survey shows 28.56% intellectual immigrants is "margined", and 44.21% is "balanced", 22.66% felt they are "slightly short", and 3.86% found themselves "in debt". Therefore, the majority of the intellectual immigrants do not have a margined income (see Fig. 3.14).

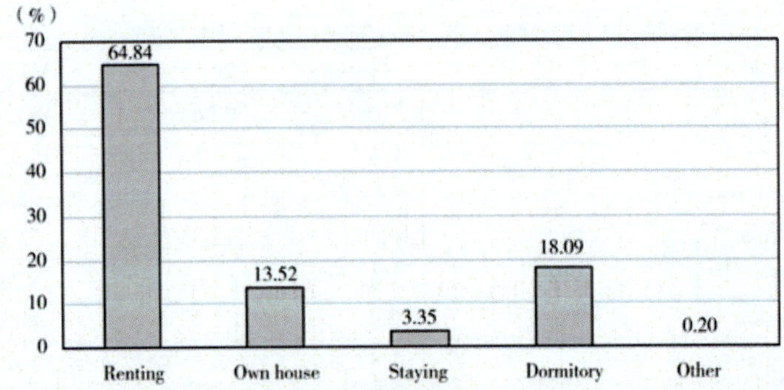


**Fig. 3.15** Housing condition for intellectual immigrants

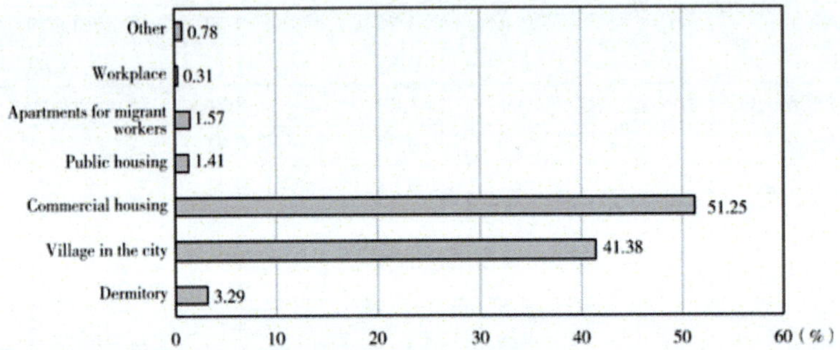


**Fig. 3.16** Renting condition for intellectual immigrants

### 3.2.4 *Housing*

The housing situation is an important factor for living conditions. The survey shows 64.84% of the intellectual immigrants renting, and 18.09% of them living in dormitory, only 13.52% owning their house (see Fig. 3.15).

Among the intellectual immigrants who rent for housing, more than half (51.25%) living in commercial residence, 41.38% living in urban village. Therefore, most of the intellectual immigrants are not "dwelling" in the urban villages (see Fig. 3.16).

### 3.2.5 *Social Intercourse*

The survey shows that the leisure companions for intellectual immigrants are mainly friends, schoolmates and colleagues. 84.62% of the intellectual immigrants chose their friends as their usual leisure companions, 79.33% chose their schoolmates as their usual leisure partners, and 78.92% chose their colleagues as their usual leisure friends (see Fig. 3.17).

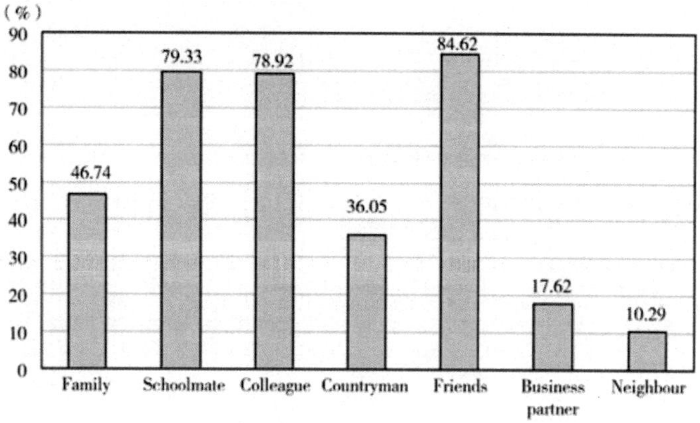


**Fig. 3.17** Companions leisure for intellectual immigrants

**Table 3.3** Three best friends for intellectual immigrants (Unit: %)

| Relationship | 1st best friend | 2nd best friend | 3rd best friend |
|---|---|---|---|
| Business partner | 1.32 | 2.66 | 2.86 |
| Schoolmate | 77.35 | 71.99 | 64.46 |
| Countryman | 18.54 | 23.81 | 24.40 |
| Neighbour | 2.78 | 2.52 | 6.93 |
| Friend | 13.77 | 10.08 | 11.60 |
| Others | 0.40 | 1.68 | 1.05 |

When asked to list their three best friends, most intellectual immigrants chose their schoolmates as their best friends. Among them, 77.35% of the intellectual immigrants chose their schoolmates as their first friends, 71.99% of the intellectual immigrants chose their schoolmates as their second friends, and 64.46% of the intellectual immigrants chose their schoolmates as their third friends. In addition to schoolmates, the second-best friend was fellow countrymen. 18.54% of the intellectual immigrants chose fellow countrymen as their first best friend, 23.81% chose fellow countrymen as their second-best friend, and 24.40% chose fellow countrymen as their third best friend (see Table 3.3).

### 3.2.6 Recreation

The survey shows most intellectual immigrants take their leisure time online with PC, at 90.7%, 80.4% choose to watch television or video for leisure, and 72.8%

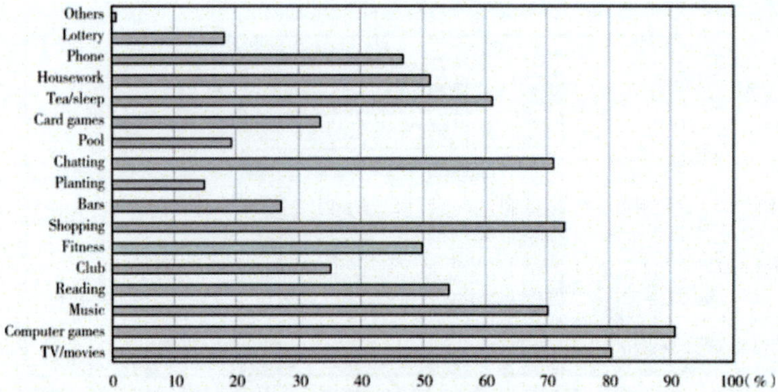


Fig. 3.18 Recreational activities for intellectual immigrants

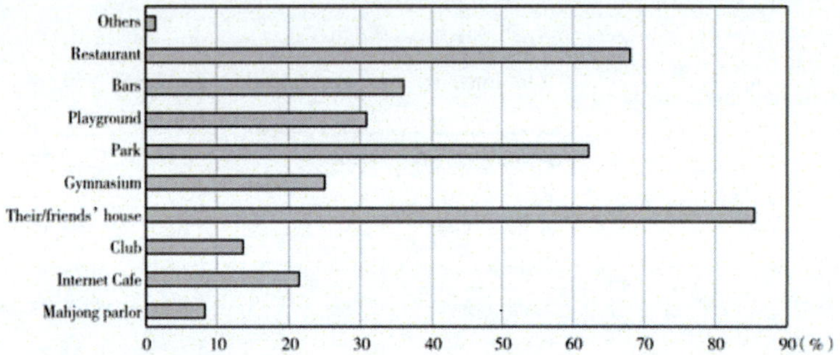


Fig. 3.19 Occasions for recreation

take shopping as their primary option, while 71.0% visit friends and chatting, 70.1% choose listening to music or radio (see Fig. 3.18).

As for the occasions for recreation, 85.6% of intellectual immigrants stay at their house or friends' house for leisure, and 68.0% choose a public place like restaurants for recreational occasion, and 62.2% go to parks for leisure (see Fig. 3.19).

### *3.2.7 Social Welfare and Public Insurance*

Survey shows that in terms of social welfare and public insurance, intellectual immigrants pay the most attention on housing welfare, 61.89% of intellectual immigrants take housing welfare as their priority, 17.68% of intellectual immigrants take health welfare most seriously, 14.94% believe that education is the most concerned, only 3.8% take recreational welfare as the their priority (see Fig. 3.20).

In compare to locals, 11.92% of the intellectual immigrants think the coverage for locals and immigrants is absolutely different, and 35.65% of whom think the

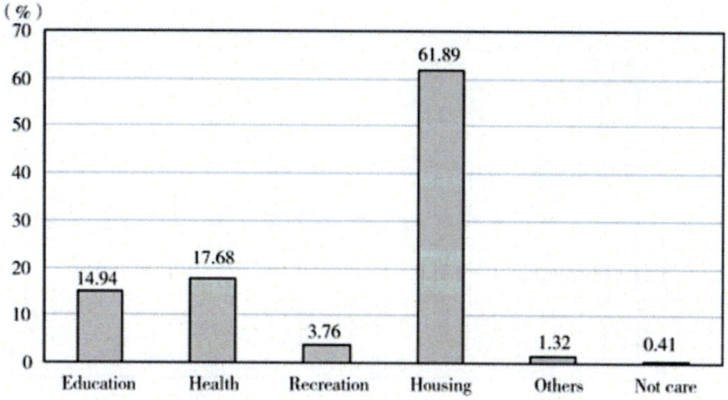


**Fig. 3.20** Social welfare and public insurance weighting for intellectual immigrants

**Table 3.4** The inequality on social welfare and public insurance for intellectual immigrants (Unit: %)

| Level | Education | Health | Recreation | Housing | Overall |
|---|---|---|---|---|---|
| Most inequality | 27.06 | 18.39 | 23.53 | 40.72 | 11.92 |
| Somehow inequality | 21.72 | 19.38 | 17.53 | 23.96 | 35.65 |
| Somehow equality | 30.55 | 32.71 | 30.32 | 19.57 | 42.38 |
| Most equality | 20.67 | 29.52 | 28.62 | 15.75 | 10.04 |

coverage is slightly different, over half of the intellectual immigrants consider the coverage for locals and immigrants is almost the same. As for the fields of coverage, intellectual immigrants feel most inequality on housing, and most equality on health (see Table 3.4).

As for public insurance, 77.44% of the intellectual immigrant was covered by health insurance, 66.87% was covered by pension insurance, 54.37% was cover by work injury insurance, 50.30% was covered by unemployment insurance, 31.40% was covered by maternity insurance, 14.13% was covered by commercial insurance, and 16.87% of intellectual immigrant was not covered by any insurance (see Fig. 3.21).

### *3.2.8 Community Participation*

The survey shows 55.39% of the intellectual immigrant never reach or reach by any community cadres, only 2.85% of them contact with the cadres regularly. 73.07% the of the intellectual immigrant feels the community cadres do not concern with them, moreover, 3.46% of the intellectual immigrant feels the community cadres

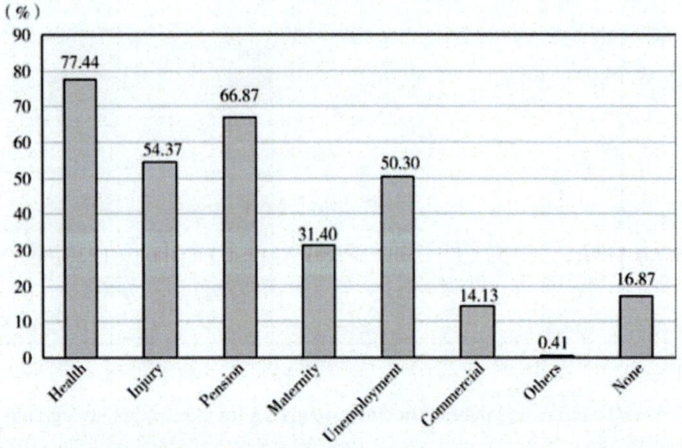


**Fig. 3.21** Public insurance coverage for intellectual immigrants

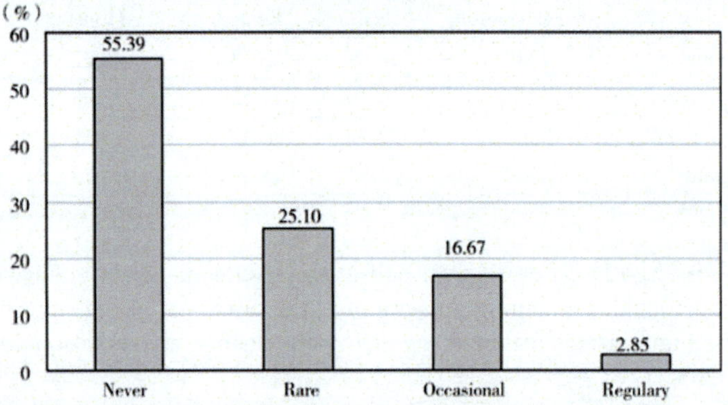


**Fig. 3.22** Contact with the community cadres for intellectual immigrants

discriminate against them, only 22.26% of the respondent consider the community cadres concern with them (see Figs. 3.22 and 3.23).

As for the willingness to participation on community affairs, nearly half (47.26%) of the intellectual immigrant said they would like to take part in the election for community committee, 30.08% said they would not, and 22.66% said the election does concern me at all (see Fig. 3.24).

For the neighbourliness, more than a third (33.84%) of the intellectual immigrant does not know their neighbour, 48.07% said they only have the meet-and-greet relation with their neighbour, only 7.62% said they often visit their neighbour, and 10.47% said they would help or seek help from their neighbour (see Fig. 3.25).

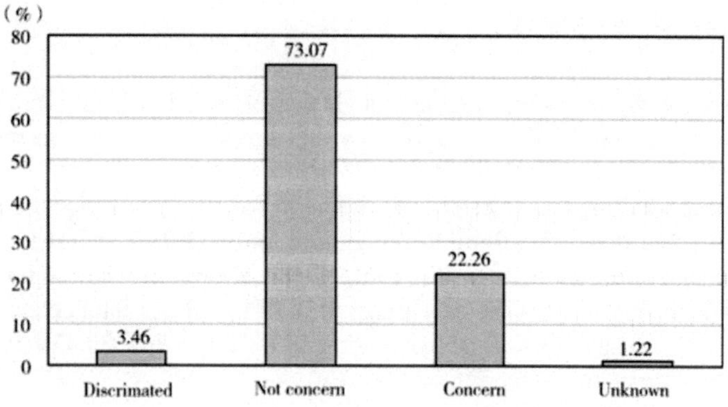


**Fig. 3.23** Community cadres' attitude towards intellectual immigrants

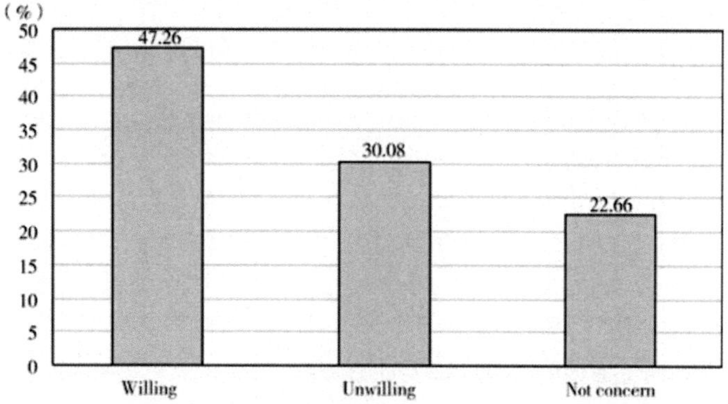


**Fig. 3.24** Willingness to participate in the election for community committee

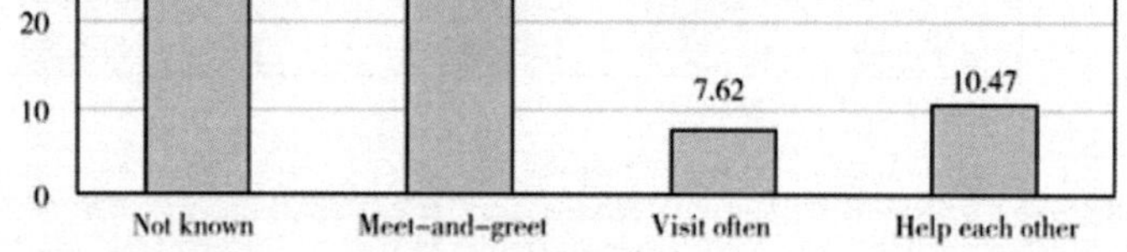


**Fig. 3.25** Neighbourliness for intellectual immigrants

### 3.2.9 Belongingness

45.53% of the intellectual immigrant thought they are non-local, only 16.87% thought they are local, and 32.42% thought they are temporary resident in the city (see Fig. 3.26).

As for the willingness to establish a home in the city, 63.81% of the intellectual immigrant said they are willing to do so, with only 12.93% said they not willing to establish a home in the city. However, those who thought they capable to do so is much less than those who are willing to, 47.41% of the intellectual immigrant thought they are capable to establish a home in the city, while 20.47% thought they are not capable to do so (see Fig. 3.27).

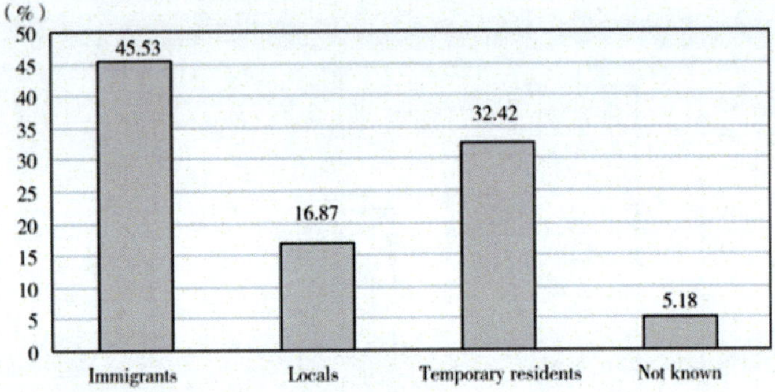


**Fig. 3.26** Belongingness for intellectual immigrants

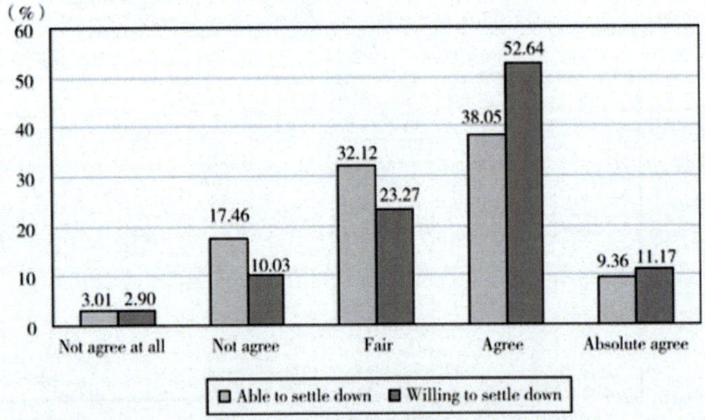


**Fig. 3.27** Willingness and capability to establishment in the city

**Table 3.5** Education level for labour immigrants

| Education level | Primary or below | Middle school | High school | Tertiary school | College or above |
|---|---|---|---|---|---|
| Percentage (%) | 7.9 | 40.2 | 49.6 | 1.7 | 0.6 |

## 3.3 Group Characteristics of Intellectual Immigrants

The above section generally described the livelihood for intellectual immigrants, compare to traditional rural migrants, the intellectual immigrants have some significant characteristics.

### *3.3.1 The Emergence of a New Group of Highly Qualified Immigrants*

As we can see from the timeline, the emergence of rural migrant was beginning at the 1980s, and cause a wide attention from the whole society. Compare to traditional labour immigrants, intellectual immigrants were emerging more recently. As the author discussed in previous section, intellectual immigrants were emerging on the background of the expansion of higher education enrolment, and the reform of employment policy for graduate. The critical reform of the employment and deployment policy for higher education graduate was began after mid-1990s, and expansion of higher education enrolment started from 1999. Therefore, the intellectual immigrants were emerged after the end of 1990s, and became more prominent at the early 2000s.

Rural migrants have relatively low education. In the national census 2005, 11.6% of rural migrants at primary education level, 48.6% at middle school level, 35.9% at high school level, only 3.9% at the college level or above.[6] Thus, rural migrant is low on education, our survey also proves the census, in our rural migrant respondents, 48.1% are at middle school level or below, 49.6% are at high school level, the sum is 97.7%, less than 3% are at the college level or above. The intellectual immigrants are at less at the college education level, relatively well educated for immigrants (see Table 3.5).

[6] Zheng Gongcheng, Wong Linda et al., ed., *Immigrant Workers' Problems and Social Protection in China*, Beijing: People's Publishing House, 2007, pp. 9–10.

**Table 3.6** Comparison on age for intellectual and labour immigrants (Unit: %)

| Age | Intellectual immigrants | Labour immigrants |
|---|---|---|
| 20 or below | 1.1 | 17.9 |
| 21–25 | 47.3 | 37.1 |
| 26–30 | 44.6 | 20.6 |
| 31–35 | 4.8 | 7.6 |
| 36–40 | 1.4 | 6.7 |
| 41 and above | 0.8 | 10.2 |

### *3.3.2 Immigrant in the Prime of Life*

The labour immigrant has a relatively even distribution of age, our survey shows labour immigrants age 21 or below take 17.9%, age 21–25 take 27.6%, age 26–30 take 20.7%, age 31–35 take 12.9%, age 35 or above take 20.9%, average age is 28.6, the oldest sample is age 65, and the youngest sample is age 15. Therefore, labour immigrant is fairly even distributed in ages.

Intellectual immigrants are a relatively young group compared with traditional labour immigrants, because the average age of college graduates is about 20 years old, and the average age of undergraduate graduates is about 22 years old. After drifting around the city for a few years, this group can generally find a stable job in the city, buy a house in the city, and finally integrate into the city and become the residents of the city.[7] Intellectual immigrants just "float" around the city after college and before they enter the local city, becoming a special immigrant group. Generally speaking, after working for a few years, these college graduates can go through the local entry formalities by purchasing real-estate.[8] Therefore, compared with the traditional labour immigrants, the age of this group is mostly concentrated between age 20 and 30. Our survey data confirm this, according to the survey, the age distribution of labour immigrants is relatively dispersed, while the age distribution of intellectual immigrants is relatively concentrated, 91.87% of intellectual immigrants are between age 20 and 30 (see Table 3.6).

[7] Ibid., p. 9.

[8] Besides Beijing, Shanghai and a few other big cities, as long as people can buy a fixed property in other cities, they can go through the local entry procedures with the property certificate. As a result, these intellectual immigrants can only buy real estate. Generally speaking, after working in the city for a few years, most of these intellectual immigrants can buy real estate in the city and obtain the household registration in the city where they live. In other words, after a few years in the city, these intellectual immigrants can eventually become true city resident.

### 3.3.3 A Group Emphasis on Schoolmate Relation

The labour immigrant has a traditional social network, most of their social network covers their countrymen and relatives. The intellectual immigrant's social network emphasis on schoolmate relation. In our survey, the responds to "leisure companion" (multiple choice question) is differed for intellectual and labour immigrants, 79.3% of intellectual immigrants chose schoolmates, 46.7% chose family or relatives, 36.1% chose countrymen. The percentage for labour immigrants is 38.8%, 53.7%, 50.4% respectively, apparently, intellectual immigrants tend to spend time with their schoolmate more than labour immigrants (see Table 3.7).

In our survey, the question of "three best friends beside kinship" was asked, the result shows intellectual immigrants chose schoolmate much more than labour immigrants who tend to choose countryman (see Table 3.8).

**Table 3.7** Comparison on social network for intellectual and labour immigrants (Unit: %)

| Companion for leisure | Intellectual immigrants | Labour immigrants |
|---|---|---|
| Family and relatives | 46.7 | 53.7 |
| Schoolmate | 79.3 | 38.8 |
| Colleagues | 78.9 | 74.8 |
| Countryman | 36.1 | 50.4 |
| Friend | 84.6 | 76.8 |
| Business partner | 17.6 | 10.2 |
| Neighbour | 10.3 | 18.7 |
| Others | | 0.1 |
| None | | 0.5 |

**Table 3.8** Comparison on "three best friends" for intellectual and labour immigrants (Unit: %)

| Order | Intellectual immigrants | | Labour immigrants | |
|---|---|---|---|---|
| | Alternative | Percentage | Alternative | Percentage |
| The 1st place best friend | Schoolmate | 77.4 | Schoolmate | 48.6 |
| | Countryman | 18.5 | Countryman | 40.2 |
| The 2nd place best friend | Schoolmate | 72.0 | Schoolmate | 43.3 |
| | Countryman | 23.8 | Countryman | 41.0 |
| The 3rd place best friend | Schoolmate | 64.5 | Schoolmate | 39.8 |
| | Countryman | 24.4 | Countryman | 38.7 |

**Table 3.9** Comparison on recreational activities for intellectual and labour immigrants (Unit: %)

| Recreational activities | Intellectual immigrants | Labour immigrants |
|---|---|---|
| Watch TV or video | 80.4 | 77.4 |
| Online with PC | 90.7 | 61.4 |
| Listen to radio or music | 70.1 | 64.6 |
| Reading, photography or calligraphy | 54.2 | 34.2 |
| Night club, Karaoke | 35.1 | 21.7 |
| Sports, exercise | 49.9 | 30.2 |
| Shopping | 72.8 | 63.3 |
| Go to barber shop or beauty salon | 27.1 | 18.7 |
| Gardening, flower or pet | 14.9 | 12.8 |
| Visit friends and chatting | 71.0 | 62.4 |
| Billiard | 19.3 | 17.2 |
| Play chess, cards or mah-jong | 33.4 | 31.4 |
| Drinking tea or eating, sleeping or just being idle | 61.1 | 50.6 |
| Chore | 51.1 | 49.5 |
| Smartphone | 46.9 | 54.2 |

### 3.3.4 A Fashionable Group of Immigrants

The intellectual immigrant is a group of high educated and young people, their lifestyle is relatively fashionable compare to labour immigrants. The intellectual immigrant generally chose more options for recreational activity than labour immigrants, 90.7% of intellectual immigrants chose to "online with PC", 54.2% of them chose "reading, photography or calligraphy", 35.1% chose to go to "night club, karaoke", 49.9% chose "sport, exercise", 72.8% chose "shopping" (see Table 3.9).

The fashionable lifestyle for intellectual immigrants can be seen from the occasions for recreation, compare to labour immigrants, intellectual immigrants are more likely to spend their leisure time in parks, cafe or teahouse, playground, restaurant, and gymnasium (see Table 3.10).

### 3.3.5 A Group of People with High Pressure

In our survey, intellectual immigrants were asked "What are the miseries in your live?" The answers were mostly "low income", "insufficient knowledge", "drab life", "unstable job". Compare to labour immigrants, intellectual immigrants were more troubled by "high pressure", "drab life", "lack of accomplishment", "no goal" (see Table 3.11).

**Table 3.10** Comparison on recreational occasions for intellectual and labour immigrants (Unit: %)

| Recreational occasions | Intellectual immigrants | Labour immigrants |
|---|---|---|
| Mah-jong house | 8.3 | 7.4 |
| Internet café | 21.4 | 31.1 |
| Ball room | 13.7 | 12.4 |
| Home, own or friends | 85.6 | 80.2 |
| Gymnasium | 25.1 | 9.5 |
| Park | 62.2 | 56.7 |
| Playground | 31.0 | 26.3 |
| Café or teahouse | 36.2 | 16.1 |
| Restaurant | 68.0 | 48.1 |

**Table 3.11** Comparison on stressful things for intellectual and labour immigrants (Unit: %)

| Stressful things | Percentage | |
|---|---|---|
| | Intellectual immigrants | Labour immigrants |
| Low income | 56.8 | 56.7 |
| Unstable job | 27.4 | 28.9 |
| High pressure | 47.2 | 36.4 |
| Lack of skills | 15.1 | 14.5 |
| Insufficient knowledge | 50.5 | 43.2 |
| Lack of accomplishment | 14.9 | 8.3 |
| Drab life | 40.0 | 29.5 |
| No goal | 16.9 | 5.7 |

In general, since the beginning of the twenty-first century, with the expansion of college enrolment, China's higher education has gradually entered the stage of popularization, and people have more opportunities to enter and receive university education. But today's college students are nothing like what they used to be. In the past, college students were rare and treasured by many. The "deployment and allocation" policy means one enrolled in a university means to have a stable job and to have an "iron rice bowl". However, today's college students, due to the reform of the employment system of higher education and the general employment pressure in the society, it is increasingly difficult for college graduates to find a stable and decent job. "Graduation is unemployment" has become a popular phrase in the society. This is especially true for undergraduates who have graduated from the second batch and third batch of universities, as well as junior college students from local universities.

In recent years, with the increase of employment pressure in colleges and universities, the survival status of this group has been widely concerned by the media. Chinese Academy of Social Sciences published the *Green Book of Population and Labor* (*2009*) points out, for the same of age of 21 and 22, college graduates and young immigrant workers salary is roughly the same at "one thousand yuan" level, the average wage for young labour immigrant even slightly higher than that of college graduates, in some fields, labour immigrants and young graduates are competitors for position.

Upon these circumstances, intellectual immigrants are "floating" in the cities, compromised to some low income, unstable and temporary jobs, many of them working in the informal sectors. Compare to local residents, intellectual immigrants are similar to labour immigrants, they do not have urban household registration, lack of social security, insufficient welfare coverage. These new immigrants, the intellectual immigrant, even they have received higher education, their immigrant status is no differed from traditional labour immigrants. Although the academics and medias have begun to pay attention to the livelihood for graduates, the in-depth studies on these particular immigrants were still absent. This research based on relevant data, analyses the causes of this group, and briefly describes the living conditions and group characteristics of this group. The group characteristics, identity, social participation and integration process of intelligent immigrants need further in-depth analysis.

# Chapter 4
# Comparison on Labour Immigrants, Intellectual Immigrants and Entrepreneurial Immigrants

As stated in Chapter 1, urban new immigrants can be divided into labour, intellectual and entrepreneurial, three types of immigration. In Chapters 2 and 3, we have outlined the sketch of labour immigrants and intellectual immigrants. In this chapter, we will take the labour and intellectual immigrant as the reference groups, make a comparative analysis on the three types of immigrants, so as to highlight the characteristics of various types of immigrants.

## 4.1 Discrepancy on Demographic Characteristic

The average age of the labour immigrant is 27.45, much older than the intellectual immigrant, and the entrepreneurial immigrant is the oldest among three, the average age 31.52, and 43.87% of them is over 31 (see Table 4.1). As for marriage, 68.52% of entrepreneurial immigrants are married, which also means more likely with child; intellectual immigrants are mostly unmarried, only 19.21% of them are married. The labour immigrant can be further divided into senior and junior group, junior labour immigrants have the lowest marriage rate of 16.99%, and senior labour immigrants have the highest marriage rate of 89.05%, only 10.24% of the senior labour immigrant without any child (see Tables 4.2 and 4.3). The gender ratio for labour and intellectual immigrants are fairly even, entrepreneurial immigrants are lean to male. As for education, most of labour immigrants are middle school or high school level, and the intellectual immigrant is above tertiary education, the entrepreneurial immigrant is dispersedly distributed in all levels of education (see Table 4.4).


D. Zhou, *Urban Migrants in China*,
https://doi.org/10.1007/978-981-99-3114-9_4

**Table 4.1** Age distribution for urban new immigrants (Unit: %)

| Age distribution | Intellectual immigrants | Labour immigrants | Entrepreneurial immigrants |
|---|---|---|---|
| 20 or below | 1.12 | 17.88 | 3.01 |
| 21–25 | 47.26 | 37.05 | 19.56 |
| 26–30 | 44.61 | 20.61 | 33.56 |
| 31–35 | 4.78 | 7.58 | 17.13 |
| 36–40 | 1.42 | 6.67 | 14.00 |
| 41 or above | 0.81 | 10.23 | 12.73 |
| Total (N) | 984 | 1,320 | 864 |
| Average | 26.11 | 27.45 | 31.52 |

**Table 4.2** Marriage status for urban new immigrants (Unit: %)

| Marriage status | Intellectual immigrants | Labour immigrants | Entrepreneurial immigrants |
|---|---|---|---|
| Single | 80.08 | 60.38 | 30.56 |
| Married | 19.21 | 38.94 | 68.52 |
| Divorced/widowed | 0.71 | 0.68 | 0.93 |
| Total (N) | 984 | 1,320 | 864 |

**Table 4.3** Filial status for urban new immigrants (Unit: %)

| Filial status | Intellectual immigrants | Labour immigrants | Entrepreneurial immigrants |
|---|---|---|---|
| None | 90.75 | 66.52 | 41.20 |
| 1 | 8.64 | 21.14 | 37.04 |
| 2 | 0.61 | 9.77 | 17.82 |
| 3 or above | – | 2.58 | 3.94 |
| Total (N) | 984 | 1,320 | 864 |
| Average | 1.07 | 1.45 | 1.46 |

**Table 4.4** Education level for urban new immigrants (Unit: %)

| Education level | Intellectual immigrants | Labour immigrants | Entrepreneurial immigrants |
|---|---|---|---|
| Primary or below | – | 7.88 | 7.29 |
| Middle school | – | 40.15 | 34.49 |
| High school | – | 49.62 | 34.72 |
| Tertiary | 49.29 | 1.74 | 14.70 |
| College or above | 50.71 | 0.61 | 8.80 |
| Total (N) | 984 | 1,320 | 864 |

## 4.2 Comparison of Willingness to Settle Down

As the author explicated before, the urban new immigration is different from the "old", because the "new" prefer to settle down after a period of time. However, the internal diversity among new immigrants is significant, intellectual and entrepreneurial immigrants are more likely to settle down than labour immigrants, a large proportion of labour immigrants answered "willing to, but cannot afford to" (see Table 4.5), which mean there is a big difference between "willing to" and "capable of", see Tables 4.6 and 4.7, intellectual and entrepreneurial immigrants are significant "willing to" and "capable of" settle down in cities; while we found out in the survey, the willingness and capability for intellectual and entrepreneurial immigrants are consistent, but intellectual immigrants have stronger intention to settle down than entrepreneurial immigrants.

**Table 4.5** Willingness to settle down for urban new immigrants (Unit: %)

| Willingness | Intellectual immigrants | Labour immigrants | Entrepreneurial immigrants |
|---|---|---|---|
| Willing to | 77.13 | 54.77 | 66.32 |
| Willing to, but cannot afford to | 11.69 | 23.03 | 16.67 |
| Not willing to | 7.22 | 12.42 | 9.84 |
| No necessary | 3.86 | 9.62 | 6.94 |
| Others | 0.10 | 0.15 | 0.23 |
| Total (N) | 984 | 1,320 | 864 |

**Table 4.6** Scale of willingness to settle down for urban new immigrants (Unit: %)

| Scale of willingness | Intellectual immigrants | Labour immigrants | Entrepreneurial immigrants |
|---|---|---|---|
| Strongly agree, agree | 62.70 | 52.20 | 64.47 |
| Strongly disagree, disagree | 12.70 | 23.71 | 17.82 |
| Fair | 22.87 | 20.00 | 14.47 |
| No idea | 1.73 | 4.09 | 3.24 |
| Total (N) | 984 | 1,320 | 864 |

**Table 4.7** Scale of capability to settle down for urban new immigrants (Unit: %)

| Scale of capability | Intellectual immigrants | Labour immigrants | Entrepreneurial immigrants |
|---|---|---|---|
| Strongly agree, agree | 46.34 | 23.48 | 49.19 |
| Strongly disagree, disagree | 20.02 | 45.30 | 26.04 |
| Fair | 31.40 | 25.61 | 20.25 |
| No idea | 2.24 | 5.61 | 4.51 |
| Total (N) | 984 | 1,320 | 864 |

## 4.3 Comparison of Mobility and Housing Conditions

In this section, we will compare the mobility and housing condition for labour, intellectual and entrepreneurial immigrants. As shown in Table 4.8, intellectual immigrants tend to stay in the same city they first immigrated to, the "no switch" option was selected by 50.61% of the intellectual immigrant, with the average of number of city switch lower than 1, entrepreneurial immigrants have the highest number of city switch, 1.58, but with a high degree of dispersion, similar to labour immigrants, some individuals of entrepreneurial immigrants long stay in one city, and some switched many, and the key variable for city switch is gender. As shown in Table 4.9, the gender difference for city switch among entrepreneurial immigrants is significant, similar to the pattern of senior labour immigrants; number of city switch for female entrepreneurial immigrants is 1, for male is 2. As for the resident time length, labour immigrants do not have a gender difference, entrepreneurial immigrants are the longest among three, with the average year of 3.77, also similar to senior labour immigrants, intellectual immigrants are similar to junior labour immigrants as shown in Table 4.10. Therefore, the length of resident time is closely related to the age of immigrants, male entrepreneurial immigrants have higher number of city switch, intellectual immigrants have the lowest number city switch, but the length of resident time is also exceeding 3 years.

As shown in Fig. 4.1, most intellectual immigrants live in this city because of "workplace" or "where the university graduated from", while the entrepreneurial immigrants want "more opportunities for career development". Besides, the entrepreneurial immigrants are older, and like the senior labor immigrants, they will consider "the future development of children" and "increase income".

In the survey, we categorized the living together companionship into three types, first is family and kinship, say "primary relationship"; second is friends, schoolmate, colleagues, lovers, cohabitant, say "secondary relationship"; third is living alone, say "independency". Intellectual immigrants are the least for "primary relationship", entrepreneurial immigrants are the most for "primary relationship", labour immigrants are in between. Meanwhile, intellectual immigrants are the most for "independency", and entrepreneurial immigrants are the least for "independency", labour immigrants are in between (see Table 4.11). No matter which type the immigrants are

**Table 4.8** City switch for urban new immigrants (Unit: %)

| Number of switch | Intellectual immigrants | Labour immigrants | Entrepreneurial immigrants |
|---|---|---|---|
| No switch | 50.61 | 40.94 | 35.53 |
| Switched once | 27.85 | 25.32 | 24.88 |
| Switched twice | 15.24 | 17.36 | 20.37 |
| Switched thrice | 4.07 | 8.57 | 9.26 |
| Switched 4 times or more | 2.24 | 7.81 | 9.95 |
| Total (N) | 984 | 1,319 | 864 |
| Average | 0.84 | 1.34 | 1.58 |
| Standard deviation | 1.30 | 1.89 | 2.19 |

**Table 4.9** Gender difference for city switch of entrepreneurial immigrants (Unit: %)

| Number of switch | Entrepreneurial immigrants | |
|---|---|---|
| | Male | Female |
| No switch | 29.13 | 42.82 |
| Switched once | 21.52 | 28.71 |
| Switched twice | 22.61 | 17.82 |
| Switched thrice | 11.30 | 6.93 |
| Switched 4 times or more | 15.43 | 3.71 |
| Total (N) | 460 | 404 |
| Average | 2.07 | 1.03 |

**Table 4.10** Length of resident time for urban new immigrants (Unit: %)

| Times | Intellectual immigrants | Labour immigrants | Entrepreneurial immigrants |
|---|---|---|---|
| 3 years or below | 35.37 | 29.24 | 18.63 |
| 3 years or above | 64.63 | 70.76 | 81.37 |
| Total (N) | 984 | 1,320 | 864 |
| Average (year) | 3.17 | 3.38 | 3.77 |

in, the willingness for residency does not change with this variable (see Table 4.12, sig. = 0.217 > 0.05), which mean the cohabitee of "primary relationship" does not lead to stronger willingness for settle down.

Table 4.13 shows that renting is the main form of housing for urban new immigrants, especially for entrepreneurial immigrants. The proportion of intellectual immigrants and entrepreneurial immigrants to buy their own houses is significantly higher than that of labour immigrants, who mostly live in dormitories. 77.31% of entrepreneurial immigrants choose to live in rented houses. The reasons (see Table 4.14) are "close to workplace" and "convenient living", which is "convenient

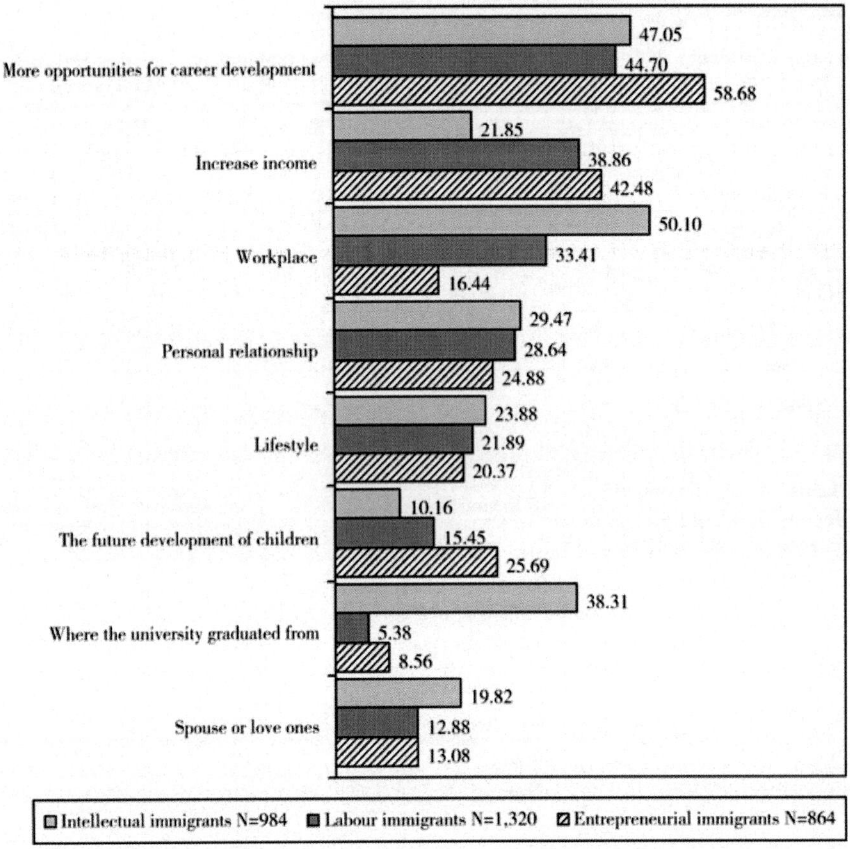


**Fig. 4.1** Reasons for stay in the presented city (%)

**Table 4.11** Housing companionship for urban new immigrants (Unit: %)

| Items | Intellectual immigrants | Labour immigrants | Entrepreneurial immigrants |
|---|---|---|---|
| Primary relationship | 25.41 | 40.46 | 74.19 |
| Secondary relationship | 50.71 | 46.24 | 16.67 |
| Independency | 25.20 | 16.35 | 12.15 |
| Total (N) | 984 | 1,320 | 864 |
| Average housing population | 2.50 | 3.59 | 3.00 |

**Table 4.12** Willingness for settle down for three types of housing companionship (Unit: %)

| Items | Primary relationship | Secondary relationship | Independency |
|---|---|---|---|
| Strongly disagree | 3.11 | 3.36 | 3.27 |
| Disagree | 16.86 | 15.92 | 14.34 |
| Moderate | 17.87 | 22.28 | 20.87 |
| Agree | 52.89 | 48.98 | 50.82 |
| Strongly agree | 9.26 | 9.46 | 10.71 |
| Total (N) | 1,382 | 1,131 | 551 |
| Chi-square = 10.733, df = 8, Sig. = 0.217 | | | |

to do business" according to the interviewees' words. The reasons for renting are "close to workplace" and "convenient living". They prefer to "living together with family or friends", but their housing is the most crowded. More intellectual immigrants go out to rent a house because "the workplace does not provide dormitory". In fact, intellectual immigrants are more likely to find the "dormitory conditions provided by the workplace are too poor" and turn to rent a house by themselves. Their per capita housing area is much more spacious than that of labour immigrants and entrepreneurial immigrants.

**Table 4.13** Housing types for urban new immigrants (Unit: %)

| Items | Intellectual immigrants | Labour immigrants | Entrepreneurial immigrants |
|---|---|---|---|
| Rent | 64.84 | 57.50 | 77.31 |
| Dormitories | 18.09 | 31.59 | 1.97 |
| Own house | 13.52 | 6.06 | 17.25 |
| Living with someone for free | 3.35 | 3.03 | 2.20 |
| Others | 0.20 | 1.82 | 1.27 |
| Total (N) | 984 | 1,320 | 864 |
| Housing area per capita ($m^2$/person) | 24.87 | 16.26 | 22.71 |

**Table 4.14** Reasons of renting for urban new immigrants (Unit: %)

| Items | Intellectual immigrants | Labour immigrants | Entrepreneurial immigrants |
|---|---|---|---|
| Close to workplace | 42.63 | 41.90 | 56.22 |
| Convenient living | 40.75 | 38.34 | 48.13 |
| Cheaper | 28.21 | 30.96 | 33.43 |
| Living together with family or friends | 27.90 | 34.39 | 27.74 |
| Workplace does not provide dormitory | 42.95 | 30.43 | 5.40 |
| Dormitory conditions provided by the workplace are too poor | 5.49 | 3.16 | 0.90 |
| Convenient to do business | 0.16 | – | 1.50 |
| Total (N) | 638 | 759 | 667 |

**Table 4.15** Job change for urban new immigrants (Unit: %)

| Items | Intellectual immigrants | Labour immigrants | Entrepreneurial immigrants |
|---|---|---|---|
| Never | 31.91 | 31.67 | 39.81 |
| Once | 24.09 | 15.68 | 15.39 |
| Twice | 23.17 | 24.09 | 20.83 |
| Thrice | 14.94 | 15.15 | 11.69 |
| Four times or above | 5.89 | 13.41 | 12.27 |
| Total (N) | 984 | 1,320 | 864 |
| Average | 1.45 | 1.91 | 1.60 |

## 4.4 Comparison of Working Conditions

Labour immigrants change their job more often than intellectual and entrepreneurial immigrants, those who never change their job as labour immigrants, significantly lower than those who as entrepreneurial immigrants. At present, intellectual immigrants have the shortest job tenure, 10 months less than the average of entrepreneurial immigrants, and labour immigrants is 8 months less than the average of entrepreneurial immigrants. Therefore, the entrepreneurial immigrant has the highest stability for job, while the labour immigrant has the lowest, the intellectual immigrant is the least experienced for work, they are younger and fresh to workplace (see Tables 4.15 and 4.16).

Figure 4.2 shows the labour immigrant emphasis income, personal interests and working environment, while the intellectual immigrant take career prospect and professional consistency more seriously, intellectual immigrants are relatively satisfied by their job prospect, promotion and social status, therefore, the job stability for intellectual immigrants are significantly high. Table 4.17 shows the working intensity for urban new immigrants, 70.07% of intellectual immigrants have weekend

**Table 4.16** Job tenure (present job) for urban new immigrants (Unit: %)

| Items | Intellectual immigrants | Labour immigrants | Entrepreneurial immigrants |
|---|---|---|---|
| ≤12 months | 33.84 | 33.48 | 17.71 |
| 13–24 months | 31.50 | 28.56 | 28.70 |
| 25–36 months | 22.05 | 20.45 | 22.92 |
| 37–48 months | 7.42 | 8.64 | 14.12 |
| ≥49 months | 5.18 | 8.86 | 16.55 |
| Total (N) | 984 | 1,320 | 864 |
| Average (months) | 22.25 | 24.19 | 32.18 |

leaves, while only 26.61% of labour immigrants enjoy weekend leaves, moreover, daily working hours for intellectual immigrants are lower than that for labour immigrants. However, the subjective satisfactory on job is almost the on the same scale for labour and intellectual immigrants, as shown in Fig. 4.3, both at 3.37. This reminds the author of a kind of "social rationality", which is different from economic rationality. It is not based on "optimal" or "maximum" criteria, but only on the principle of "satisfaction" and "rationality".[1] It presented as the fact that labour immigrants maintain a high degree of satisfaction despite their poor job stability and high work intensity.

## 4.5 Comparison of Living Conditions

The monthly disposable income of labour immigrants was mostly below RMB3,000, while 24.90% of intellectual immigrants had a monthly disposable income of more than RMB3,000, and 49.00% of entrepreneurial immigrants had a higher monthly disposable income (see Table 4.18). The average savings of intellectual immigrants and entrepreneurial immigrants were more than that of labour immigrants. However, 26.52% of intellectual immigrants had tight income and expenditure balance and sometimes could not make ends meet, which was much higher than that of entrepreneurial immigrants (see Table 4.19). If one wants to assure income and expenses balance, the monthly salary of the labour immigrant should be RMB2,486.85, the intellectual immigrant is RMB3,501.93, the entrepreneurial immigrant is as high as RMB5,406.93. It is clear that high earners have high salary expectations and low earners have low salary expectations.

[1] Wen Jun, From Survival Rational Choice to Social Rational Choice: Analysis on the Motivation of Farmers' Outgoing for Employment in Contemporary China, *Sociological Studies*, Vol. 6, 2001, pp. 19–30.

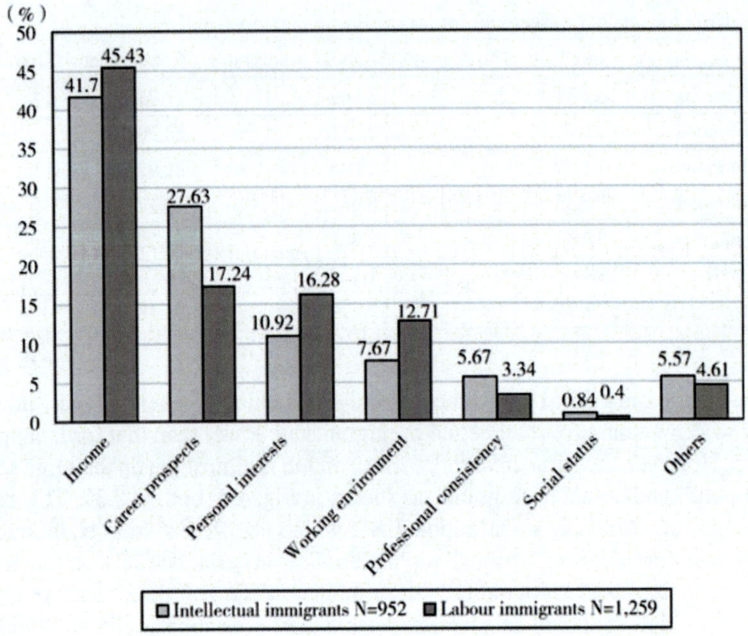


**Fig. 4.2** Important factors in job choosing for intellectual and labour immigrants

**Table 4.17** Working intensity for intellectual and labour immigrants (Unit: %)

| Items | Intellectual immigrants | Labour immigrants |
|---|---|---|
| Leave on weekend | 50.11 | 11.68 |
| Leave on Sunday | 19.96 | 14.93 |
| Leave on a set day in a month or a week | 4.83 | 8.34 |
| Shift leave | 10.71 | 22.56 |
| No fixed leave | 11.87 | 28.67 |
| No leave | 2.52 | 13.82 |
| Total (N) | 952 | 1,259 |
| Average working days in a week | 5.46 | 6.11 |
| Average working hours in a day | 8.23 | 9.16 |

Financial balance should be observed at the family level, especially at the structure of family expenditure; the average family expenditure of the labour immigrant is RMB1,712.24 (see Table 4.20), in which the proportion of food expense takes 0.31, and the housing takes 0.15, food and shelter combine takes nearly 0.5; the intellectual immigrant and the entrepreneurial immigrant has higher housing expenditure, both above 0.2 (see Fig. 4.4), 67.78% of the intellectual immigrant listed housing expenses

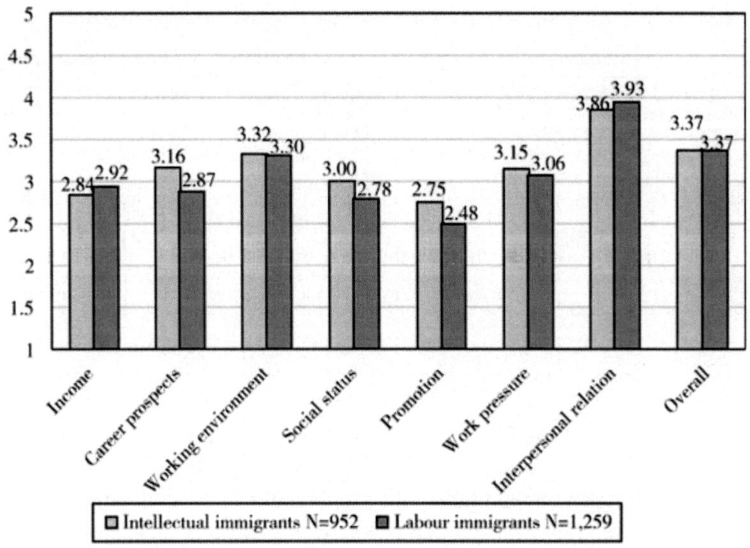


**Fig. 4.3** Satisfactory on present job for intellectual and labour immigrants

**Table 4.18** Monthly disposable income urban new immigrants (Unit: %)

| Items | Intellectual immigrants | Labour immigrants | Entrepreneurial immigrants |
|---|---|---|---|
| ≤RMB1,000 | 8.13 | 23.86 | 8.10 |
| RMB1,001–2,000 | 42.89 | 61.82 | 30.90 |
| RMB2,001–3,000 | 24.09 | 11.67 | 21.99 |
| RMB3,001–4,000 | 13.31 | 1.59 | 14.93 |
| ≥RMB4,001 | 11.59 | 1.06 | 24.07 |
| Total (N) | 984 | 1,320 | 864 |

**Table 4.19** Financial balance for urban new immigrants (Unit: %)

| Items | Intellectual immigrants | Labour immigrants | Entrepreneurial immigrants |
|---|---|---|---|
| Margined | 28.56 | 21.74 | 29.28 |
| Balanced | 44.21 | 52.05 | 50.12 |
| Slightly short | 22.66 | 21.44 | 16.55 |
| In debt | 3.86 | 3.71 | 2.78 |
| Total (N) | 984 | 1,320 | 864 |
| Income makes ends meet (RMB) | 3,501.93 | 2,486.85 | 5,406.93 |

**Table 4.20** Family expenditure structure for urban new immigrants (Unit: RMB)

| Items | Intellectual immigrants | Labour immigrants | Entrepreneurial immigrants |
|---|---|---|---|
| Food | 618.73 | 499.05 | 1,065.78 |
| Housing | 541.59 | 280.55 | 1,136.60 |
| Communication | 153.91 | 114.43 | 224.88 |
| Commute | 155.79 | 78.20 | 271.56 |
| Recreation | 320.34 | 175.52 | 357.11 |
| Social | 342.57 | 194.30 | 523.33 |
| Remittance | 300.39 | 370.19 | 414.43 |
| Total | 2,433.32 | 1,712.24 | 3,993.69 |

as their main expenditure, 77.55% of the entrepreneurial immigrant listed housing expenses as their main expenditure, and only 48.18% of the labour immigrant take housing as their main expenditure. The proportion of remittance for labour immigrants is significantly higher than that of intellectual immigrants and entrepreneurial immigrants. 18.03% of labour immigrants spend most of their income on supporting their parents, which is significantly higher than that of intellectual immigrants and entrepreneurial immigrants. Entrepreneurial immigrants usually bring their families to live with them, so the proportion of food expenses is relatively large. 40.86% of them listed raising children as their main expenditure. Intellectual immigrants, on the other hand, tend to spend more on interpersonal communication and entertainment (see Table 4.21). The expenditure structure of labour immigrants tends to be basic needs, and the proportion of household expenditure is relatively large, while entrepreneurial immigrants need to support the family. Intellectual immigrants prefer personal enjoyment, their family burden is relatively small.

Psychological burden is also an important aspect in our investigation on the living condition of urban new immigrants. In the survey, we adopt eight indicators for psychological distress, and we use 4 points scale to measure the degree of these distress, the 4 points scale are "never (1 point)", "seldom (2 point)", "often (3 point)", "severe (4 point)". Most of the results (see Fig. 4.5) are between "never (1 point)" and "seldom (2 point)", while the "feeling exhausted physically and mentally" was the severest problem among immigrants, as for intellectual immigrants, the scale is 2.13 point. "irritable", "vanish prospect", "feeling that life is hard" followed, "crying easily or prone to cry" is the least severe problem. Besides "insomnia" and "feeling that life is hard", intellectual immigrants are generally more troubled by psychological burden, labour immigrant are severer than entrepreneurial immigrants on the aspects of "vanish prospect", "feeling useless", "crying easily or prone to cry" and "feeling lonely".

Psychological distress is caused by the gap between ideal and reality. This chapter also tries to judge or explain the psychological status of new immigrants by the "main criteria of life happiness" and "life distress". Intellectual immigrants place more importance on career success, achieving their dreams and living a well-off life.

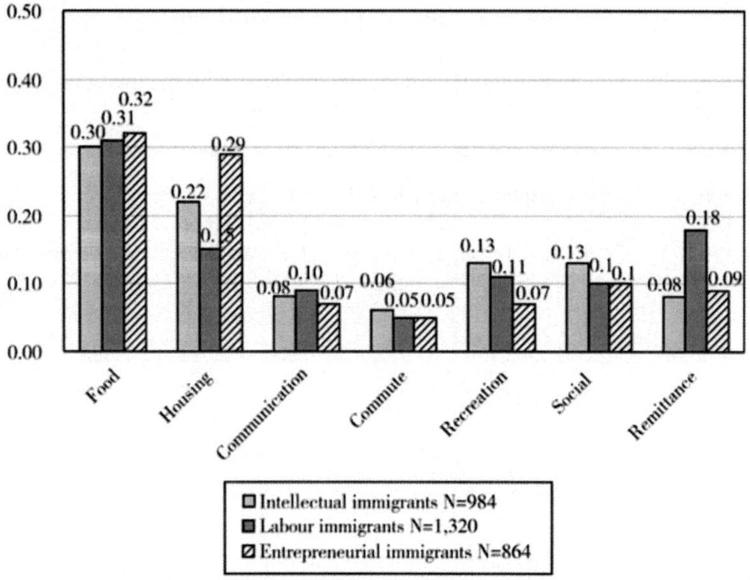


**Fig. 4.4** Family expenditure proportion for urban new immigrants

**Table 4.21** Top 3 expenditures for urban new immigrants (Unit: %)

| Items | Intellectual immigrants | Labour immigrants | Entrepreneurial immigrants |
|---|---|---|---|
| Food | 67.07 | 61.59 | 66.90 |
| Rent | 52.74 | 42.80 | 65.51 |
| Child support | 7.42 | 24.24 | 40.86 |
| Clothing | 39.33 | 44.32 | 20.60 |
| Social | 30.49 | 20.08 | 20.25 |
| Communication | 17.48 | 29.39 | 16.44 |
| Recreation | 22.87 | 19.17 | 12.62 |
| Mortgage | 15.04 | 5.38 | 12.04 |
| Parent support | 9.76 | 18.03 | 11.92 |
| Commute | 10.57 | 8.56 | 11.34 |
| Health | 3.46 | 7.50 | 7.64 |
| Insurance | 2.85 | 2.42 | 3.94 |
| Cosmetics | 7.11 | 6.36 | 3.70 |
| Books\training | 12.20 | 7.42 | 3.13 |
| Others | 1.42 | 2.12 | 3.01 |
| Total (N) | 984 | 1,320 | 864 |

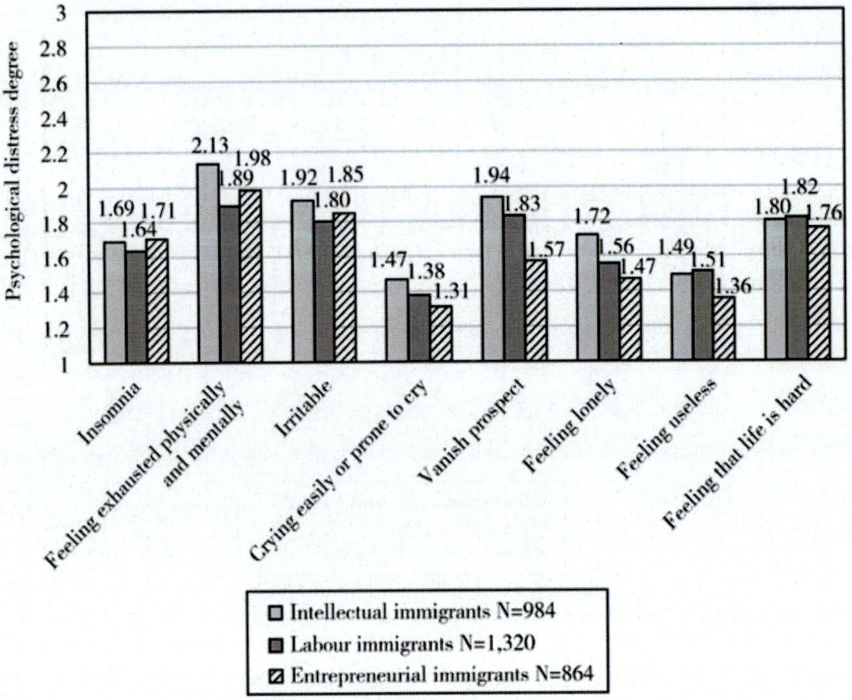


**Fig. 4.5** Psychological distress for urban new immigrants

However, they complain about lack of achievement, drab life, high pressure of work and low income. Labour immigrants prefer to earn a lot of money and gain the respect of others. However, labour immigrants suffer more from low income, unstable jobs, lack of professional skills and lack of knowledge (see Figs. 4.6 and 4.7).

Entrepreneurial immigrants are more likely to think that "a happy family", "filial piety", "successful career" and "good health" are more important, while they are less likely to complain about "low income" and "lack of achievement". Moreover, 74.19% of entrepreneurial immigrants are able to live with their families and relatives. This can explain why the degree of psychological distress of entrepreneurial immigrants relatively less severe, because they have a career, a higher income, and have relatives around; similarly, the expectations of intellectual immigrants are in conflict with the reality, and they have no goals to strive for, so their psychological feelings are very poor.

In the above studies, we found that the willingness of intellectual immigrants and entrepreneurial immigrants to settle down in cities are similar and both higher than that of labour immigrants. However, in some ways, intellectual immigrants and junior labour immigrants share some traits, while entrepreneurial immigrants are similar to senior labour immigrants. Firstly, the age is similar, junior labour immigrants and intellectual immigrants are younger, while the senior labour immigrants and

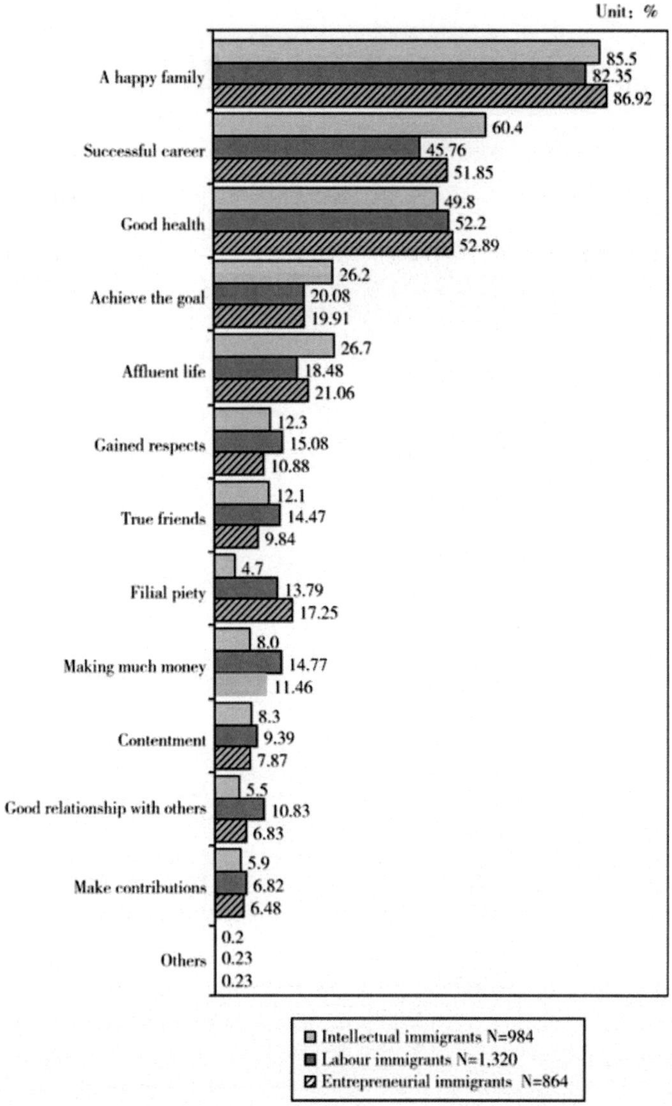


**Fig. 4.6** The main criteria of life happiness for urban new immigrants

entrepreneurial immigrants are older. However, they happen to fall into four age groups (22, 26, 32, 38), and these are influential in defining the characteristics of intellectual and entrepreneurial immigrants. The group characteristics of intellectual immigrants and entrepreneurial immigrants are summarized as follows:

1. The vast majority of intellectual migrants have college degrees or above, and many of them stay in cities after graduation. Entrepreneurial immigrants mainly

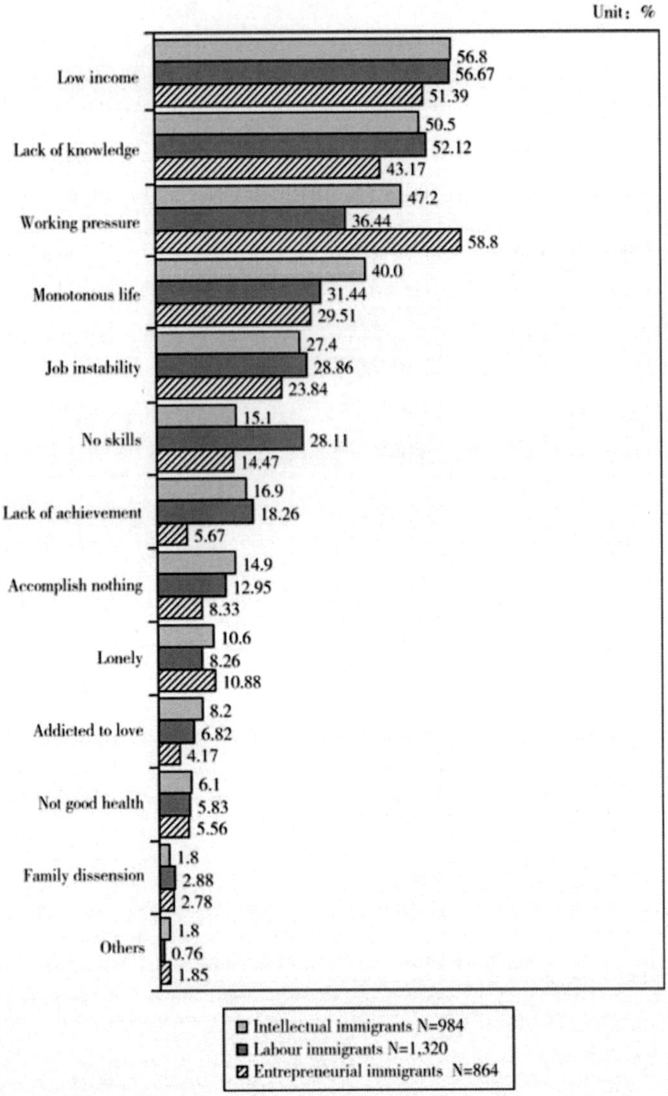


**Fig. 4.7** The life distress for urban new immigrants

focus on "more career development opportunities" in cities, but they also take "children's future development" and "increasing income" as important reasons, just like senior labour immigrants.

2. Due to the younger age, the time from graduation to present employment for intellectual immigrants is relatively short, and entrepreneurial immigrants have longer residential time and working experience.

3. Intellectual immigrants are similar to junior labour immigrants, mainly living with friends or colleagues, some living alone, which are defined as secondary and independence living types in the research. Most of entrepreneurial immigrants bring their families to live with them, which is a primary relationship household.
4. Labour immigrants take income, personal interest and working environment as important factors in job finding. In contrast, intellectual immigrants value job prospects and professions more.
5. The monthly disposable income of intellectual immigrants is higher than that of labour immigrants, and that of entrepreneurial immigrants is even higher. Intellectual immigrants are similar to junior labour immigrants and prefer individual enjoyment, while most of entrepreneurial immigrants must support their families.
6. Intellectual immigrants attach more importance to career success, realization of goals and a prosperous life. However, they often complain that they lack of achievement, drab life, work under great pressure, low income and have poor psychological feelings. Entrepreneurial immigrants generally under less psychological distress, because of entrepreneurial immigrants tend to take "family happiness", "filial piety", "business success" and "good health" as their goals of life, and less complain about "low income" and "lack of achievement", 74.19% of entrepreneurial immigrants are able to live with their families, which improve greatly on their psychological condition.

# Chapter 5
# The Immigration and Establishment of Urban Immigrants

From the perspective of urban new immigrants, the immigration process is not only a change of location but also a change of social context and norms. Immigration means being away from the original social context, including the social network and social support built up in long-term life practice, and the familiar social values and norms. Meanwhile, a series of emergencies in the new environment must be coped with, so as to achieve social adaption and integration after continuous adjustment. In the next few chapters, the authors will demonstrate the process of social adaptation and integration of urban immigrants with quantitative data. From a general perspective, it is a process of immigrating and establishing, in specific, it is a complex process involving the community, social networks, social participation, social security, and so on. For the sake of analysis, the authors will expand them separately, although they corresponding to each other and together constitute the process of social adaption of urban immigrants.

In this chapter, the authors will first describe and analyse the processes and methods of different types of urban new immigrants immigrating and establishing in cities. The first concern of academic studies on population migration is the motivation of migration, and the most influential model in this field is the "push-pull" theory, that is, the motivation of population migration comes from two main sources, the first is push factors from the original place, the second is pull factors from the destination.[1] In each place, there are some factors that attract some people, as well as some factors that exclude some people and those that are irrelevant to some people. The individual immigrant should make decisions on whether to migrate after subjectively and rationally weighing the positive and negative social-cultural factors. But in reality, this individualistic choice of migration is almost impossible. We are not only rational individuals but also embedded in a multi-structured social network, which

[1] Lee, Everett S., *Demography*, Vol. 3, No. 1 (1966), pp. 47–57.



D. Zhou, *Urban Migrants in China*,
https://doi.org/10.1007/978-981-99-3114-9_5

makes the decisions we make are often influenced by others, sometimes even determined by others. For instance, the migration of children with their parents and wives with their husbands[2] is a family-centred collective migration rather than a rational choice of individualism.

Immigration means rebuilding social relationships and social networks in the urban new environment and embedding daily life into the structure of social relationships. It's not only about the integration of social relationships, but also the adaptation of social systems, social psychology, and other multiple aspects. Generally, urban immigrants will go through three stages: cultural shock, anxiety, and adaptation, and their specific adaptation process will be affected by personal and social factors. The cases the authors obtained through research showed that different factors have different degrees of influence on different types of urban immigrants.

## 5.1 The Immigration and Establishment of Intellectual Immigrants

As the authors mentioned earlier, the main characteristics of intellectual immigrants are high academic qualifications and professional knowledge. For example, among the ten intellectual immigrants in the age range of 26–35 years old interviewed in Shenyang, the under-30s are all undergraduates or above. Most of the interviewees are from surrounding cities of Shenyang and the major cities of Liaoning province, Jilin province, and Heilongjiang province. After receiving higher education in Shenyang, most of them chose to work and settle there. Their occupation types include office clerks, teachers, chief editors, civil engineers, department managers, business secretaries, and freelance. One of them was unemployed and was an office clerk before. However, when our interviewer paid a return visit a month later, she had found a satisfactory job in a famous software company. It can be seen that intellectual immigrants mostly engage in occupations with high social recognition and are more active and competitive in career acquisition. It is generally believed that the higher the education level, the easier it is to adapt to the new environment and vice versa, and the authors' research is indeed consistent with this assertion.

The 49 typical cases of intellectual immigrants obtained through investigations and in-depth interviews in six major cities reveal some common problems of intellectual immigrants, and also show their regional differences in specific urban scenarios. The intellectual immigrants have received less attention in previous academic studies on immigrants, however, with the increasing demand for intellectual labour in urban development, intellectual immigrants have now become an important force. For intellectual immigrants, immigration is not only the process of spatial transfer from rural

[2] Liao Zhenghong, *Population Migration*, Taipei: San Min Book Co., Ltd., 1985.

or second- and third-tier cities to metropolises, but also the process of identity transformation from college students to urban professionals and settlers. In addition to the advantages of intellectual capital, intellectual immigrants also have a stronger desire for urbanisation and a clearer plan for immigrating to cities.

### *5.1.1 From Graduation to Job Hunting: The Immigration Mode for Intellectual Immigrants*

As the number of college students increases year by year, their employment has become a widespread concern of society. In China, the number of college graduates, including undergraduates and graduates, exceeds 6 million every year. An employment survey for all undergraduates of China Agricultural University (CAU) conducted by the Rural Development Research Association of CAU in 2010 found that more than 90% of undergraduates want to work in big cities, and nearly two-thirds want to work in first-tier cities like Beijing and Shanghai, only 2.3% are willing to work in rural areas,[3] which shows that big cities are very attractive to intellectual immigrants. However, these non-local college students may not be able to find a stable job or have local household registration, so they can't enjoy equal welfare as city dwellers, thus forming a group of intellectual migrants. The formation of this group is closely related to the expansion of college enrolment and the reform of the employment system for college students. With the reform of the job assignment system for college students, college students, especially those from rural areas, face more employment and urbanisation pressures due to the lack of necessary social capital.

As can be seen form cases of 10 intellectual immigrants interviewed in Shenyang. After graduating from college, the majority of those intellectual immigrants choose to stay in Shenyang by applying for jobs. According to the "push-pull" theory, the socioeconomic conditions of the original place and destination will both be the push and pull factors to the decision-makers of migration. At the same time, the judgment of the positive and negative factors of the original place and destination will also vary depending on decision-makers' subjective feelings and stages of life cycle. For most people, the original place is where they grew up, and the migrant destination is where they started their business and struggled for adaptation. Under the influence of the above subjective and objective factors, decision-makers of migration will make a decision based on his or her judgment. Nowadays, the group of college graduates has become an important component of urban immigrants, and the group's willingness to settle in cities is also the strongest among urban immigrants.

[3] Wang Miao, Employment of College Students: Unable to Escape the "Black Hole" of Big Cities, *China Reform Daily*, 24 June 2010.

Case: Shenyang-Intellectual-001

Ms Wang comes from a second-tier city near Shenyang. After passing the college entrance examination, she entered a junior college in Shenyang to study computer science, and later obtained a bachelor's degree through Top-up.[4] After graduation, she chose to stay and work in Shenyang. She explained, "as far as my major is concerned, staying in Shenyang should have more opportunities than staying in my hometown, and even if I go back to my hometown, there are few relatives." This statement is typical of intellectual immigrants when selecting occupations. First of all, Ms Wang believes that her future career development should be consistent with her major, so as to apply what she has learned. Secondly, she believes that Shenyang, as a big city, will have more opportunities than her small hometown, this yearning for opportunities is a vaguely perceptual understanding to a certain extent, and it reflects her expectations for her bright future. Finally, "there are few relatives" is also one of the important reasons why she doesn't want to return to her hometown. It is very important to rely on the social network to live in a small hometown, but in big cities, she believes that she can create a better future with her professional expertise.

Shenyang is located in eastern China, and this geographical advantage is also one of the considerations for some intellectual immigrants, especially for people from the border areas of China. Western and eastern China is not only a division of geographical space but also a division of economic strength, the latter has more convenient transportation and broader development prospect than the former. Although the Chinese government has been vigorously implementing the strategy of developing the western region,[5] it is still lagging behind the eastern region due to poor environmental conditions and weak infrastructure. Therefore, many college students from the west choose not to return hometown after graduation but stay in the city where they studied or other large cities in the east.

Case: Shenyang-Intellectual-008

Ms Li comes from Xinjiang and worked in Shenyang after graduating from college in Xi'an. When asked why she came to work in Shenyang, she replied, "because there are relatives in Shenyang, and there are many job opportunities, transportation here is also more convenient."

Compared with college graduates from urban areas, college graduates from rural areas tend to have a stronger desire to work and settle in cities. For example, Ms Miao (Case: Shenyang-Intellectual-002) comes from a village in a second-tier city near Shenyang. Since she came to Shenyang to study at university, she has established a clear life direction to settle in Shenyang. After graduation, she and her boyfriend found jobs in Shenyang. Although she experienced many difficulties and setbacks, she finally bought a house and got married in Shenyang, fulfilling her wish to settle in Shenyang. She currently works for a scientific research instrument company and believes that the company has a "good humanistic environment", she explained,

[4] Junior college students can upgrade to university students through Top-up, it's an examination system for people who want to promote their education degree.

[5] "The development of the western region in China" is a policy of the Central Government of the People's Republic of China. The purpose is to "use the sufficient economic development capacity of the eastern coastal areas to improve the economic and social development of the western regions and consolidate national defence" (by translator).

"most of the colleagues in the company are from outside Shenyang, and they often have dinner and chat together, so their relationship is very harmonious." Compared with other types of urban immigrants, intellectual immigrants tend to encounter less social exclusion during the process of immigrating to cities. As the authors mentioned earlier, the occupations of intellectual immigrants are often distributed in fields with high social recognition. In addition, homogeneous people are more likely to gather in similar occupational fields. As we can see in Ms Miao's case, most of her company's colleagues are from outside Shenyang, so they have similar experiences of immigrating to Shenyang and living in a different city, making this group more likely to resonate and form an identity. In Ms Miao's opinion, it is a lucky thing to be able to work in an office building from 9 to 5 every day with a satisfactory salary and a boyfriend to rely on. Indeed, it is the desire of some college graduates from rural areas to become a white-collar worker with a high degree of social recognition and live an ordinary but stable life in cities. When this desire is fulfilled, their happiness index will increase significantly.

Among intellectual immigrants the authors interviewed in Shenyang, some have higher self-positioning and career goals. For example, Ms Lei (Case: Shenyang-Intellectual-007) has applied for many companies before joining her current company, including TV stations, newspapers, IT companies, securities firms, insurance companies, web editors, photography production, and so on. In the beginning, she felt that jobs were easy to find and also worked in several small companies, but she still felt that these small companies did not meet her needs for higher-level development, so she quit. In her view, with her ability and knowledge, she can enter a large enterprise and obtain a position with a higher social status. This kind of confidence based on academic qualifications and abilities is common among intellectual immigrants, and sometimes confidence even becomes a kind of arrogance. Ms Lei believes that "compared to the sales staff with limited education and knowledge in her company, she is quite excellent, so she sometimes looks down upon them." Many intellectual immigrants believe that knowledge and ability are above everything when they first start working, and they think they can realise their own values with their abilities. However, after accumulating a certain amount of work experience, they will begin to realise the importance of social relationships and social networks, which sometimes even determine their personal development space. As a department manager, Ms Lei believes, "in order for the work to proceed smoothly, you need to be friends with your colleagues first, so that they will not easily refuse the work you entrust to them." Knowing the importance of social networks, many intellectual immigrants also began to build their own social networks based on the principle of instrumental rationality to help achieve their goals. Mr Wang (Case: Shenyang-Intellectual-005) was a victim of the company's unspoken rules, and he summed up his experience and said, "I failed because I ignored the interpersonal relationships in the department. In fact, as newcomers, apart from working hard, we should also pay attention to social relationships in order to quickly integrate into the department. I used to work hard in obscurity, and I was too humble in front of my leaders and colleagues, so they doubted my ability to deal with social relationships and looked down upon me. Now I will pay special attention to overcome my weakness in dealing with social relationships." It can be seen from the above cases that how to strike a balance between doing a good job by virtue of ability and dealing with social networks has become the first test of whether these intellectual immigrants who have just started to work can realise their dreams of settling in the city.

### 5.1.2 From Employment to Establishment: Key Steps for Urban Adaptation

After graduating from college and starting to work, intellectual immigrants begin to move towards different life paths, some of them may have a successful career, and some of them may feel difficult to adapt to changes because of frustrating work. In general, the step from employment to the establishment have become a watershed for the integration of intellectual immigrants into urban life.

From college graduates to professional workers, intellectual immigrants have to undergo a process of transformation and adaptation of social roles. After graduating from college, Ms Gao (Case: Shenyang-Intellectual-003) entered a training institution to work as a primary language teacher. However, she did not have any teaching experience, so she suffered great setbacks and failures at the beginning. The training institution allows students' parents to listen to the course, but Ms Gao's performance in the first lecture is very poor, which caused harsh criticism and complaints from students' parents. This lecture experience severely damaged Ms Gao's confidence and self-esteem, and she did not know how to face her students and their parents in the future. However, Ms Gao did not treat this failure negatively but actively chose to gain recognition from students' parents and her leaders through her own efforts. Finally, hard work pays off, Ms Gao was later promoted to be the head of the language discipline, but at the same time, she needs to face the pressure and challenges of adapting to this new role. Some senior teachers do not cooperate with Ms Gao's work, which brings difficulties to her management, but Ms Gao finally won everyone's respect through her excellent work and meticulous care for colleagues. From college graduates to professional workers who continue to make progress and promotion, Ms Gao has been working hard to adapt to new social roles, and her ability and social status have also been improved during this adaptation process, laying a solid foundation for the establishment. However, Ms Gao is not satisfied with the status quo, and she is also planning to pursue a master's degree to increase her competitiveness in the city.

Nevertheless, not all intellectual immigrants can be successfully integrated into the city. Some intellectual immigrants will be hindered in the process of career development, and they may have physical and psychological problems due to huge pressure and anxiety. Ms Li (Case: Shenyang-Intellectual-008) said, "when I was working as an accountant in an overseas employment service company, I was particularly stressed out and had depression for some time. Sometimes I suffered from insomnia, and sometimes even I fell asleep quickly, I still woke up at 2 AM due to anxiety, then I began to think carefully about the work I need to do during the day, for fear of making a little mistake in work." Ms Li's pressure comes not only from work but also from her strict requirements for self-improvement. In order to improve her professional competitiveness, she also prepares for a test of assistant accountant certificate in her free time. She has to go to Northeastern University for self-study on weekends, and she always reads books at night until she is tired and falls asleep, feeling particularly nervous. Essentially, Ms Li's pressure comes from the competitive pressure and the sense of instability that she faces in living in the city. Ms Li said, "I have a strong sense of crisis. If I am a native of Shenyang, even if I do not bother to take the accountant certificate, I can still rely on the local social networks to find a decent job. But I am an outsider in Shenyang, so I must obtain an accountant certificate in case I am unemployed and there is no social network to rely on, I can still use this certificate to find a job."

Under this kind of pressure and sense of crisis, Ms Li's physical condition got worse and worse, she was not only suffering from insomnia but also a lot of hair loss. She said, "I lost a lot of my hair in one month. For this reason, I went to more than six hospitals. It cost a total of RMB40,000-50,000 and took me more than four years to recover. In those few years, I didn't have chances to go to the hair salon because there was no need to trim my hair." In consideration of her poor physical and psychological conditions, Ms Li finally chose a more relaxed internal work position and began to focus on the quality of life. But deep down, she still yearns for a more stable life and believes that the current company's instability is a big problem. In fact, Ms Li's depression is a reflection of her inner panic about the uncertainties in urban life. As outsiders in cities, intellectual immigrants need to deal with the reconstruction of social networks and huge pressure of survival, and they will be inferred by various kinds of negative emotions such as anxiety, fear, depression, etc. Physically, it manifests itself in the form of diseases.

### *5.1.3 From Establishment to Getting Married: Immigration and Settlement*

Generally, intellectual immigrants have a strong desire to settle in cities. After graduating from university and having a stable job, intellectual immigrants usually consider starting a family in the city to achieve the transition from non-locals to locals.

But in reality, not all intellectual immigrants will have a solid material basis to achieve the goal of settling in cities, so some of them will choose to rely on their partners to achieve this goal. Ms Wang (Case: Shenyang-Intellectual-004) is a typical case. Ms Wang desperately wants to get rid of rural household registration and become a fashionable urban citizen. To achieve this goal, she had done many jobs and even entered the MLM (multi-level marketing) organisation to make money before she came to Shenyang. Besides, she even cancelled her marriage contract with her former fiancé. After coming to Shenyang, Ms Wang still made every effort to make money, but her career suffered many setbacks. Finally, Ms Wang married a postgraduate student majoring in international finance at a local college in Shenyang. Her husband bought a house and the two settles in Shenyang. In the end, Ms Wang achieved her goal of settling in the city through marriage.

The case mentioned above only accounts for a few of the cases of intellectual immigrants. Actually, the vast majority of intellectual immigrants realise the dream of settling in cities through the joint efforts of husband and wife. As housing prices increase year by year, buying a house becomes a threshold for many young intellectual immigrants who want to settle down and live a stable life in cities. Usually, it is difficult to buy a house in the city based on the economic strength of the young people who have just started working, and the savings of their parents of both sides play an important role. When Ms Miao (Case: Shenyang-Intellectual-002) married her boyfriend in 2008, she used the money borrowed by the parents of both sides to pay the down payment and bought a second-house with a loan near her workplace. When Ms Gao (Case: Shenyang-Intellectual-003) and her husband got married in 2007, they also relied on the help of their parents and parents-in-law to pay the down payment of the house. In 2008, Ms Gao and her husband moved into a new house and also began to work hard to create better material conditions for their upcoming children, as their parents did for them.

Changes in intimacy and marital status can also affect the choice of cities for intellectual immigrants to settle in. After graduation, intellectual immigrants usually work alone in big cities, so their social circle is relatively small and single, and the main objects of interaction are their classmates, teachers, and colleagues. In this case, the intimacy of lovers is usually at the core of their social networks and is one of the key considerations for them to stay in the city. When this kind of intimate relationship is in crisis, intellectual immigrants will no longer consider the factors of lovers or partners as they did before when choosing the city to settle in but will choose a city that is more suitable for their development or has relatives and friends to settle in. Ms Lei (Case: Shenyang-Intellectual-007) and her ex-boyfriend are from the same hometown, they were going to study for a master's degree together, but only Z passed the postgraduate exam. Later, when Z went to school, Ms Lei continued her busy work. After some time, Z felt that he was ignored by Ms Lei, so he found a new girlfriend. After breaking up with Z, Ms Lei felt that there was no reason to stay in Shenyang. Nonetheless, some intellectual immigrants do not attach importance to love relationships and families but put personal development and future in the first place, which is also a prominent feature of intellectual immigrants when compared with other types of urban immigrants. And compared to women, highly educated men are more inclined to put their personal development above love relationships and families. Yang (Case: Shenyang-Intellectual-006) frankly said, "I stayed in Shenyang purely for the development of my career, and it has nothing to do with my emotional life, I will leave here eventually." Similar to Yang, many intellectual immigrants take personal development as the ultimate goal of life, which is closely related to their self-confidence based on their high education degree.

The case analysis of intellectual immigration is consistent with its quantitative data analysis, showing the importance of high education degree and intellectual capital for the immigration and adaptation of intellectual immigrants to cities. Also, although intellectual immigrants must undergo a stage of social role adjustment and adaptation to cities, on the whole, they will gradually find a suitable position in the city. Therefore, intellectual immigrants not only have a relatively strong desire to settle in cities but also have a deep desire to start a family in cities.

## 5.2 The Immigration and Establishment of Labour Immigrants

Labour immigration is a type of immigration with a large scale and long duration among urban immigrants. Previously, there was no lack of academic studies on labour immigration, and the concepts of "rural migrant", "migrant workers", and "migrant population" were mostly used. Large-scale urban-rural migrants emerged in the late 1980s, expanded dramatically in the 1990s, and peaked in the mid-1990s. At present, there is more than 80 million migrant population across China throughout the year, accounting for 15–20% of the rural labour force. Large-scale migrant population flocking to cities not only brings problems to both cities and countryside but also brings difficulties of adapting to cities to the rural labour.[6]

At present, the academic study on labour immigration focuses on three aspects. The first is from the perspective of the destination of immigration, studying the source of rural labour, as well as their purposes of immigration, methods of employment, occupations, incomes, work and living environment, protection of rights, social networks, adaptation processes, impact on destinations of immigration, etc. The second is from the perspective of the source of immigration, studying the reasons for rural labour's choice to go out for employment, as well as their sources of information, characteristics, occupations, impact on their family and hometown after going out and individual's acquisition of modernity.[7] The third is to explore and construct the theory of migration of migrant labour in rural areas from a macro perspective, and to study the significance of rural migrants going out for work and the path of rural urbanisation in China.[8] With China's socioeconomic development and changes in social structure, especially with the emergence of a new generation of rural migrants, the problems of urban integration and urbanization of labour immigrants have become increasingly prominent.

---

[6] Guo Zhenglin, Zhou Daming, Going Out for Work and the Cultivation of Modernity of the Farmers, *Journal of Sun Yat-sen University (Social Science Edition)*, Vol. 5, 1996.

[7] Du Ying, *Out of the Village, Empirical Research of Rural Labour Migration in China*, Beijing: Economic Science Press, 1997.

[8] Huang Ping ed., *Seeking Survival*, Kunming: Yunnan People's Publishing House, 1997.

The 10 labour immigrants interviewed in Shenyang are all from rural areas, their occupations are characterised by manual labour, including leaders of building sites, cleaners, bricklayers, dormitory cleaners, mineral water workers, salesmen, etc., some of them are even informally employed, their work is not only unstable but also largely unacceptable to society. One hypothesis is that the lower the education degree, the more difficult it is to integrate into the city. Among the 10 labour immigrants the authors interviewed in Shenyang, only one mineral water worker has a bachelor's degree, and the rest of them only have a junior high school education degree or high school education degree. Besides, except that Mr Lu (Case: Shenyang-Labour-001), who has been working for 16 years, has obtained a Shenyang household registration through a loan house, the other 9 people are still non-Shenyang household registration and cannot enjoy the same social welfare as local citizens. Most labour immigrants face social injustice and social segregation, which seems to confirm the hypothesis mentioned above. In the following, the authors will analyse the process of immigrating and establishing of labour immigrants in cities through cases.

### 5.2.1 Emigration from Rural Areas to Cities

The labour immigrants the authors interviewed in Shenyang all came from northern provinces, mainly in areas with low economic development outside Shenyang or outside Liaoning province, and most of them came from rural areas in Northeast China. In addition, some labour immigrants are laid-off workers from the former old industrial base in Northeast China. Ms Wu (Case: Shenyang-Labour-003) and her husband were originally employees of a state-owned enterprise in Anshan city, Liaoning province. They were laid off after the reform of state-owned enterprises[9] in 2000. After that, Ms Wu has been working in Shenyang for four years and is now a cleaner in a dormitory apartment in a college. The case of Ms Wu also represents the living conditions of many laid-off workers in the old industrial base in Northeast China.

From the perspective of the motivation of migration, some scholars believe that the current phenomenon of rural labour going out for employment is the result of the combined effect of "survival pressure" and "rational choice". And the specific manifestation of the advanced "rational choice" is that after the basic material needs are met, the rural labour will pursue a higher level of material and spiritual satisfaction.[10] A quantitative survey of the new generation of rural labour in 2001 revealed that the initial purpose of rural labour going out for employment in the 1990s has shown a diversified trend. When the new generation of rural labour went out in the 1990s,

[9] The reform of state-owned enterprises is a reform carried out by traditional state-owned enterprises in order to adapt to the socialist market economic system in terms of systems, mechanisms, and management systems. Its core content is to establish a modern enterprise system, enhance the vitality of state-owned enterprises, and improve the economic benefits of state-owned enterprises (by translator).

[10] Huang Ping ed., *Seeking Survival*, Kunming: Yunnan People's Publishing House, 1997.

while considering making money, they also considered going out for employment as a way to change their living conditions and purse urban lifestyles. Compared with the first generation of rural labour, the motivation of the new generation of rural labour has changed from purely economic-oriented to the coexistence of economic-oriented and modern lifestyle-oriented or purely modern lifestyle-oriented. However, the motivation of most of the new generation of rural labour is still in a coexistence state of economic-oriented and modern lifestyle-oriented, they cannot yet change from rural migrants to urban residents.[11] Ten years later, the authors' research found that although the younger generation of labour immigrants pays more attention to the pursuit of modern life or urban life, economic-oriented motivation still accounts for a considerable proportion of labour immigrants. In a society where the gap between rich and poor is widening, there is still a group of people struggling to survive at the bottom of society. In the following, the authors will show several possible reasons for going out for employment through specific cases.

Case: Shenyang-Labour-001

After the failure of the college entrance examination, Mr Lu was recommended by his physical education teacher to practice gymnastics and study for another year to prepare for the second college entrance examination because of his good physical coordination and high comprehension. At that time, it was common to prepare for the second entrance examination. Basically, people who passed the college entrance examination were all students in the town or those who chose the second college entrance examination. However, considering the family's economic situation and his grades of cultural classes, Mr Lu gave up studying for another year to prepare for the second college entrance examination. After giving up studying and returning home, the responsibility of supporting the family fell on Mr Lu.

Case: Shenyang-Labour-002

Mr Zhang came from Yangjiazhangzi Township, Xingcheng City, which is far away from the county seat. He has three brothers and an elder sister, and his family's economic situation is relatively poor, so he and his brothers dropped out of school before graduating from junior high school and helped their parents do some farm work at home. Although the economic situation is better than before, life is still very tight. Several of Mr Zhang's neighbours who grew up together went out for employment, after a few years, they saved some money and married their wives. However, Mr Zhang has no economic basis to find a girlfriend or even get married, and his brothers have not been married yet, making it even harder for him to get married. Therefore, Mr Zhang wants to go out to work and change the poor economic situation through his own efforts. His basic idea is to earn some money, and then return to hometown to marry a wife. So, he packed up his luggage, brought only a few clothes and RMB200, and came to Shenyang alone.

Case: Shenyang-Labour-003

Ms Wu and her husband were both employees of a state-owned enterprise in Anshan City, Liaoning Province. They were laid off after the reform of state-owned enterprises in 2000. In order to provide for her daughter's tuition and maintain the family's daily expenses, her husband went to a construction site in Shenyang to work as a skilled worker, and Ms Wu was caught in the difficulty of finding a job. Ms Wu told the authors that she couldn't remember how many jobs had changed since she was laid off, only remembering that it was a particularly difficult time. Because of her old age and low education degree, it was really

[11] Wang Chunguang, Social Identity of the New Generation of Rural Migrant and Merger of Urban and Rural, *Sociological Studies*, Vol. 3, 2001.

difficult for her to find a job, even the job of washing dishes in a restaurant. It can be seen how hard life is, but even so, Ms Wu still emphasise the importance of children's education. For the sake of the children of the family, Ms Wu rushed to do several jobs every day. She used to be a nanny during the day and went to the night markets to sell things at night. Three years later, Ms Wu's daughter was admitted to a college in Shenyang and chose to major in architecture.

Case: Shenyang-Labour-004

Mr Lin was born in a peasant family in Jilin Province. He is the second child in the family, and his brother has already established a family and works in his hometown. In 1993, Mr Lin lived in Shenyang for 3 years with his trading father. After graduating from junior high school, he returned to his hometown in Jilin and studied mechanical machining at a local technical school for 2 years. In 2002, Mr Lin's father set up a factory in the far suburbs of Shenyang. For the family's business and helping the family make money, Mr Lin once again came to Shenyang to temporarily help in his father's factory. During the 7 years in Shenyang, Mr Lin changed a total of three formal jobs. The first two jobs were found with the persuasion and help of his father, and the third job was chosen by Mr Lin independently.

Case: Shenyang-Labour-005

Ms Wang is from Inner Mongolia. She came to Shenyang in 2001 and has lived in Shenyang for 8 years. Ms Wang now rents a house in Shenyang with her 18-year-old daughter and elderly parents. In her hometown, her family lived mainly on land, and her husband was the main labour force in the family. However, after the death of her husband in 2001, the family lost the main labour force, and the family lacked mechanical farming, so it was impossible to continue living on land. In order to make a living for the family, Ms Wang had to go out to work. However, at that time, her parents were old and her daughter was still in elementary school, Ms Wang could not leave them at home, so she took them to Shenyang together.

Case: Shenyang-Labour-006

Ms Li's daughter is born with disabilities and cannot take care of herself. Despite being physically disabled, Ms Li's daughter is particularly bright and hard-working, and it has been her dream to go to college. Ms Li devoted a lot of love for her daughter. When Ms Li's daughter was in high school in the county, the family also moved to the county for the convenience of taking care of their daughter. While taking care of her daughter, Ms Li also worked in a restaurant near her daughter's high school. After her daughter went to college, Ms Li followed her daughter to live near the university, taking care of her daughter while working in the cafeteria of her daughter's university.

Case: Shenyang-Labour-007

Mr Jia said that he is a person who is not satisfied with the ease and likes to challenge. Since he was a poor student, he dropped out of school after graduating from junior high school. Later, he was assigned to be a correspondent at the police station. In rural areas, working in the police station is very decent, but Mr Jia still chose to resign from the police station. He is unwilling to stay in the countryside all his life and wants to go out and experience the outside world. Mr Jia's mother opens a restaurant in the countryside, and his father is reselling goods elsewhere. Probably influenced by his parents, Mr Jia feels that he had the talent to do business, so he wants to go out and give it a try.

Case: Shenyang-Labour-008

Yang said that his mother longed for big cities. In order to stay in the big city and let his parents come to the city in the future, he willingly struggled in Shenyang alone. Although he suffered many setbacks and pains along the way, he still did not regret it and would not give up.

As can be seen from the above cases, family economic factors are still the main reason why people go out for employment (such as Case 001,002,003,005). Poor family economic conditions not only affect children's school education but also affect partner selection and family formation, all of which prompt people to go to the city, hoping to return to the countryside to complete their unfinished things after making money in cities. Some of the above cases also show that some people move from the countryside to the city because of their personal desire of living and developing in the city (such as Case 007,008).

### 5.2.2 *Surviving in the Flow*

Unlike intellectual immigrants who take the initiative to move to a larger development space with a clear plan, labour immigrants are more mobile and passive when changing jobs. The occupational mobility of labour immigrants can be divided into initial occupational mobility and secondary occupational mobility. The initial occupational mobility is divided into two types, one is the flow from agricultural workers to non-agricultural workers, and the other is the flow in the field of non-agricultural workers. To sum up the characteristics of the occupational mobility of labour immigrants, there are the following six points. First, the occupational mobility of labour immigrants is mainly at the horizontal level, and the occupational status tends not to rise or fall significantly; second, due to the informality and impermanence of works, the occupational mobility of intellectual immigrants is becoming more and more frequent; third, intellectual immigrants lack accumulated social resources; fourth, the process of occupational mobility of labour immigrants lacks a good management system and service mechanism; fifth, men are more likely to flow in occupations than women, which is related to gender discrimination in employment and the female physiological characteristics; sixth, most of the labour immigrants are engaged in manual work.[12] Contrary to the rise in occupational status caused by the occupational mobility of urban residents, the correlation between occupational status before and after the occupational mobility of labour immigrants is weak, so the occupation of rural labour has always been maintained at a low social level, and they generally engage in low reputation, low-skilled, and low-paying occupations that urban residents are not willing to engage in.[13] Specifically, the authors once classified labour immigrants into four categories: enterprise workers, agricultural workers, casual workers, and professional workers. These four categories can also be subdivided, and different occupational types have different situations. Among them, casual workers and workers in individual enterprises have the worst survival situation. The labour immigrants the authors interviewed in Shenyang are mainly

[12] Li Qiang, Occupational Mobility of Rural Labour in China's Mainland, *Sociological Studies*, Vol. 3, 1999.

[13] Yin Xiaoqing, The Obtain and the Maintain of Mobile Employment, *Journal of Nanjing Normal University (Social Science Edition)*, Vol. 5, 2001.

enterprise workers and casual workers, and these two types of urban immigrants are the most marginal and bottom-level groups in cities.

It can be seen from the cases that frequent occupational mobility that does not cause a rise or fall in position is the normal state of survival of labour immigrants. Labour immigrants often achieve employment through informal channels because they cannot meet the employment requirements of the formal employment sector. Some scholars believe that in the informal economy there are also conflicts between exploitative, abusive, and aggressive business owners and workers who lack defensive capabilities. In addition, there are also capitalist relations of production and reproduction. Workers in the informal economy are in a special situation. They are lower-ranked workers, with low wages and benefits, and poor working conditions, which makes them more vulnerable. However, it is also true that the informal employment sector is easier for labour immigrants to enter.[14]

According to the labour market segmentation theory, the labour market can be divided into primary segments and secondary segments. And the dual urban-rural labour market formed under the constrictions of China's household registration system and the bipolar labour market based on factors such as the requirements for occupational skills have jointly constructed a pattern of multiple labour market segmentation. Due to the constrictions of their own qualities and institutional factors, labour immigrants are mostly in the marginal position in the pattern of multiple labour market segmentation, and they can hardly affect the social structure with individual actions. The cases demonstrate the weakness of labour immigrants and their frequent occupational mobility.

Case: Shenyang-Labour-009

Ms Liu finished her first year of junior high school in 2002 and dropped out of school to come to Shenyang. She had a cousin who opened a barber shop in Shenyang, so Ms Liu came to Shenyang to study hairdressing with her cousin. Soon after, Ms Liu returned to her hometown to work in another barber shop. But it didn't last long, Ms Liu decided to resign and went to Shenyang again. After coming to Shenyang for the second time, Ms Liu initially worked as an assistant in a barber shop and was later entrusted by her friend to look after a barber shop. But her friend didn't want to continue to run the barber shop, so Ms Liu returned to her hometown after taking care of the barber shop for two months and stayed in her hometown for another month. Later, Ms Liu found a waiter job in Fushun, and she didn't come to Shenyang because of the cold weather. After that, she worked in the barber shop in Fushun and Shenyang for three months each. So far, Ms Liu has been changing jobs frequently, on average every two or three months, except for working in her cousin's barber shop for more than a year. In addition to barber shop staff, she also worked as waiters and cashiers. Three years ago, Ms Liu went to a company to do business on Sanhao Street under the introduction of her friend.

Case: Chengdu-Labour-007

Mr Zhang comes from a remote village in Dazhou City, Sichuan Province. After graduating from junior high school, he dropped out of school because of family economic conditions. But he didn't want to live a poor life in the village like his parents, so he went out to work with his fellow villagers at the age of 16. At first, Mr Zhang could only do odd jobs

[14] Cai He, ed., *Rural Migrant in the Process of Urbanization: A Study from the Pearl River Delta*, Beijing: Social Sciences Academic Press (China), 2009.

everywhere because of his young age and lack of skills, such as transporting coal briquettes, water, garbage, and working as temporary workers on construction sites. Although he is very tired and often looked down upon by others, he still feels very satisfied by being able to earn his own living.

Case: Zhengzhou-Labour-005

Since 1992, Mr Liu has worked in southern cities such as Guangzhou, Zhuzhou, and Liuzhou. He has done a lot of different jobs, selling vegetables, working as a construction worker, and also working in factories, living a completely different life than before. Mr Liu said, "When I used to sell chicks, it was very easy to make money and I did not feel tired. Now, on the contrary, making money is particularly difficult and tiring." Mr Liu had bought a second-hand tricycle before, and he had to get up at 4 o'clock every morning, rode an hour to buy vegetables at the wholesale market, and then rode another hour to go to the market to sell vegetables. Mr Liu said, "Selling vegetables is particularly hard and earns very little money. On average, you can only earn RMB50 a day." During his time working on the construction sites, he mainly did simple work such as painting, or maintenance work, including repairing water pipes, windows, floor tiles, and so on. Since 1992, Mr Liu has been living such a wandering life, wandering between various cities and industries. Currently, he works on construction sites in Zhengzhou. Probably because of his older age, Mr Liu returned hometown more frequently than other rural migrants. Unlike other rural migrants who does not go home all the year-round, Mr Liu will go home as long as he has no work to do, and then go out again when he wants to work outside. When returning to his hometown, Mr Liu still goes to the county to sell chickens and chicks, which is his main source of income.

As mentioned earlier, Mr Lu (Case: Shenyang-Labour-001) went out to work after failing the college entrance examination. He has changed more than ten jobs and has worked in Hebei Province and Shenyang, Dalian, Guangzhou, and other cities. During this long and arduous working life, Mr Lu's ability has been improved, and his work gradually shifted from manual work to technical work. In terms of marriage, after experiencing a failed marriage, Mr Lu finally remarried and bought a house to settle in Shenyang, realising his dream of taking root in the city, which is also the only case of all Shenyang cases to buy a house and obtain Shenyang household registration. After getting married, Mr Lu ended his wandering life and began to take the family as his main responsibility. When looking for a job, Mr Lu will also consider family as the key factor. However, most young labour immigrants find it difficult to find a partner in cities, not to mention getting married.

Case: Shenyang-Labour-008

The single problem currently haunts Yang. Yang had a blind date once, and the date started just fine until the girl heard about his job and directly refused to continue the date. The girl said, "people say that my future husband is career-oriented." When the authors asked Yang what else he wanted to say at the end of the interview, he repeatedly asked us to help him pay attention to the right person who can live together for a lifetime. Yang said, "I don't want my partner to be the kind of person who always asks for various valuable gifts, but someone who can really live together."

Case: Shenyang-Labour-009

When talking about her nearly seven years of social experience and feelings of living in Shenyang, Ms Liu said, "now people are too realistic and material". Ms Liu's conclusion was drawn through her brother's love experience. Her brother's girlfriend broke up with her brother because of the money and marriage delay, and the seven-year relationship ended. Ms

Liu believes that people now value money more than love. In fact, Ms Liu does not completely distrust love, but no longer blindly puts love first. After Ms Liu's first relationship ended, she no longer valued love as much as before. As for work, Ms Liu feels that work is everyone's own business, so she never asks the performance of her colleagues, but just diligently does her job.

In contemporary China, the life of labour immigrants should be understood in a broader social context. China's social development is driven by two interacting forces. One is the state's governance of society and the transformation of political projects. The other is the capitalisation and marketization of socialist countries, which is named "pursuing modernity" and "for globalisation" by the country, and the country even put forward the slogan of "connecting with the world".[15] In the most energetic youth, the labour capital of labour immigrants was requisitioned by global capitalism and national socialism to promote the development of cities and industries, jointly completing socialist capital accumulation.

Labour immigrants, as participants in China's socioeconomic transformation today, have promoted the development of urbanization and national modernization. However, under the dual power of the political and economic structure of national socialism and the logic of capital operation, these promoters of social progress did not undergo a simultaneous transformation of social identity but were isolated in cities. Labour immigrants are not fully integrated into cities but are limited to the marginal space of cities under the pattern of multiple labour market segmentation. Most of labour immigrants are engaged in temporary, labour-intensive, and low-status occupations and become the main component of informal employment. Furthermore, labour immigrants were further divided and segregated under the influence of the urban-rural structure and regional differences. The rural emigrant who migrates from rural areas to cities or economically developed areas to work are not only restricted by rural household registration, but also discriminated against because of their identity of the immigrant population. Besides, they will be treated differently by the divided industrial sectors within the city, such as the public sector, professional and technical departments, and government monopolies, etc., all of which have no access to rural migrants due to access restrictions.[16] Although labour immigrants are in cities, they cannot enjoy the same treatment as urban residents, they are only introduced into the urban production space as the most marginal labour force under the labour market segmentation pattern.

The daily living space of labour immigrants is also relatively segregated from the urban living space. There are three patterns of locale distribution for labour immigrants. The first is the immigrant population gathering area formed in the city based on the geographical connection, the second is villages within cities formed by the gathering of the immigrant population during the stage of urban expansion, the third is the factory dormitory commonly found in urban industrial zones or

[15] Pan Yi, *Chinese Women Workers: The Call of the Emerging Working Class*, Hong Kong: Ming Bao Publishing House Co., Ltd., 2007.

[16] Cai He, ed., *Rural Migrant in the Process of Urbanization: A Study from the Pearl River Delta*, Beijing: Social Sciences Academic Press (China), 2009.

economic development zones, which usually form a binary community with the local community.[17] No matter where labour immigrants live in the city, they only have the status of urban labour, but not the status of urban residents. Their social attributes are divided into rural areas by the household registration system, while in cities they are just passers-by, which makes them trapped in a dilemma. And the temporary residence permit system currently implemented in most cities is one of the institutional obstacles for the situation of labour immigrants being segregated in cities.

Case: Hangzhou-Labour-009

I think the temporary residence permit system is very unreasonable. The temporary residence license needs to be renewed once a year and charges more than RMB40 each time, which is particularly troublesome and expensive. I originally wanted to bring my children together, so that I can take care of them myself, but now I give up this idea. If the immigrant population want to study here, they will have to pay temporary schooling fee, not to mention the cost of living here. Therefore, only my child who has gone to college is studying in the city, while my other two children are studying in their hometown. However, the quality of education in my hometown is not as good as in the city, and the knowledge level of rural children is lower than that of urban children, which is also a social phenomenon of inequality between urban and rural areas.

Mr Lu (Case: Shenyang-Labour-001), who was a bricklayer porter in the early days, and Mr Zhang (Case: Shenyang-Labour-002), who is a bricklayer, are typical of casual workers among labour immigrants. These casual workers are not employed by formal institutions, and their rights and interests are not guaranteed by the formal employment system.

The authors once conducted a survey of casual workers in Guangzhou and found that 87.7% of casual workers did not sign labour contracts with their employers.[18] The wages, working hours, and wage payment time of casual workers among labour immigrants are mostly in the hands of employees, so the phenomenon of arrears of casual workers' wages may occur from time to time. What's more, these casual workers' medical and health insurance are not guaranteed, especially for casual workers who are engaged in some high-risk occupations, such as construction workers, who have a relatively high injury rate with a low possibility of receiving injury compensation. Therefore, this part of the casual workers engaged in high-risk occupations is the most vulnerable among labour immigrants. Their material and life security in the city cannot be guaranteed, let alone integrate into the city and settle in the city.

Case: Shenyang-Labour-001

"I initially worked in a brick factory. The brick factory emphasises efficiency, and workers have to synchronise with the machine, so I feel particularly tired and unfit for the job. After working for a month, I lost a lot of weight and weighed only 121.25 pounds, and had thick calluses on my hands. Working as a porter in a brick factory is acceptable to people who have been doing it all year round, but it is difficult for the novice. We work like machines

---

[17] Xiang Biao, *Communities Crossing the Boundary: The Life History of "Zhejiang Village" in Beijing*, Beijing: SDX Joint Publishing Company, 2000.

[18] Zhou Daming, *"Free" Urban Marginal People: A Study on Casual Workers in China's Southeast Coast*, Guangzhou: Sun Yat-sen University Press, 2007.

without thought, what we need is just physical strength." Later, he felt that he could not bear this kind of work and resigned after a period of time. After changing a job, he was satisfied with the salary, but the new employer wanted to keep his identity document in case he left, and he would not even be allowed to ask for leave, so he finally decided to quit.

Case: Shenyang-Labour-002

In 1996, Mr Zhang went to work in a shoe factory, but he resigned again after only one month. Mr Zhang told the authors that the pungent chemical smell of the glue in the shoe factory is unbearable, and a thick layer of glue was always stuck on his hands. In addition, although the shoe factory said that it covered food and accommodation for employees, the actual food and accommodation conditions were extremely bad. When the shoe factory was busy, Mr Zhang needed to work overtime, but there was no overtime pay. Finally, Mr Zhang left the shoe factory.

……

Relying on his skills, Mr Zhang finally settled in Shenyang. Mr Zhang is not too satisfied with his current state of life. He feels that his life and work are not very stable. In his words, he does not know what will happen tomorrow, and he is also unable to change the current state of life.

Case: Zhengzhou-Labour-004

"I have no ambitious thoughts; I just want to live a stable and healthy life. What I think about every day is that I can earn RBM70 a day, and how much I have spent, and whether the remaining money can make up for the expenses of my family."

Case: Zhengzhou-Labour-005

"Once I installed an air conditioner for a family, when I came down, I didn't see the cutting machine beside me, and my leg was slipped along by the cutting machine. At that time, there was blood flow, and I finally got five stitches on my leg, and now there are still obvious scars on my leg." As he told the authors, he lifted his trousers to show the authors the scars on his leg. He also worked in a glue factory in the south. The glue in the factory was toxic, and he often choked on the pungent chemical smell of the glue, making him have no appetite. He did not dare to stay here after working in this factory for two months.

As job recruitment raises the requirements for the academic qualifications of the employees, some people who still rely on manual labour to survive also hope to improve their education degree to change their destiny, and some people seek benefits for themselves through informal channels and unspoken rules. Although labour immigrants cannot directly resist the entire social system and structure, they can still use the "weapon of the weak" such as "counterfeiting" and "being opportunistic" in their daily life practices to conduct hidden text-style struggles. This kind of negative individual action that exploits the loopholes of the existing social structure and rules cannot trigger an organised collective action, nor can it change the current life situation of labour immigrants. Moreover, most of labour immigrants are silent on the existing social system and rules, and will not take practical action, which makes the change less likely to occur. However, the emergence of labour immigrants' resistance to the existing social system and structure also enlightens us that the structural reform of social distribution and labour system cannot simply rely on the reform of the top-down political and economic system, but also on labour immigrants mobilise from the bottom up.

Case: Shenyang-Labour-002

Mr. Lu only has a high school degree. Although he took the self-study examination for higher education, he only passed two subjects and did not get a higher education degree. With the popularity of higher education, junior college or college education degree has become the minimum threshold for employment.

Case: Shenyang-Labour-009

"It's not necessary to rely on this education degree to find a job, but I still feel that at least I need a decent education degree, now I don't even have a junior high school education degree." When registering at Radio and Television University, she truthfully told the person in charge that she had no diploma. She has already rented a house next to the school and is ready to use her spare time to go to classes at night.

The above analysis shows that labour immigrants face more challenges and difficulties than intellectual immigrants in terms of immigrating into cities and establishing in cities. Although labour immigrants have the desire to settle in cities, their income and work still keep them in a difficult state of life in cities. Therefore, returning home has become their ultimate plan after they spent their youth in the city. In the end, they are still passengers who have contributed to the city.

## 5.3 The Immigration and Establishment of Entrepreneurial Immigrants

From the perspective of economic nature, the entrepreneurial immigration economy belongs to the private economy and the individual economy. After Deng Xiaoping's south tour in 1992, the status of the private owner's economy in the socialist market economic system with Chinese characteristics was recognised and legalised, thus accounting for an increasing proportion and playing a more important role in China's economic development. Although the emergence of the private sector has opened the door to new opportunities for the Chinese, its social significance is different for urban residents and rural residents. Some studies have pointed out that under the conditions of different opportunities brought about by the dual urban-rural structure based on the household system, self-employed workers in urban and rural areas have very different personal backgrounds. Early field investigations revealed that the owners of private enterprises in the countryside were mainly leaders of production teams, educated youths, and veterans who remained in the countryside. Compared with ordinary farmers, these people have more abundant human capital and political capital. In cities, most self-employed and private owners come from marginalised groups of society, such as rural migrants, unemployed youth, laid-off workers, and retirees. For these people, there is basically nothing to lose, and it is better to participate in the private economy, from which they might profit more.[19]

[19] Wu Xiaogang, Plunging into the Private Business: Self-employment and Social Stratification in the Transformation of China's Urban and Rural Labour Market (1978–1996), *Sociological Studies*, Vol. 6, 2006.

As rural labour continues to migrate to cities on a large scale, the rural labour in the city is also constantly changing, and its internal differentiation is taking place. The most notable one is the accelerated differentiation of occupations. People transform from the pursuit of survival to the pursuit of higher development and transform from pure manual labour to the coexistence of technical, capital, intellectual labour, and manual labour. In the course of these holistic transformations, some manual workers have gradually transformed into entrepreneurial immigrants who hold means of production and even hired workers through the accumulation of social and monetary capital, realising the change of social roles.

Entrepreneurial immigrants generally work for a longer period of time and have their own business projects. For example, the authors interviewed 11 entrepreneurial immigrants in Shenyang, and their business projects include barber shops, repair shops, fish stalls, snack bars, clothing stores, beauty shops, and so on. As can be seen from the occupational distribution, entrepreneurial immigrants mostly operate self-employed shops. Compared with intellectual immigrants who are engaged in occupations with high social recognition and respect, the social status and recognition of entrepreneurial immigrants are relatively low but higher than labour immigrants who rely on manual work. Besides, the development of entrepreneurial immigrants is carried out under the logic of market capital, which is not substantially comparable to the logic of the development of intellectual immigrants based on intellectual capital. Mr Ji's (Case: Shenyang-Entrepreneurial-001) ideal is to choose a good location and open a larger barber shop in Shenyang, preparing to develop his career in hairdressing. He said, "after all, this is my former occupation." The combination of skills and capital is the basis for entrepreneurial immigrants to establish in cities, and the education degree as a variable has little effect on the behaviour of entrepreneurial immigrants.

In the following, the authors will analyse the immigration and establishment of entrepreneurial immigrants in Shenyang through case analysis.

As mentioned earlier, since the reform and opening up in 1978, most self-employed and private owners in cities come from marginalised groups of society, such as rural migrants, unemployed youth, laid-off workers, and retirees. These people are not included in the formal employment system and are marginalised by society. Compared with people in the formal employment sector in cities, they are confined to informal jobs in the secondary labour market and survive through self-employment. In the early stage of development, some entrepreneurial immigrants and labour immigrants have no essential difference, but the specific occupation distribution is different. Actually, a considerable part of entrepreneurial immigrants was originally working for small owners engaged in small-scale production, operation, and service businesses. Later, they gradually developed into small owners or self-employed vendors with no labour relationships, which belonged to self-employment.

Case: Shenyang-Entrepreneurial-001

Mr Ji comes from Fushun City, Liaoning Province. He has been diligent since he was a child and often helps his family to do farm work, deeply understanding the difficulty and hardship of peasants. After graduating from junior high school, Mr Ji went to Shanghai with his parents to do medicinal materials business. But the business was not easy to do, so he followed his fellow villagers to work in Shenyang and washed dished in a restaurant for two months. Later, he felt that if he kept doing such work, there would be no bright future. So he decided to go back to Fushun to learn a skill, after careful consideration, he chose to learn haircut. After graduating from the school of hairdressing in Fushun, he confidently returned to the county town where his hometown was located and opened a barber shop. However, he was not skilled enough, and there were few customers, and it didn't take long for the barber shop to close. Therefore, his family thought that he was not suitable for the hairdressing industry at all, and relatives and friends also opposed him to engage in the hairdressing industry again. Mr Ji, who is introverted, said nothing, and still insisted on coming to Shenyang to continue studying hairdressing. After working in salon A and salon B, he acquired a small barber shop to operate by himself, and he planned to find a better location to open a branch after he had sufficient funds, realising his dream step by step.

Case: Shenyang-Entrepreneurial-004

In the summer of 2000, Mr Li graduated. At that time, he had not yet found a job matching his major, and he was unwilling to continue to let his family pay for his living. But he was also unable to live in Shenyang alone without any financial resources. To make ends meet, he found his first job selling phone cards. Later, Mr Li found a second job as a salesman in an online company. Mr Li's third job was a purchaser of a company, after four or five years in the company, he found the fourth job in real estate sales in 2005. In 2007, Mr Li saved RMB60,000-70,000, and he borrowed another RMB100,000, from a friend, and carefully opened a charcoal-fired hotpot restaurant.

Case: Shenyang-Entrepreneurial-003

Ms Duan's daughter went to high school, and her son also went to junior high school. The tuition for the two children was a heavy burden, and the family's economic situation became a little bit tight. Ms Duan's younger brother did business in Shenyang. In the winter of 2004, his business was too busy, so he asked Ms Duan's husband for help. Therefore, Ms Duan stayed at home to take care of the farmland and her family while her husband went to Shenyang to do business with her younger brother. In 2006, her husband's business was particularly prosperous, and he also acquired the stalls of the owners selling tobacco and tea next to it. Then the business was too busy, so Ms Duan rented out her farmland and came to Shenyang to do business with her husband.

Case: Shenyang-Entrepreneurial-005

Yanhong comes from Tieling. She went out to work at the age of 18. By the time she was 32 years old, she had been working, getting married, having children, setting up stalls, and opening small shops. Unlike most urban immigrants who want to settle in Shenyang, she is desperate to make money every day just to save enough money to return to her hometown. Actually, she does not deny that living in the city is better than living in the countryside, but the reality does not allow her to settle in the city, and she still longs for the familiar land and people in her hometown. She came from her hometown, and strongly hoped that she could return to her hometown one day. She worked hard every day for the purpose of living a simple but happy life. The life she hopes is that there is a big house in her hometown, where parents, daughters, husband and herself live in, and then raise a few pigs and a group of chickens and ducks with her family. In her own words, "a simple and happy life is enough."

The above cases collectively reflect the relationship between entrepreneurial immigrants and labour immigrants, that is, most entrepreneurial immigrants are transformed from labour immigrants. Compared with labour immigrants who have been working for others, entrepreneurial immigrants use the money earned by early work to run some large or small business and become self-employed owners or even hire other people to work. Previous academic studies have used the "push-pull" theory to analyse the social phenomenon of rural migrants immigrating to cities, but this analysis ignores the social psychology of rural migrants, thereby affecting insights into the future direction of this group of people. We can use the anti-driving mechanism to explain the behaviour logic of entrepreneurial immigrants represented by Ms Duan (Case: Shenyang-Entrepreneurial-003) and Yanhong (Case: Shenyang-Entrepreneurial-005). From Ms Duan's words, we can see that the change of the family's consumption structure forced them to come to the city, and only then could they fulfil their responsibility of providing tuition for their children. In Yanhong's view, what matters most is not living in the city or in the countryside, but living with her family. Whether it is for the consideration of children's education or the quality of family life, these considerations show the profound tradition of entrepreneurial immigrants, that is, the family is the primary consideration at all times. The family-centred collective immigration or family-unionise trend has become the most prominent feature of entrepreneurial immigrants.

However, we also need to note that the family-unionise of immigrants is not consistent, so family-centred collective immigration inevitably brings certain risks and increase immigration costs. From the perspective of the evolution of population migration around the world, the historical process of population migration is roughly divided into three closely related and progressive stages. The first stage is the pioneer stage of population migration, the second stage is the family-unionise stage of population migration, and the third stage is the popularise stage of population migration.[20] Previous academic studies have found that in the late 1980s, especially in the early 1990s, the migration of rural labour in China has entered the second stage of population migration, which is the family-unionise stage of population migration. Nonetheless, in this trend of family-oriented immigration, we should also pay attention to the rural migrant who does not have the will to settle in cities. In fact, not every rural migrant can fulfil the desire of family-centred collective immigration. In a society where the upward mobility opportunities lack, not everyone can afford the high costs of settling in cities. What's more, within the framework of China's household registration management system, choosing to settle in cities means giving up rural household registration and its corresponding land and collective income. A 2006 survey of the rural migrant in the Pearl River Delta also showed that 56% of the rural migrant was willing to give up land, while 44% were unwilling to give up land, and 60.2% were unwilling to convert rural household registration to urban household registration.[21] Therefore, a considerable portion of entrepreneurial immigrants hopes to benefit from both businesses in cities and land in rural areas. In fact, the willingness of rural migrants to settle in cities is not as high as we thought.

In addition, there are some entrepreneurial immigrants who have a strong desire to settle in cities. They try to expand their small business through their own efforts and wealth accumulation, and eventually achieve upward social mobility. Due to life

[20] Yu Xianzhong, The Floating Position of Chinese Population, *Journal of University of Jinan (Social Science Edition)*, Vol. 6, 2004.

[21] Cai He, ed., *Rural Migrant in the Process of Urbanization: A Study from the Pearl River Delta*, Beijing: Social Sciences Academic Press (China), 2009.

cycle limitations, some middle-aged entrepreneurial immigrants often see themselves as stepping stones for future generations to settle in cities. However, they themselves are still in a state of transition. They must not only consider the business in the city but also the land and their elderly parents in the countryside.

Case: Shenyang-Entrepreneurial-006

The authors can feel their family's love for life in big cities. She said, "every aspect of urban life is good, and shopping is particularly convenient, but living in rural areas is not as convenient as in cities." She is willing to live in the city, and hopes that her grain and oil store can continue to operate until the day she can't do it. But she also said, "It depends on the situation to go and stay in the end. There are elderly parents in my hometown, so we may still have to go back to our hometown." She and her husband are in their forties, and it is unlikely that they will have any greater achievements. Moreover, due to the economic crisis, fewer people went out to eat, and the restaurant has bought less rice and oil from their stores, so their earnings are worse than in previous years. In the future, they may hand over the grain and oil shop to a son, but she still hopes that her two sons will be able to find jobs in Shenyang, and finally settle in Shenyang and get married.

Case: Shenyang-Entrepreneurial-009

Due to busy work, Ms Jiang only returned home for the Spring Festival during the 14 years since she left her hometown, and also occasionally went home to see her family during the off-season of work. The authors asked her, "Do you feel homesick when you can't go back during the Spring Festival?" She said, "Of course." In order to cope with her homesickness, Ms Jiang spent every Spring Festival with her fellow villagers in Shenyang. It is also this nostalgia that made Ms Jiang always have a stronger sense of identification with her hometown. Even if she will settle in Shenyang in the future, even if she says that she feels the same everywhere, she still thinks she is from Sichuan. After she retires, she still hopes to go home to spend the rest of her life. Then the authors asked Ms Jiang's son, "where are you from?" Her son replied without hesitation, "Shenyang." This year, Ms Jiang intends to take her son and husband to return home for the Spring Festival. Because their parents are old, they must cherish the time they can get along with their parents.

It has been decades since the large-scale urban-rural migration, and we should also note the differences in the willingness of settling in cities of the new generation of rural migrants due to age differentiation. Compared with the first generation of rural migrants, the new generation of rural migrants already have a strong sense of identification and willingness to settle in cities when they go out of the countryside, and they work hard to achieve upward social mobility.

Case: Shenyang-Entrepreneurial-008

Chengcheng's plan for the future is to expand the size of the beauty salon, and then open the branch to the urban area. It is good for her to have this kind of idea. After all, she needs an ideal to be motivated to realise it. In addition, she wants to obtain a management degree through the self-study examination. The authors also briefly introduced her to the process and method of self-study examination. There are two reasons why she is ready to study management. First of all, management can help her manage the beauty salon. If she opens a chain store in the future, management will help a lot. Secondly, even if she no longer works in the beauty industry in the future, she can still rely on the management degree to find a good job. She said, "a stable life with a fixed salary is also very good. The income from running a beauty salon is not stable. If there is no income this month, I can only rely on the principal to maintain business and life, which is huge pressure for me. What's more, running a beauty salon alone can be particularly tiring, and I need to manage a lot of miscellaneous things. Of course, there are also good things about doing business by myself, that is, I will feel free and there will not be too many restrictions." In any case, enriching herself with knowledge is a good choice. And she hopes that whether she is opening a chain store or studying management in the future, she will succeed.

Previous academic studies on urban immigrants have paid little attention to the impact of ethnic identity and ethnic relationships on the immigration and establishment of urban immigrants. But in reality, ethnic identity and ethnic relationships are important factors that affect the immigration and adaptation of urban immigrants to cities. Besides, one of the common issues in the urbanization process around the world includes the issue of ethnic and social disputes. There are people of different races, and ethnic groups in the city. They may be scattered or gathered, and they may also conflict with each other due to misunderstanding and competition in the process of interaction. In a word, ethnic issues are usually intertwined with other urban issues, affecting the willingness of urban immigrants to immigrate and settle in cities.

Therefore, when we are discussing the issue of integration of urban immigrants, we should not only focus on the problems faced by individual immigrants but also the relationship between ethnic groups. Due to historical reasons, some ethnic groups have been stereotyped as difficult to reach. Hence, the elimination of geographical discrimination and ethnic discrimination is still our focus in the process of urbanization.

In this chapter, we analysed the process of immigration and the establishment of three types of urban immigrants through cases. Among these three types of urban immigrants, intellectual immigrants usually have high education degree and relatively clear intentions and strategies of urbanization. For the intellectual immigrants, immigration is not only the process of spatial transfer from rural or second- and third-tier cities to metropolises, but also the process of identity transformation from college students to urban professionals and settlers. In this process, intellectual capital is one of the most prominent advantages and characteristics of intellectual immigrants. Besides, although intellectual immigrants must undergo a stage of social role adjustment and adaptation to cities, on the whole, they will gradually find a suitable position in the city with their high education degree and intellectual capital. Therefore, intellectual immigrants not only have a relatively strong desire to settle in cities but also have a deep desire to start a family in cities. In contrast, although labour immigrants are also active participants in China's socioeconomic construction, they are still isolated in cities and can hardly enjoy the same welfare as urban residents under the dual power of the political and economic structure of national socialism and the logic of capital operation. In addition, although labour immigrants have the desire to settle in cities, their income and work still keep them in a difficult state of life in cities. Therefore, returning home has become their ultimate plan after they spent their youth in the city. In the end, they are still passengers who have contributed to the city. To some extent, most entrepreneurial immigrants are developed from labour immigrants. They started their own business through the accumulation of capital in their early labour work. Compared with labour immigrants, entrepreneurial immigrants have a stronger desire to settle in cities and show a tendency of family-unionise of immigration. However, considering the high cost of family living in the city and the income of rural land, they will rationally hope to get benefits from both businesses in cities and land in rural areas. Therefore, their desire to settle in cities will also be affected by many factors.

# Chapter 6
# The Social Network of Urban Immigrants

The social network is one of the core concepts in the field of social sciences, and its research path has been elaborated from different perspectives by sociologists such as Granovetter, Bourdieu, Coleman, Burt, Putnam, Lin Nan, et al. Among them, the most influential theories include Granovetter and Bian Yanjie's assumptions about the strength of strong ties and weak ties,[1] Lin Nan's theory of social resources,[2] Bourdieu and Coleman's social capital theory,[3] and Burt's theory of structural holes.[4] The social support network in the field of social sciences is explained from the perspective of the interaction between individuals and the structural society. Scholars such as Zhang Wenhong believe that the difference in viewpoints in social support networks stems from scholars with different research orientations focusing their analysis on the location of networks, the strength of ties, or embedded resources.[5] In fact, the earliest measurement standard proposed by Granovetter to measure the strength of ties and support from four dimensions of interaction frequency, emotional strength, intimacy, and reciprocal exchange is still guiding the research of social support networks.

From the perspective of the relationship between social-psychological stimulation and individual mental health, social support can be regarded as the positive effect that individuals can obtain through social connections to mitigate psychological stress reactions, relieve mental stress, and promote social adaptation. Social connections refer to spiritual and material support and help from families, relatives, friends,

---

[1] Granovetter, M. S., "The Strength of Weak Ties", *The American Journal of Sociology*, 1973, Vol. 78, No. 6, pp. 1360–1380. Bian Yanjie, Zhang Wenhong, Economic Systems, Social Networks and Occupational Mobility, *Social Sciences in China*, Vol. 2, 2001.

[2] Lin Nan, Ensel, V. M., Vaughan, J. C., Zheng Lu, eds. Trans, Social Resources and Strength of Ties, Structural Factors in Occupational Status Attainment, *Social Science Abroad*, Vol. 3, 2001.

[3] Coleman, James, "Social Capital in the Creation of Human Capital", *American Journal of Sociology*, Supplement 94, 1988, pp. S95–S120.

[4] Burt, Ronald S., *Structural Holes, The Social Structure of Competition*, Cambridge, MA: Harvard University Press, 1992.

[5] Zhang Wenhong, Social Capital: Theoretical Discussion and Empirical Research, *Sociological Studies*, Vol. 4, 2003.



D. Zhou, *Urban Migrants in China*,
https://doi.org/10.1007/978-981-99-3114-9_6

colleagues, groups, organisations, and communities. Besides, there are also some academic studies that equate social support networks with social connections. It is generally believed that social support is premised on the existence and interaction of social networks. Typically, social support networks can be divided into formal social support and informal social support. Among them, formal social support includes the support system that can be provided in the social structure, such as medical and social services, while informal support includes interpersonal interactions in daily life such as families, friends, neighbours, relatives, churches, clubs, and so on. Xiao Shuiyuan divides social support into two categories, one is the visible and practical support, including direct material support and the existence and participation of social networks and group relationships, the other is subjective, experiential, and emotional support refers to the emotional experience and satisfaction of individuals being respected, supported, and understood. Specifically, in the field of urban immigrants, seeking relationships and constructing social networks is an adaptive strategy for urban immigrants, and it is also an important social capital for them to deal with institutional barriers, social barriers, and cultural barriers.

After analysing the sample questionnaire survey conducted in Jinan city, Shandong province, China in 1995, Li Peilin found that rural migrants mainly relied on their traditional kinship and geographical relationships in the process of migrating from rural areas to cities. Just as rural migrants migrated their kinship and geographical relationships to the enterprise when they entered the township enterprise, now they migrate this social network to their life circle in the city. Surprisingly, in the process of China's market transformation, this traditional social network, which is incompatible with the principles of modernity, plays an important role in saving the cost of urban-rural labour migration and effectively allocating resources as an informal institution.[6] Zhang Jijiao also noticed the phenomenon of rural social networks being migrated to the city, and he called the social network of rural migrants who immigrated into the city "the urban version of the difference sequence pattern". Zhang Jijiao believes that for urban immigrants from rural areas or small towns, the city is an unknown and uncertain new world. On the one hand, the social network based on kinship and geographical relationships is rural migrants' most intimate and reliable social foundation, which provides them with the basic conditions to survive in the city. On the other hand, they cannot survive only in the acquaintance circle. In order to survive and develop in the city, they must deal with strangers and establish new social networks of work, investment, and management, as well as relationships such as friends, marriages, and exchanges, in accordance with city rules.[7] It can be seen that Zhang Jijiao not only paid attention to the application of rural social networks in cities but also noticed the construction of new social networks in cities. The authors also believe that it is necessary to explore the current status and changes of urban immigrants' social networks in combination with the two aspects mentioned above. Based on the distinction of nature, Cai He and other scholars divided social networks into work support networks, life help networks,

---

[6] Li Peilin, Social Networks and Social Status of Rural Migrants, *Sociological Studies*, Vol. 4, 1996.

[7] Zhang Jijiao, *Adaptation to the City*, Beijing: The Commercial Press, 2004.

and emergency help networks and conducted a quantitative statistical study in the Pearl River Delta. The study found that consanguinity, geographical relationships and kinship support for rural migrants' work mainly in the provision of preliminary work information, and it has little effect on their follow-up work development, but it is very important for their sickness care, emotional comfort and decision-making on key issues.[8] The authors' survey of urban immigrants in major cities across China shows that the social network of different types of urban immigrants has different characteristics. In addition, the study of changes of social networks itself is constantly changing, so the authors also explored the impact of occupational mobility and life cycle on social networks.

## 6.1 The Social Network of Urban Immigrants: A Quantitative Description Perspective

On the basis of classifying the social network of urban immigrants into three categories, namely primary relationships including families and relatives, secondary relationships including classmates, colleagues, fellow villagers and friends, and formal relationships including government, banks, communities committee and so on, the authors found that the quantitative data clearly showed the difference of the three types of urban immigrants on social networks.

First of all, the authors will explore the social network of urban immigrants from recreational occasions and recreational partners. Typically, intellectual immigrants have plenty of recreations, and they also have a lot of recreational occasions to go to. Besides, intellectual immigrants have a relatively large number of classmates, colleagues and friends. In terms of recreations, labour immigrants are more involved in activities such as "playing with mobile phones", "shopping", and "listening to music or radio", while entrepreneurial immigrants choose to "watching TV/DVD", "playing chess/cards/ mah-jong" and other activities (see Fig. 6.1). Correspondingly, in terms of recreational occasions, entrepreneurial immigrants usually stay at home or go to mah-jong club, while labour immigrants often go to the internet cafe or parks (see Fig. 6.2). In terms of recreational partners, the main recreational partners of entrepreneurial immigrants are family members, relatives, and friends, and their business partners are also more than those of intellectual immigrants and labour immigrants. Compared with intellectual immigrants, the recreational partner of labour immigrants is mainly family members, relatives, fellow villagers, and neighbours, showing relatively strong ties of kinship and geographical relationships (see Fig. 6.3).

The social support network of intellectual immigrants has the largest scale, with an average size of 15.89 people, and almost every sub-type of their social support network is larger than that of labour immigrants and entrepreneurial immigrants.

[8] Cai He ed., *Rural Migrant in the Process of Urbanization: A Study from the Pearl River Delta*, Beijing: Social Sciences Academic Press (China), 2009.

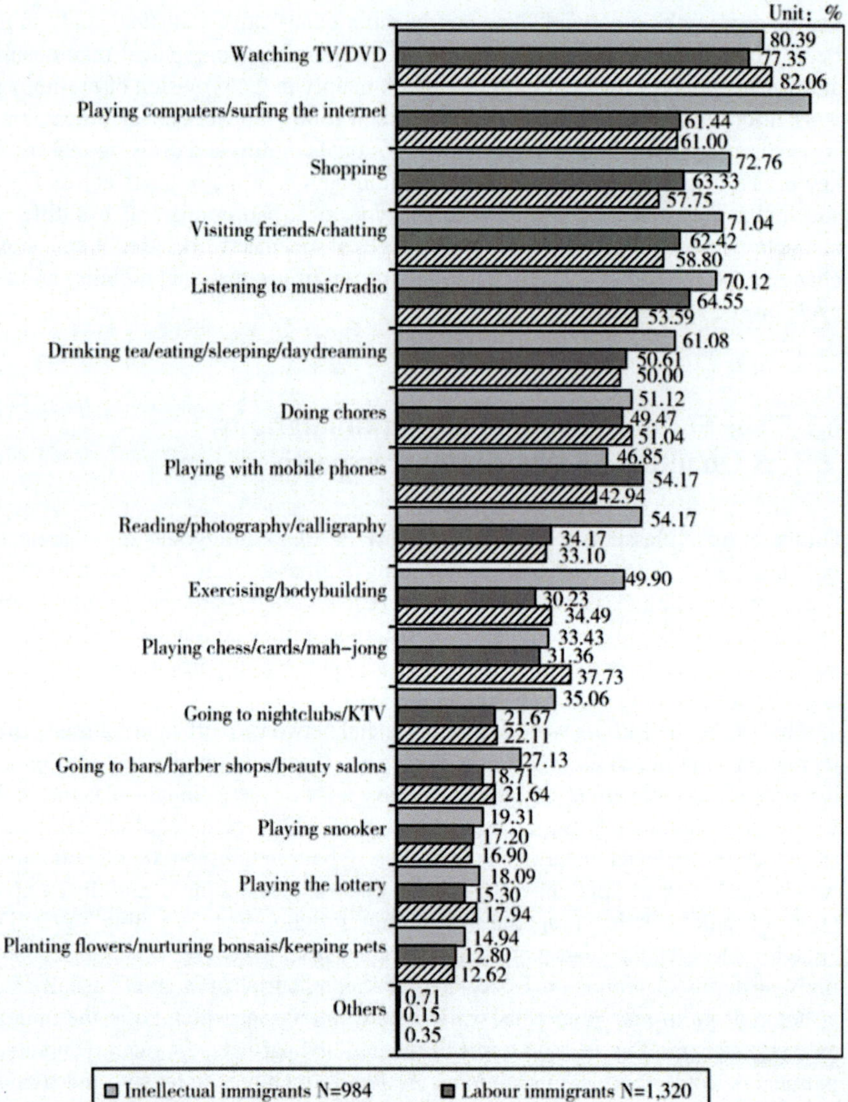


**Fig. 6.1** The recreation of urban immigrants

Among them, the average size of support network for borrowing money is 2–33 people, which is much higher than other social networks, the average size of the support network for job hunting of intellectual immigrants is 1.74 people, and the average size of the support network for dispute resolution is 1.76 people (see Table 6.1). The relationship types of urban immigrants' social support network

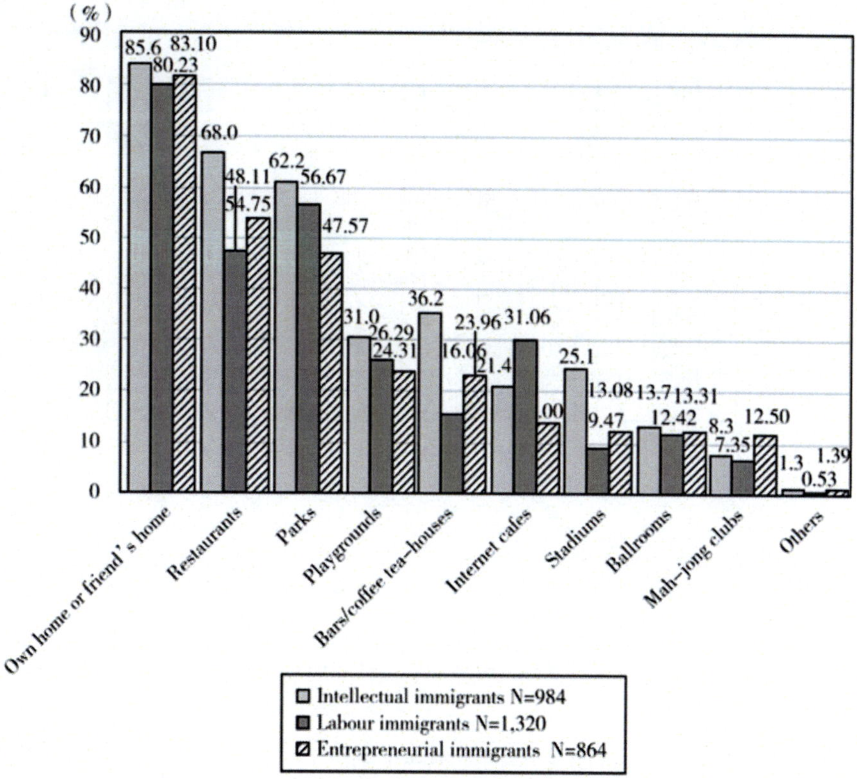


**Fig. 6.2** The recreational occasion of urban immigrants

include formal relationships such as the government in their hometown, the relevant departments of municipal government, insurance companies, banks, work units, social assistance, and primary relationships such as family members, relatives, and secondary relationships such as classmates, colleagues, fellow villagers, and friends. For intellectual immigrants, in the social support network for borrowing money, formal relationships account for 10.94%, primary relationships account for 66.60%, and secondary relationships account for 79.67%. Similar to intellectual immigrants, the social support network of labour immigrants is also dominated by secondary relationships, but the proportion of secondary relationships in the social support network of labour immigrants is lower than that of intellectual immigrants. On the contrary, primary relationships account for the largest proportion of the social support network of entrepreneurial immigrants. The composition of the relationship in social support networks for job hunting and dispute solution is similar to the above. However, formal relationships account for a large proportion of the social support network for dispute solution of three types of urban immigrants, especially for entrepreneurial immigrants (see Table 6.2).

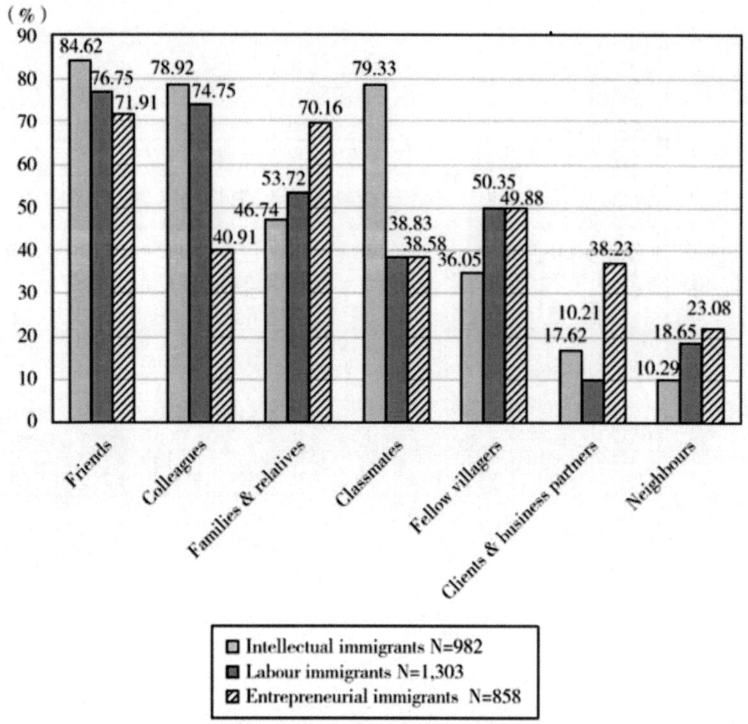


**Fig. 6.3** The recreational partner of urban immigrants

From the above analysis, we can find that the intellectual immigrant has plenty of recreations, and they also have a lot of recreational occasions to go to. Besides, intellectual immigrants have a relatively large number of classmates, colleagues and friends. As for entrepreneurial immigrants, their main recreational partners are family members, relatives, friends, clients, and business partners. In terms of social support networks, intellectual immigrants mainly rely on secondary relationships, while labour immigrants and entrepreneurial immigrants rely more on primary relationships. In the following, the authors will present the social network and its changes of three types of urban immigrants through cases.

**Table 6.1** The scale of social networks of urban new immigrants (Unit: %)

| Items | Intellectual immigrants | Labour immigrants | Entrepreneurial immigrants |
|---|---|---|---|
| Borrowing money | 2.33 | 2.24 | 2.16 |
| Getting medical services | 1.87 | 1.72 | 1.68 |
| Being cheated out of money | 1.60 | 1.47 | 1.34 |
| Being out of employment | 1.74 | 1.68 | 1.51 |
| Lack of accommodation | 1.80 | 1.60 | 1.51 |
| Children's schooling | 1.04 | 1.96 | 1.07 |
| Getting into disputes with others | 1.76 | 1.51 | 1.54 |
| Having traffic accidents | 2.00 | 1.62 | 1.50 |
| Occupational injury | 1.74 | 1.44 | 1.19 |
| Overall average number | 15.89 | 16.24 | 13.51 |
| Total (N) | 984 | 1,320 | 864 |

**Table 6.2** The composition of relationships of social networks of urban new immigrants (Unit: %)

| Items | | Intellectual immigrants | Labour immigrants | Entrepreneurial immigrants |
|---|---|---|---|---|
| Social support networks for borrowing money | Primary relationships | 66.60 | 66.90 | 81.32 |
| | Secondary relationships | 79.67 | 72.74 | 65.37 |
| | Formal relationships | 10.94 | 11.84 | 5.67 |
| Social support networks for job hunting | Primary relationships | 60.52 | 62.38 | 73.86 |
| | Secondary relationships | 71.09 | 69.66 | 59.72 |
| | Formal relationships | 8.20 | 6.31 | 10.10 |
| Social support networks for disputes solution | Primary relationships | 38.52 | 48.99 | 60.19 |
| | Secondary relationships | 72.96 | 58.68 | 52.83 |
| | Formal relationships | 37.55 | 36.48 | 45.84 |
| Total (N) | | 984 | 1,320 | 864 |

## 6.2 The Social Network of Labour Immigrants

### 6.2.1 *Reconstruction of Traditional Relationships*

Zhai Xuewei once pointed out that the core of rural migrants' job hunting is the trustworthiness of information acquisition and assistance. The so-called trustworthiness does not refer to the authenticity and reliability of the information itself but refers to the factor by which the information receiver judges the authenticity of the information. Traditional Chinese people trust their kinship and geographical relationships more. Therefore, instead of accepting authentic information provided by strangers, labour immigrants may accept false information introduced by their relatives and friends when applying for jobs. This practice of social networks is essentially a manifestation of the lack of modernity in the city as a whole. A sample survey of 13 cities in 6 provinces across the China in 2000 showed that 60.55% of rural migrants were employed through social networks and self-employment. According to the questionnaire survey conducted by the research group of the Institute of Economics of Wuhan University in 2005, rural migrants who achieved employment through social networks of acquaintances and self-employment accounted for 86.02% of the total questionnaire sample that year, and rural migrants who achieved employment without the help of social networks of acquaintances accounted for only 11.02%,[9] which further confirmed the above-mentioned characteristics of social networks.

The authors once pointed out in the study of casual labour belonging to labour immigrants that the social circle of work-study and the social circle of spare life of the ordinary labour force are mostly relatives and fellow villagers. Originally, the term "friend" is often used to refer to the modern relationship formed in the context of work and study. However, in the expression of labour immigrants, it is usually used to refer to the traditional kinship and geographical relationships. In the social network of labour immigrants, friend relationships usually overlap with those of relatives and fellow villagers. However, as individual agents, a considerable portion of labour immigrants are also actively constructing new social networks based on instrumental rationality, the process of which is also the process of labour immigrants adapting to the city.

Case: Zhengzhou-Labour-005

In 1983, Mr Liu came to work in the city from the countryside at the age of 30. Before he came to the city, in addition to farming at home, he also travelled between his home and the local county to sell chicks, known as "chick sellers". Because of his small business of selling chicks, his family's economic situation is not bad, and his family built a multi-storey building earlier. After expanding his business networks through selling chicks, he met more and more businessman doing related businesses, so he came to Beijing to continue his business of selling chicks in 1983. However, with the increasing number of chick hatcheries and competition in the business, he was unable to continue his business in Beijing, so he returned to his hometown to continue the business of selling chicks in 1992. And when his

[9] Gao Jun, Promotion of Employment and Citizenization of Rural Migrants, *Theory Monthly*, Vol. 10, 2008.

business in his hometown was bleak, he went to work in other cities. Since then, he has started a mobile life, having worked in southern cities such as Guangzhou, Zhuzhou, and Liuzhou, and also engaged in many different occupations, including selling vegetables, working as a construction worker, working in factories, and so on. For him, as long as he can make money, there is no work he will not do. Later, he felt that he was older and no longer suitable for a mobile life, so he returned to work in Zhengzhou, which was closer to his home. He can go back to help when the farm work is busy at home, and he can also come to Zhengzhou to find a job at a construction site when he has nothing to do at home. In this way, he began to travel between Zhengzhou and his home. He is now 56 years old, and he says he can work for another ten years.

In the early nine years of selling chicks in Beijing, Mr Liu's social networks were relatively fixed, mainly including his fellow villagers, hatchery owners, farm owners, and some peasants. In addition, playing mah-jong basically took up most of his spare time. Whether playing cards with the owner of chicken farms or playing mah-jong at the mah-jong club, he basically plays mah-jong in his spare time. He admitted that playing mah-jong is not a good hobby, but he was addicted to it. Since 1992, he began to migrate to various cities. Not only did he not have a stable job, but his social networks were also not fixed. Currently, he mainly works in Zhengzhou. Instead of renting a house in Zhengzhou, he brings two sets of clothes and a thin set of bedding with him. If he finds a job at the construction site, he will sleep at the construction site at night. And if he does not find a job, he will sleep on the street or go home to sell chicks. He thinks such a life is very free and comfortable. Whether he speaks or does things, he is very straightforward, so he is not very close to the foreman and his fellow workers. He said, "I am very straightforward, and I will not approach others intentionally and try to establish close relationships with others." The main component of his social networks is his fellow villagers. For example, he was introduced to the authors by one of his fellow villagers, who is now a foreman and has already established a family in Zhengzhou. Mr Liu said, "this fellow villager is particularly willing to help us, and whenever there is a job available, he will always introduce it to me." This fellow villager also commented on Mr Liu, saying, "he is a straightforward and stiff person, and he always speaks very straightforwardly. But he works very hard, and now there are not many rural migrants like him who work very hard." Mr Liu said that he is getting older, and playing mah-jong is not as frequent as before. Only if he really wants to play mah-jong, will he play mah-jong with his co-workers.

In the above case, a large part of Mr Liu's social networks is recreational partners, but his social networks are still dominated by his fellow villagers based on geographical relationships.

### *6.2.2 The Outsider in the City*

According to the theory of modernity, the migration of population means the establishment of modern society and modern social relationships. However, Frank proposed that in the process of population migration, developed regions and undeveloped regions are connected, and the distinction between the two and the latter's attachment to the former is strengthened by the structure of "centre-periphery". In the study of African labour immigrants in France, Mellisoux pointed out that the real division of labour between the central area and the peripheral area is that the peripheral area produces and reproduces labour, while the central area uses the labour according to the life cycle of labour, and then expelled labour who lost the ability to work back to the peripheral area. It can be seen that the separation of the production and use of labour is the secret of the continuous economic surplus in the central area.

In a sense, the social interaction objects of labour immigrants in cities can be divided into locals and non-locals. Moreover, interactions between colleagues who are also non-locals are usually easier than those with colleagues who are locals. For labour immigrants, the social segregation and social isolation caused by the two opposing labels of locals and non-locals make these two groups both become the others in the city, and they are almost irrelevant to each other. The authors once proposed the concept of "dual community" to analyse this social phenomenon. The so-called "dual community" means that under the existing household registration system, locals and non-locals form different systems in terms of distribution, employment, social status, locales in the same community, and even cannot accept each other psychologically. As Li Qiang said, the household registration system is essentially a "social closure" system, that is, it separates some people in the society from sharing the social resources in the city, which further strengthens the segregation of dual communities in cities. According to the theory of social interaction, the identification of the identity symbol is the basis for the interaction between individuals. The social network of labour immigrants is shaped by multiple aspects. First of all, the occupational types of labour immigrants are characterised by long working hours and high intensity, which will affect the rhythm and frequency of their social interaction. In addition, the distinction of social identity formed on the basis of the urban-rural structure has also resulted in the segregation between urban and rural areas, locals and non-locals in social space and social interaction. At last, labour immigrants are usually limited by the sense of security brought about by homogeneous cultural identity, so that they create an internal interaction barrier that causes them to alienate from urban culture. In reality, labour immigrants have become "the others in cities", and their social networks are embedded in the logic of urban society, which is also a way for them to live in the city.

Case: Shenyang-Labour-003

Since being laid off, Ms Wu has been working in Shenyang for four years, and she is now a cleaner in a dormitory apartment in a college in Shenyang. Ms. Wu and her husband were both employees of a state-owned enterprise in Anshan City, Liaoning Province. They were laid off after the reform of state-owned enterprises in 2000. In order to provide for her daughter's tuition and maintain the family's daily expenses, her husband went to a construction site in Shenyang to work as a skilled worker, and Ms. Wu was caught in the difficulty of finding a job. Ms. Wu told the authors that she couldn't remember how many jobs had changed since she was laid off, only remembering that it was a particularly difficult time. Because of her old age and low education degree, it was really difficult for her to find a job, even the job of washing dishes in a restaurant. It can be seen how hard life is. Later, she became a cleaner in a dormitory apartment in a college in Shenyang under the introduction of her husband's boss.

During these years in Shenyang, apart from her family, she had the most contacts with her colleagues. When she first came to Shenyang, she felt very uncomfortable. The prices here were relatively high and most of her colleagues were locals, who would look down on non-locals. In order not to be underestimated by locals, she worked hard to prove her ability and passed all of the clean checks. Besides, she is very diligent and always takes the initiative to do dirty work and tiring work that others do not want to do. After a long time, everyone praises her for her ability and begins to enjoy getting along with her, and their relationship has become very harmonious. But she rarely contacts other locals except for her colleagues. And the most worrying thing for her in Shenyang is the frequent traffic accidents. She is also very simple in eating and wearing in daily life. She has to get up at 4:30 every morning to prepare breakfast for her family, and then goes out at 6:00 AM and takes an hour and a half bus to the school department to work, and eats the lunch she brought from home at noon. After such a day, she often feels exhausted. However, when she returned home at night and saw the family eating dinner together in happiness, she felt very satisfied. In addition to the tiring work every day, Ms Wu also has to do a lot of housework, so she rarely has spare time. She said, "I have to worry about the lives and health of my family, so I have neither the mood nor the money to buy clothes, let alone cosmetics and health products."

Case: Shenyang-Labour-004

Mr. Lin was born in a peasant family in Jilin Province. He is the second child in the family, and his brother has already established a family and works in his hometown. In 1993, Mr. Lin lived in Shenyang for three years with his trading father. After graduating from junior high school, he returned to his hometown in Jilin and studied mechanical machining at a local technical school for two years. In 2002, Mr. Lin's father set up a factory in the far suburbs of Shenyang. For the family's business and helping the family make money, Mr. Lin once again came to Shenyang to temporarily help in his father's factory. During the seven years in Shenyang, Mr. Lin changed a total of three formal jobs. The first two jobs were found with the persuasion and help of his father, and the third job was chosen by Mr. Lin independently.

During these years in Shenyang, he has had the most contacts with his colleagues. He said, "I will not offend anyone, and I have a good relationship with my friends and colleagues." Spending dinner with friends accounted for a large part of his daily expenses. Although he and his friends will go Dutch on dinner, the average cost is still tens of yuan a day. Besides, most of his friends are locals, and he feels that these locals are not easy to get along with. In his view, these locals are too conceited and will look down on non-locals. He and these locals have different values. These young colleagues always spend a lot of money every time they go to the mall, and they also borrow money from Mr. Lin, ranging from RMB300 to RMB1,000, and then usually forget about having borrowed money from Mr. Lin. However, after spending time with these locals for a while, Mr. Lin said, "good people are still the majority. People get to know each other gradually in the process of getting along with each other, and only when people think you are good, will they continue to get along with you." He thinks he is not a non-local. The first reason is that he has stayed in Shenyang for a long time, and the second reason is that people cannot tell from his accent that he is a non-local. However, he still believes that in order to further develop and settle in Shenyang, he must have a certain amount of capitals and a house. Otherwise, Shenyang is still not suitable for his long stay.

### 6.2.3 *Social Networks in Changing*

Affected by the type of work, most labour immigrants have high occupational mobility, so the social network built on which is also temporary. Compared with workers who enter the factory through formal employment channels, the working and living conditions of workers who achieve employment through informal channels and mainly rely on manual labour are worse. However, they are not restricted by the strict rules and regulations of enterprises, and therefore have more freedom of occupational mobility, even if this freedom is based on continuous migration for survival. In addition, their social networks changes with the change of workplace, and the relatively stable relationship in their social networks is still consanguinity and kinship.

Case: Shenyang-Labour-001

After several job changes, Mr. Lu is now the site construction manager of an engineering company in Shenyang. He hasn't been home for two years. Last winter his mother went to see her granddaughter at his mother-in-law's house and stayed there for more than a month. Her mother also wanted to spend more time with her granddaughter, but she couldn't get away because she had farm work at home. So, she could only take advantage of winter when there was no farm work to help take care of her granddaughter. When she left, she left RMB500 to her daughter-in-law, who used the money to buy tickets and clothes for her mother-in-law.

After having children, Mr Lu's life became very tight. He has to spend a lot of money every time he goes home, including round-trip fees, and gifts for family members and relatives, and the total cost of each trip to home is at least RMB2,000, so returning home is a very extravagant thing for him. When Mr Lu is homesick, he will make a phone call to his parents, and the frequency of phone calls is usually once a month.

Kinship is the main component of Mr Lu's daily social networks. His wife's home is in Liaoning Province, and there are also some relatives in Shenyang. These family members and relatives took care of them a lot, and Mr Lu also often visit their relatives' houses. Besides, whenever Mr Lu encountered financial difficulties, he could borrow some money from these relatives. For example, Mr Lu suffered from waist disease after the Spring Festival this year. He was unable to go to work, and his income was greatly affected. Payment of mortgages and maintenance of living expenses became a major problem. At this time, Mr Lu mostly relied on the help of relatives to go through this difficult time.

Every holiday, Mr Lu will go back to his mother-in-law's house with his wife, which is a three-hour train ride from Shenyang. Mr Lu's child is also at his mother-in-law's house, and he will call the child every two or three days. Mr Lu's income is relatively low, so he rarely gives money to his mother-in-law, and he wants to wait until his income is higher.

Mr Lu's interpersonal circle is relatively small, mainly based on the friends he met at work. Mr Lu has some close friends that he met in different jobs. However, he and these friends only occasionally talked on the phone about their current situations, and rarely ate dinner or drink together. Most of his friends and colleagues he knew in the chain enterprises were in other places, and their whereabouts were not fixed. These friends are distributed in various provinces and cities, some of them can no longer be reached, and only a few of them can still keep in touch by phone.

As we can see, the main component of Mr. Lu's social networks has changed from geographical relationships to kinship, which was mainly affected by marital relationships. In addition to kinship, Mr. Lu also has work relationship networks and geographical relationship networks. But in general, Mr. Lu's social circle is still relatively narrow, which is also related to his long-term work in various cities. Mr. Lu's social network is usually impacted or even broken due to the transfer of work relationship and the change of workplace.

## 6.3 The Social Network of Entrepreneurial Immigrants

The social network of entrepreneurial immigrants has two major characteristics. One is the family member and relatives as the main component, and the second is that the social network built on the basis of business relationship also plays an important role. Besides, the authors also noticed that the social network built on the basis of business relationships are mostly social interactions out of instrumental rationality. Once one of the two interacting parties of this relationship engages in behaviors other than those expected by the other, this social network will face a crisis of disintegration. Relatively speaking, the interaction based on kinship and geographical relationships is at the core of the social network of entrepreneurial immigrants. Entrepreneurial immigrants usually use and integrate new and stable social networks built on kinship and geographical relationships growing in urban spaces, and share and reciprocate benefits and information through such social networks, which can be seen from the authors' interviews with entrepreneurial immigrants in southern cities such as Guangzhou.

### *6.3.1 "Chains of Immigration": Immigration with Relatives as Mediums*

Guangzhou, in the hinterland of the Pearl River Delta, the world's manufacturing centre, has more than 500 wholesale and retail markets in all walks of life, of which the small and medium-sized market is the majority, and there are nearly 150 professional markets with larger scale, including products of clothing, books, cosmetics, toys, electronic products, tea, hotel supplies, and so on. In these wholesale trade markets, there is an extremely dynamic group of Chaoshan merchants who practice the tradition of immigration and business culture in Chaoshan.[10] In this group, people help each other and are consistent with each other based on kinship and geographical relationships, and they frequently reciprocate on the basis of morality, forming a sales alliance of the same type of product or different products in the upstream and downstream of the industry. Besides, with the development of new markets, the expansion of Chaoshan merchants has also covered multiple markets, forming a strong economic and social cohesion. At the same time, the aggregation of entrepreneurial immigrants also provides a space for later urban immigrants to adapt to cities, effectively helping the urban immigrant achieve stable and upward mobility of social status in the city.

The Chaoshan merchants in the Guangzhou wholesale and retail market have close ties with each other, and they will guide and help each other in the process of immigration, thus presenting the form of "chains of immigration". They cooperate in business and interact frequently in life, forming a very close social network.

[10] Zhou Daming, Immigration Culture: A Hypothesis, *Jiangsu Social Sciences*, Vol. 5, 2005.

Specifically, there are two types of social networks. One is the social network based on the relationship of relatives and friends, mainly including relatives and peer groups in the same village. The other is the social network based on business relationships, mainly including clients and business partners. And the boundary of these two types of social networks is indistinct, they may overlap each other, and also can shift in different scenarios. Chaoshan young people who have just arrived in Guangzhou usually choose to turn to relatives or friends, and they normally can find relatives in nearly every trade market. Typically, when an urban immigrant from Chaoshan is established in Guangzhou, he will guide some more people to Guangzhou, and those who have been guided to Guangzhou will continue to guide more people after establishing in Guangzhou, thus forming a social network woven by immigration chains connected to each other. Chaoshan merchants are connected through these overlapped social networks of different sizes and blurred boundaries. The aggregation of several entrepreneurial immigrants interviewed by the authors in Guangzhou is the result of the social network woven by the immigration chain (see Fig. 6.4).

The kinship diagram above shows a social network formed by Chaoshan merchants. In this social network, Zhao G was the first to come to Guangzhou to do business. He came to Guangzhou in 1997 to do electronic capacitor trading. He chose to come to Guangzhou because of a friend doing business here and this friend introduced him to Guangzhou. In the beginning, he cooperated with his wife and his wife's younger brother Xie G to run the shop together. Two years later, Xie G decided to do business alone because he wanted to get married and needed to start and take care of a new family. Therefore, Xie X dropped out of school and replaced Xie G's position to work for Zhao G after finishing the second year of junior high school. Three years later, Xie X also decided to do business on his own. So, Xie X brought his younger brother Xie N, who had just finished the second year of junior high school, to Guangzhou to let Xie N learn how to do business, and Xie N has now become an important helper in Xie X's business. As mentioned earlier, Xie X left Zhao G to

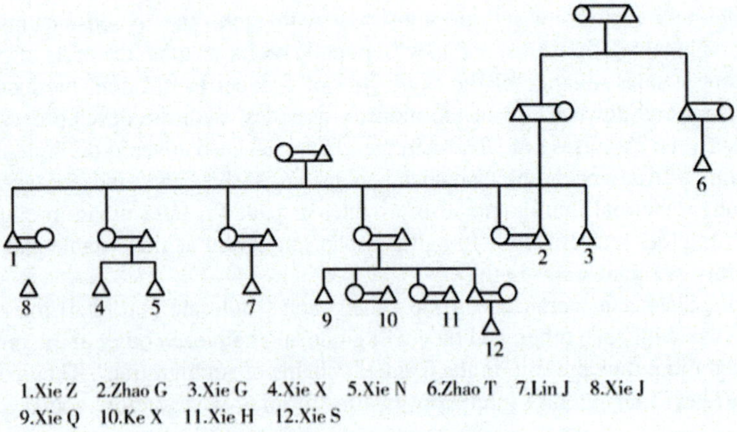


**Fig. 6.4** Diagram of social networks formed by the interviewees

do business alone. And after Xie X left, Zhao G also brought his cousin Zhao T and his fellow Chen X to Guangzhou to do business together for three years. Later, Zhao T and Chen X also started business separately, so Zhao G found his nephew Lin J as his new business helper. Besides, when Zhao G first came to Guangzhou to do business, he also brought the business information of Guangzhou back to his family, so Xie Z, who is Xie X's uncle, also came to Guangzhou in 1998 to do hardware and electrical business. Ke X came to Guangzhou in 1994 to do electronic business, and he is a senior merchant in this industry. In 2000, Xie Q came to work in Ke X's shop after graduating from junior high school. And a year later, Xie Q started his own business and made a lot of money in two years. Xie H also started his own electronic business with the help of Ke X, and soon his business was established. In 2001, Xie S, who had just graduated from junior high school, came to Guangzhou to work as an apprentice at Xie H's shop. Two years later, Xie S also started to do his own business. In addition, Xie J dropped out of junior high school and came to Guangzhou to study business with his cousin Xie Q. In 2007, Xie J also rented a counter to start his business alone.

There are several relationship chains in the above social network. The first is the relationship chain centred on Zhao G, who successfully brought out Xie G, Xie X, Xie Z, his cousin Zhao T, his fellow Chen X, and his nephew Lin J, which took a total of 12 years. And then Xie X brought out his younger brother Xie N. The second is the relationship chain based on Ke X, who brought out Xie Q and Xie H. And then Xie Q and Xie H brought out Xie J and Xie S respectively. These relationship chains are actually singed chains composed of people who are related to each other through kinship. At the same time, the person in every single chain has kinship with the person in another single chain, so these single chains are also interconnected. In this way, an intricate social network is formed under the connection of kinship. And everyone in this social network can have more resources to rely on in different cities, they can be more closely connected to integrate resources, exchange information, and develop together through this social network. The social network reconstructed in different cities was originally based on kinship, but because of the additional business connection and the emotional resonance of being in a different place, the relationship between the relative in this social network is stronger, which is actually a strengthening of the original kinship.

The social network formed by "chains of immigration" promotes the aggregation of Chaoshan people. And the gathering of stores is a major feature of Chaoshan merchants doing business. The several markets involved in this survey are Chaoshan merchants gathering areas. According to market managers, Chaoshan store owners account for more than half of these markets. The five Ye family brothers interviewed by the authors are from a village in Nanjing town, Puning city, which has the largest business in the Guangzhou wholesale market. They have dominated the business in the two major trade markets of Shaxi and Nantian. Among them, they have 9 buildings of stalls in Nantian Market, and each building has 20 shops. Besides, the member of the same branch of their family in the village also followed them to Guangzhou to do business. In Nantian Market, there are seven households brought out to do business by them. Chaoshan people like this form of store aggregation, they call it "many a little makes a mickle", and they think it will bring good luck and wealth. Through the way of store aggregation, on the one hand, they can get geographically advantageous and cheap stores, on the other hand, they can use the scale effect to enhance the business atmosphere and drive business development.

### *6.3.2 From Wage Earners to Bosses: A Platform for Qualitative Changes of Identity*

To realise the passionate dream of starting a business, people need the support from various aspects. And the social network constructed by the chain of immigration is the strongest backing for young Chaoshan merchants to establish in the whole and retail market. The authors found that young people working for relatives in the wholesale market generally do not have high incomes. Although relatives provide accommodation for young people, the wages of these young people are usually only a few hundred yuan a month, and some young people do not even have fixed monthly income.

Xie X recalled, "I was only 16 years old when I came to Guangzhou to work for my uncle in 1999. And I was not a formal employee in my uncle's shop because I had neither salary nor fixed-job responsibilities. At that time, my family just hoped that I could learn some business experience from my uncle's shop. And my uncle's responsibility was to take care of me so that I won't wander around and go astray. My uncle told me, 'if there is anything in the shop that you need to do, just do it. Although I will not give you salaries, I will cover your food and lodging. If my business is prosperous, I will also give you some benefits'".

However, these young people are not too concerned about income. In fact, what's more important is that they need to accumulate a series of elements that they need in relatives' shops as soon as possible to change from wage earners to bosses.

To start a business in the wholesale and retail market, people first need sufficient capital. Young people working in relatives' shops mainly want to gain experience in doing business, and they do not have much money. Therefore, their start-up funds usually need to rely on other people's support. Sometimes, this start-up fund comes from the support of the family. But more often, this support comes from relatives who they have worked for. These relatives will either directly invest in helping young people rent shops, or delegate young people to take care of their newly opened chain stores, and then gradually hand over the shop to young people. When young people first start a business, the scale of business is usually not large and the profit is also not high, but they finally have their own business and have achieved the leap from being a wage earner to being a boss. Xie J, who is doing electronic business in Guangzhou Jinghong Electronic City, started his own business after working for his cousin for four years.

He recalled, "In 2007, with my own deposit, plus the help of my family and my cousin, I opened a shop in Guangzhou Jinghong Electronic City, and the monthly rent was RMB3,000. I also rented an apartment with two-bedroom and one-hall that is a few hundred metres away from the shop, and the monthly rent is about RMB1,300. In addition to other expenses, I spent a total of about RMB10,000 per month in Guangzhou. However, I could only earn a few hundred or a few thousand yuan per order when I first started doing business, so I only had a slight surplus every month."

Once the store is opened, the next issue the merchants need to solve is the source of supply. Since the wholesale and retail market in Guangzhou has people who are responsible for publicity, shop owners do not need to advertise, and customers will naturally find the market to choose sellers, so customers are not crucial factors. On the contrary, the quality and price of the supply are the key factors for business success.

The supply of goods is generally concentrated in certain areas. If young people have worked in this industry, they can basically get some information about the supply of goods. Chaoshan young people should not only be familiar with the various models and functions of goods, but also master the sources and sales of various goods while working in relatives' shops. Besides, some manufacturers will also come to the store to promote their new products. For young people who have just started a business, in addition to relying on the experience they have obtained through working for relatives to purchase goods, the support and experience of their predecessors are also essential at the beginning of their business.

After working for his brother for a few years, Ye X began to start a business in the Nantian International Hotel Supplies Market. He explained to the authors his channels for obtaining the information of supply, "Purchasing is the most difficult and critical part of doing my business. There are two main channels for my purchase, one is through the help and introduction of my brother, and the other is through the experience of the recruited workers who have worked in the market before. Relying on these two channels, I solved my problem of being unfamiliar with the supply of goods. Besides, my brother and these workers also helped me find the best-selling products on the market today, which is very important for determining the type and quantity of purchases. My supply of goods is mainly from the Pearl River Delta and Chaoshan. More than twenty of the hundreds of types of goods in my store are directly purchased from the manufacturer, and the rest is purchased from the wholesale market. My customers are mainly from Guangzhou and the rest of the Pearl River Delta, and a small number of my customers are from abroad, mainly Africans."

In terms of commodity prices, young Chaoshan merchants learned to trade small losses for long-term and large profits while working for relatives, and they will use this experience flexibly in the process of doing business. They usually use some low-priced products to promote the sale of other products, thereby earning more profits. In addition, they are very willing to reach a "win-win" with their customers, and pay attention to maintaining their own customer base. Many customers have grown small businesses together with young Chaoshan merchants. Chaoshan merchants regard customers as their business partners, and believe that only when the customer' business grows can their own business grow. They will never dismiss the customer because of the small number of products that customers demand, nor will they deliberately increase the price of their products because customers are not familiar with the product. These young Chaoshan merchants have a consensus that if the quality of products they give to customers is not good and the price is high, then the customer's business will deteriorate, and finally their own business will also be difficult to continue. For some customers who have a long-term cooperative relationship, they will also actively offer low prices, which make the authors feel puzzled, and the explanation given by these young Chaoshan merchants is that this behaviour can reduce the cost of customers and make customer's products are more competitive in the market, and many customers will be willing to cooperate and even become good friends with them because of this behaviour.

Since working for his brother in the market, Ye X has learned and accumulated a lot of business experience by himself. Ye X runs a retail shop, so the price is the key to a successful business. Of all the goods sold in his shop, some are sold at a loss, some are sold at a cost price, and others are sold at a high price. Ye X told the authors that the pricing and selling methods of each kind of goods are well-founded. Some products with relatively high market transparency must be sold at a low price. Because it is easy for customers to shop around for this kind of product, and if you sell it at a high price, then customers will not choose to buy it in your shop. For example, Yangjiang Shibazi Group's knives are famous all over

China and are sold in many stores. In this case, the price you sell cannot be higher than in other stores. If another store sells RMB18 with a kitchen knife, then you have to sell it with RMB17, so that your store will be more competitive. Once customers come to your store, they may continue to buy other products because your kitchen knife is cheaper than that of other stores, and it is the other products that customers buy that actually make you money. If selling some large items, Ye X's experience is to sell it at a low or cost price, and then make money by selling small accessories of those large items. As for the specific price, it will be different for different customers. Ye X told the authors that the above-mentioned experience is accumulated in his long-term practice, and he also usually shares his own business experience and learns from others when drinking tea and chatting with others.

Although these Chaoshan merchants who are doing business in Guangzhou are young, they all have their own set of experience and guidelines for doing business. And meeting friends with tea is the main way for these young Chaoshan merchants to exchange information and experience. Usually, a set of kung fu tea set is placed in the shop of Chaoshan merchants. Once someone comes to their shop, the host will warmly invite the guest to sit down and drink a cup of tea. Then they will chat while drinking tea, and the exchange of information begins. From outside the shop of many Chaoshan merchants, we often see several people sitting around drinking kung fu tea. They exchange information about different places, opportunities to make money, and meet new and old friends. In this way, their social network has been strengthened and expanded through a pot of tea.

### *6.3.3 Morality and Reciprocity: Overall Interests and Individual Interests in Balancing*

Among the Chaoshan merchant in these wholesale markets, there is a general phenomenon of "guiding". For Chaoshan merchants, "guiding" is above all an obligation and responsibility. That is, a person who has successfully established a business in a different place has the responsibility and obligation to guide their fellow villagers, usually relatives, to learn about industry-related information and solve some initial problems of starting a business. "Guiding" is divided into two categories in terms of content. One is to convey information about different places to the fellow villager, such as the general environment of the industry and the prospect of the industry's development. And the other is to help relatives or friends enter the industry, provide them with opportunities to work and learn, or help them solve problems such as capital and experience in starting a business in a different place. Relatives and friends "guided" through these methods usually engage in business in the same industry, and even if there is no obvious competitive relationship, there may be potential competitive relationships.

In fact, competition exists when Chaoshan young people start working for others. Chaoshan people usually choose to work for their fellow villagers in the same industry, and this fellow villager maybe their relatives or friends, which is an important way for most Chaoshan entrepreneurial immigrants to enter the market. Since the young Chaoshan merchants entered the city, they were supported by their predecessors who guided them to work in the city. With the information and platform provided by this social network, Chaoshan young people gradually adapt to the city, and then start a real business.

Therefore, Chaoshan people understand that very few Chaoshan people are willing to work for others for a long time, and most Chaoshan people will change from wage earners to bosses within a few years and become potential competitors of their business. To avoid this situation, Chaoshan merchants will firstly consider guiding their family members and relatives to work and do business in the city, and secondly consider people of the same branch of the family. And they generally do not rush to accept a Chaoshan wage earner who has no social relationships with themselves, especially in some fiercely competitive markets, Chaoshan merchants will be more cautious when recruiting employees.

Yang S, 25 years old, comes from Puning City, Guangdong Province. Since 2003, Yang S has been working in the electrical appliance store opened by his father and uncle, but he is not very interested in this industry. Last year, he visited the Zhongda cloth market and Dashatou mobile phone market, and planned to choose these two industries as his future development direction. He had planned to work for others and accumulate some experience first. However, he searched for dozens of stores in the market for several days, but he could not find one willing to accept him. He originally thought that since most of the owner in the market are from Chaoshan, as Chaoshan people, he should also be taken care of by these Chaoshan merchants. However, as soon as the boss knew that he was from Chaoshan, these bosses knew that he didn't really want to work here. So, these bosses dare not hire him, worrying that he will start his own business after accumulating experience, and then become a competitor.

Since there are competitions between the same industries, why do these Chaoshan merchants still guide people to the market to do business? There are two reasons for this. One is that the Chinese value kinship, especially the Chaoshan people. In the Chaoshan area, ancestral hall of various clans can be seen in every village. Besides, Chaoshan people also pay special attention to activities of worshiping ancestors. It is this strong clan identity that drives Chaoshan people to start businesses. They hope to strengthen their clan through their own efforts and bring glory to their ancestors.

Xie X said, "I came to the city to do business, not only to let our family live a good life but also to earn the reputation for our branch of the clan. I look forward to the day when we return to the village, the other branches of the clan will say, 'the third branch of the clan is really successful.' I hope that I and other successful merchants of the same branch of the clan can have authority in the village and can influence the decision-making of major affairs in the village. In the past, major affairs in the village were settled by the sixth branch of the clan. There was only one person from our branch of the clan in our village committee, and our branch of the clan does not have any power to participate in decision-making on major affairs in the village." Then he said proudly, "However, in recent years, the people of our third branch of the clan have developed well in the city. As we all know, it costs money to build ancestral halls, roads, drainage channels, etc. And our third branch of the clan is now rich, so we are more confident when we participate in village affairs."

The other reason is that it can reduce competition and promote cooperation. In reality, the market is open to anyone, and anyone can enter the market with or without guidance from others, so competition is inevitable. In this case, it is better to form a business alliance with relatives and friends than to compete with others alone, so as to expand their market.

On the one hand, Chaoshan merchants refused to guide Chaoshan people with no social relationships into the market. On the other hand, they would guide Chaoshan people with social relationships into the market. This may seem contradictory on the surface, but in fact, it is related to the moral stratification standards of Chaoshan merchants. With this standard, Chaoshan merchants maintained their relationships and interests.

The young people who are guided by predecessors to do business also have a moral responsibility for predecessors. These young people can't successfully start a business without the platform built by their predecessors. Therefore, they must remember and be grateful for the help and favour that is given by their predecessors, and they must not damage the interests of their predecessors for their personal benefit. In order to avoid competing with their predecessors, the young Chaoshan merchants usually choose to start a business in a different market. If conditions do not allow this, they will avoid selling the same type of products as their predecessors, or choose products on the upstream and downstream of the industrial chain to avoid competing with their predecessors. For example, although merchants guided by Zhao G mentioned earlier are all engaged in the business of electronic components, they sell different products. In addition, they help each other and expand the market together. The younger generation guided by the predecessors to do business are all grateful to their predecessors. And it is a particularly unethical behaviour for them to compete with their predecessors after learning the experience of doing business from their predecessors.

From the immigration process of Chaoshan merchants, we can see how they practice social networks in the process of transitioning from wage earners to bosses. These Chaoshan people usually do not work for others for a long time. In other words, most of them have achieved their transformation of personal identity and successful entrepreneurship in a very short time. There are three distinct characteristics of their practice of social networks.

First of all, Chaoshan people are entrepreneur-oriented urban immigrants, and the goal of starting a business determines their development path from the beginning. They are reluctant and unwilling to work for others for a lifetime, so they are already brewing their own business plan when they work for others.

Furthermore, Chaoshan people pay special attention to the use of social networks. Although immigrants generally rely on members of their social network to obtain residence and work in order to successfully pass the initial stage of immigration. However, Chaoshan people have unique characteristics when using social networks. They rely on standards of social networks with strong clan identity to distinguish between close and alienated relationships, which in fact strengthen their internal relationships. In addition, information exchange platforms such as Kungfu tea party make the information resources based on social networks more fully utilised.

Last but not least, Chaoshan merchants who successfully started a business with the support of social networks all have similarities in business types, so there may be potential competition between them. However, by distinguishing between industry types and market segments, they can not only avoid internal competition but also achieve business cooperation at a higher level and form a strong family business.

Chaoshan merchants' use of their social networks shows us the powerful socioeconomic potential of cohesion based on social networks. From the authors' survey, we can see that Chaoshan merchants who are doing businesses in markets in Guangzhou have risen along with the reform and openingup, and they have continued to grow and develop as the market expands. Besides, they continuously guided young people from their hometown into the market, which become an important part of the immigration and business culture of the Chaoshan area.[11] In the social network formed by chains of immigration, people assist each other and avoid competition based on kinship and geographical relationships, and frequently reciprocate on the basis of morality. Chaoshan merchants usually form a sales alliance of the same type of products or different products in the upstream and downstream of the industry, and this sales alliance will also cover multiple different markets with the development of new markets and the expansion of Chaoshan merchants, which is a form of cohesion that has strong socioeconomic potential.

## 6.4 The Social Network of Intellectual Immigrants

Intellectual immigrants are a relatively young and highly educated group, but this group is no longer as shiny as before due to the expansion of college enrolment and the reform of the college employment system. Whether in job hunting or daily life, they will still encounter many difficulties and setbacks.

### *6.4.1 Job Hunting*

In term of job hunting, the employment channels of intellectual immigrants are relatively diversified, including posting resumes online, introductions by parents, relatives and friends, job fair on campus and talent market, and so on. Compared with other types of urban immigrants, most intellectual immigrants get jobs through their own efforts. But at the same time, social networks also play a key role in helping some intellectual immigrants find jobs.

Case: Shenyang-Intellectual-001

Ms Wang was born in a small town in Panjin City. She is the only child in the family. And her parents are both ordinary workers. In the 2002 college entrance examination, Ms Wang's score exceeded the admission score of the undergraduate university. However, she filled in the college application form inappropriately, so she finally went to a junior college in Shenyang to study computer science. Later, she obtained a bachelor's degree through Top-up. In her words, "The path of my studies is very tortuous". During college, she studied very hard. She had excellent grades and was awarded scholarships every year. Besides, she also actively participated in various school events. She joined the Communist Party of China before graduation and was awarded as an outstanding graduate in Liaoning Province.

[11] Zhou Daming, Immigration Culture: A Hypothesis, *Jiangsu Social Sciences*, Vol. 5, 2005.

After graduating from college in 2007, she chose to stay and work in Shenyang. She explained, "as far as my major is concerned, staying in Shenyang should have more opportunities than staying in my hometown, and even if I go back to my hometown, there are few relatives." Like most college graduates, she participated in several job fairs, but there were few work positions for computer majors, so job offers are not easy to get.

At that time, there was a way of employment, that is, she needed to pay a certain amount of tuition to participate in the company's pre-job training. However, she did not choose this way of employment. She said, "If I want to participate in the training of company, not only do I have to pay the training fee, I also have to spend three to five months to receive the training, which will delay me from becoming a regular employee. So, I decided at that time that I would only go to the training company if I couldn't find a job in the end." During the graduation season, the College Student Career Guidance Centre held a job fair. And she submitted her resumes to several companies. Unexpectedly, she was successfully accepted by a company at the end of June, and officially entered the company in July. The company she works for is a medium-sized private company with about 40 employees. The company provides all kinds of insurance for employees. And the salary of ordinary employees is about RMB1,400 per month, and after deduction of insurance, it is about RMB1,000 per month.

Case: Shenyang-Intellectual-002

Ms. Miao has been in Shenyang for 8 years and now works in a Shenyang office of a scientific research instrument company. In the past, Ms. Miao spent a lot of time and efforts to find a job. In the beginning, she hunted for jobs by searching for job information in the newspaper and posting resumes on various websites. At that time, she seldom went to the talent market. She thought that the talent market was too much to be targeted. However, whenever the recruitment company knows that the applicant is a fresh graduate, regardless of whether these fresh graduates have actual abilities, they will be rejected as long as they have no work experience. Ms. Miao can understand this phenomenon now. She said, "80% of recruitment companies tend to recruit people with work experience. It is difficult for fresh graduates without work experience to quickly adapt to the job. So, when hiring employees for our company, I will also tend to employees who already have work experience." Besides, while there is much online job information, there are also many risks. Once, a fraudulent gang obtained Ms. Miao's phone number online, and they called and harassed Ms. Miao every day, forcing Ms. Miao to turn off her mobile phone. Then the fraudulent gang started calling Ms. Miao's parents, claiming that Ms. Miao was involved in a car accident and urgently needed RMB4,000. After Ms. Miao turned on her phone twenty minutes later, she found that her father text her that the money had been transferred. So, Ms. Miao quickly called the police in Shenyang. Fortunately, the police in Ms. Miao's hometown finally detected the fraud and recovered the money belonging to Ms. Miao's parents. Not only did she fail to find a job, but also frightened her family, Ms. Miao almost collapsed at that time and did not want to continue hunting for jobs anymore. Luckily, with the persuasion and comfort of friends and her boyfriend, Ms. Miao finally regained her spirit. Twenty days later, Ms. Miao successfully applied for the reception work of a computer training school by chance. Although the monthly salary of this job is only RMB700, and it is also not related to her undergraduate environmental science major, Ms. Miao still believed that it is more important to have a job first. In the first year after graduation, Ms. Miao still needed help from her family to make up for her monthly living expenses of RMB900. Ms. Miao believes that in order to find a stable job with a professional counterpart, a certain level of the social network is required. And she believes that she cannot find a good job in Shenyang for two reasons, one is that she does not have a good social network in her hometown, and the other is that she is an outsider in Shenyang, where she does not have a broad social network. Later, she was fired after working in that computer training school for three months. And she later learned that the computer training school closed down the second month she left because of poor management.

Most of Ms Miao's classmates have already found jobs, and she was unemployed as soon as she graduated. Ms Miao was extremely stressed. Ms Miao wanted to find a better job than the last one, but although she has a little work experience, the work experience is still not enough for her to find a good job, so Ms Miao could only take a chance. Ms Miao has always been thankful for her boyfriend's company and encouragement so that she can persevere without giving up. A month and a half later, Ms Miao received an interview notice from a scientific instrument company's office in Shenyang. The interviewer is particularly kind and Ms Miao successfully passed the interview. The scientific instrument company is headquartered in Beijing and has offices throughout China. The company's office in Shenyang has a total of ten employees. In addition to the manager, the other nine people are salesman and travel all year round. And Ms Miao stays in the office and is responsible for all matters of the office in Shenyang. The company provides employees with "five social insurances and one housing fund". Besides, the employees' monthly basic salary plus performance bonus is at least RMB2,000, and the company will also provide RMB100 for every holiday. Although the company is not large, Ms Miao believes that the company has a good humanistic environment. She explained, "Most of the colleagues in the company are from outside Shenyang, and they often have dinner and chat together, so their relationship is very harmonious." Ms. Miao told the authors that when she first came to the company, she was unskilled in many things, and did not even know how to send a fax, and it was common for her to be criticised by her superiors for failing to complete tasks well. After a year of hard work and accumulation of experience, Ms. Miao became more skilled in her work and became the backbone of the company. Ms. Miao has been working in this company for four years, and renewed her three-year labour contract with the company a few days ago. Although the industry is fiercely competitive and suffered a financial crisis, and the company's sales have declined, Ms Miao's annual bonus has also been cancelled, and her monthly income is 30% lower than before, Ms Miao is still very satisfied with her job. Ms Miao clearly understands that there is not much room for development in her current job. And she also wants to live a different life, do something more challenging, and take a look around China. She is not afraid of changing a new job, but she is also worried that no other company will hire her, so she currently has no plans to change jobs. In Ms. Miao's opinion, it is already a lucky thing to be able to work in an office building from 9 to 5 every day with a satisfactory salary and a boyfriend to rely on.

Both of the above cases demonstrate the hardship of intellectual immigrants in finding a job. Although intellectual immigrants have a higher education degree, they may still be rejected by companies due to mismatched majors and lack of work experience. However, compared with labour immigrants and entrepreneurial immigrants, intellectual immigrants have more available social resources. They can mobilise school resources and social resources, such as job fairs organised by the school and online job hunting. Although some intellectual immigrants mentioned that they found jobs with the help of their parents and relatives, this is only a few cases. In reality, the existing social networks of most of the intellectual immigrant who immigrates to the city from rural areas or small towns to find jobs are difficult to extend to the city, so most of the intellectual immigrant can only rely on their own efforts and diversified job-hunting channels.

### 6.4.2 Daily Interaction

As mentioned above, intellectual immigrants will use multiple networks in the process of job hunting, especially online resources and school resources. In daily life, the social network of intellectual immigrants has the following characteristics. First of all, the social network of intellectual immigrants' centres on the family. Secondly, intellectual immigrants have relatively more contacts and interactions with their classmates. And if they work in a city different from the place where the university is located, then they will have more daily interactions with their fellow villagers. Finally, only a small percentage of intellectual immigrants establish good relationships with their colleagues, and their interaction with their colleagues is generally limited to work. It can be seen from this that the social network based on consanguinity, kinship, and geographical relationships still occupies an important position in the life of intellectual immigrants, which may supplement some scholars' assumption that the social network of intellectual immigrants will be more diverse and complicated, that is, the social network of intellectual immigrants has the characteristic of outward expansion. In this standard interpretation of the characteristic of the social network, some intellectual immigrants believe that "colleagues are not easy to get along with, and they all put personal interests first, and you should be very careful when dealing with them", and some intellectual immigrants believe that "once you start working, you will be in a competitive relationship with your colleagues, and it will be difficult for you to be friends with your colleagues". It can be found that instrumental rationality is the main logic among the professional interaction of intellectual immigrants, and the interaction between intellectual immigrants and their colleagues has the characteristic of temporary and mechanical combination, which has little effect on supporting their daily life.

Case: Zhengzhou-Intellectual-002

Ms Wang's university is in Xinxiang, so she doesn't have many classmates in Zhengzhou, and she feels lonely. Ms Wang works as a substitute teacher in school, but the turnover rate of substitute teachers is very high, so Ms Wang cannot establish a fixed relationship with them. She only interacts with other substitute teachers when they are in school. After they leave school, the connection between them is basically interrupted. Ms Wang's uncle is in Zhengzhou, but the aunt's mother also lives with her uncle, and the aunt's mother is very bad to Ms Wang, who will look down on Ms Wang because she is from rural areas. Ms Wang said, "When I first came to Zhengzhou to live in my uncle's house, my aunt's mother complained that I didn't save water when I was doing laundry. I understood that she said this only because she didn't like me, so I will either call my uncle or directly go to my uncle's company when I need help from my uncle." Ms Wang's uncle is willing to help her, but on the one hand, her uncle's ability is limited, on the other hand, her own ability is also not enough. Although Ms Wang has a bachelor's degree, she still can't find a formal job in school, so she can only live in an unstable state.

On 8 August, 2008, Ms Wang and her husband registered to get married. Her husband is a graduate of clinical medicine in a very famous university in China, so her husband was recruited by medical college as a teacher after graduation. And her husband is currently studying on-the-job postgraduate courses while working as a teacher in medical college. Her husband is also from rural areas, and his family's economic conditions are very poor. Ms Wang said, "My husband is very introverted. He is pessimistic about everything, but I am a very optimistic person. I think the two of us should exchange personalities." Her relationship with her husband is not very good. Her mother opposed the marriage from the beginning because of her husband's economic background. Although she had obtained a marriage certificate with her husband, her mother repeatedly emphasised that they only obtained a marriage certificate, but had not yet held a wedding, so it was not considered a true marriage. Ms Wang's mother doesn't want them to be together until now. And Ms Wang is more or less influenced by her mother and complains that her husband is not masculine and has poor economic conditions. Although Ms Wang bought a house in Xinxiang, she did not intend to live in Xinxiang for a long time. She felt that she could not find a good job in Xinxiang, and Xinxiang is not her hometown. Ms Wang is now substituting at school for three days a week. For the rest of the time, her husband will come to Zhengzhou to accompany her, or she will return to Xinxiang to accompany her husband. Although Ms Wang is very outgoing, her social status is not high, and she has no fixed social circle of life and work, so she basically has no friends in Zhengzhou and has not participated in any leisure activities. Apart from substituting at school, Ms Wang spends most of her time with her husband.

Case: Zhengzhou-Intellectual-009

Mr Yuan's social network is relatively simple. He is willing to establish close relationships with colleagues, but it is difficult to have pure friendship in the workplace, and the interaction between colleagues usually has a utilitarian purpose, so some people are unwilling to have too much contact with him. This is not the only case when interacting with colleagues. This also happens when Mr Yuan gets along with his former classmates. Ms Yuan's university is in Zhengzhou, so he keeps in touch with some classmates. But he didn't have many opportunities to meet with his classmates, basically meeting with classmates every six months. When he returns to Nanyang every year, he will organise the Nanyang fellows who work in Zhengzhou to meet on the evening of the fourth day of the Chinese New Year. He thought it is ridiculous that everyone works in Zhengzhou but did not meet in Zhengzhou. Instead, they returned to Nanyang to meet each other. Mr Yuan believes that social networks are very important. Relying on good social networks, it will be more convenient and easier to do all kinds of things. However, it takes a certain amount of time and money to maintain a social network. And Mr Yuan believes that he does not have any advantages in this regard, making him in a dilemma.

In the past few years, Mr Yuan mainly worked as a client receptionist for Zhuoyue Company. He was mainly dealing with customers from Hangzhou and Shanghai. Although he became friends with these customers, Mr Yuan thought that he would not have any contact with these friends in a year. After all, these friends were too far away from him and they were also disconnected from the business connection. Mr Yuan told the authors that when he was in Zhengzhou, those foreign friends often invited him to parties together.

Mr Yuan has an aunt working in Zhengzhou, but he has no contact with his aunt. Mr Yuan told the authors that his family are all from rural areas and has no contacts with these relatives working in the city, so he does not want to ask his aunt for help. Mr Yuan has no special close friends in Zhengzhou who can let him confide his thoughts, so he can only tell his mother when he wants to complain after quarrelling with his girlfriend. However, his mother could not give any good advice. On the contrary, sometimes things got worse after he told his mother. Within two hours of the authors' interview with Mr Yuan, Mr Yuan's mother made five calls to ask him about his argument with his girlfriend.

A quantitative survey on the social interaction between white-collar urban immigrants in Shanghai and local residents from the perspective of occupation types show that most white-collar urban immigrants have established social support relationships with local residents, but there are not many social relationships that can provide actual support, which shows that there is a temporary social groups segregation between white-collar urban immigrants and local residents, that is, the segregation and alienation of different social groups due to the social distance between social groups.[12] In addition, qualitative interviews by the authors also indicate that the process of structural integration of intellectual immigrants in primary relationships with local residents is not smooth.

In summary, quantitative and qualitative surveys both show that the three types of urban immigrants have different social networks in cities, and their ability to mobilise social network resources is also different. Research on Chaoshan entrepreneurial immigrants show that the aggregation formed on the basis of social networks provides subsequent Chaoshan young people with a way to adapt to the city and effectively help them improve their social status. Due to the limitations of the institutional structure, the most likely classes of rural migrants to achieve upward mobility are individual business households and business service personnel. The socioeconomic aggregation environment created by Chaoshan entrepreneurial immigrants, on the one hand, provides young people who have entrepreneurial aspirations with opportunities to acquire knowledge of finding supplies, pricing strategies, management and finance through short-term work, and on the other hand, it provides young people with a solid foundation for overcoming the difficulties they may face in the early stages of starting businesses. It can be said that the strong support provided by the socioeconomic aggregation of entrepreneurial immigrants is a key factor for Chaoshan young people to successfully start a business. However, labour immigrants' social network of work and life is mostly limited to their relatives and fellow villagers, so their close friends are also usually their relatives and fellow villagers. Originally, the term "friend" is often used to refer to the modern relationship formed in the context of work and study. However, in the expression of labour immigrants, it is usually used to refer to the traditional kinship and geographical relationships. In the social network of labour immigrants, friend relationships usually overlap with those of relatives and fellow villagers. However, as individual agents, a considerable portion of labour immigrants are also actively constructing new social networks based on instrumental rationality, the process of which is also the process of labour immigrants adapting to the city. When faced with the pressure of employment, compared with labour immigrants, intellectual immigrants can mobilise more social networks and resources, so that they can stabilise more quickly and smoothly in cities. However, the segregation between intellectual immigrants and local residents has not been eliminated, and the structural integration of intellectual immigrants and local residents in the primary relationship still has a long way to go.

[12] Zheng Chuangui, Lu Xiaohui, The Causes, Harms and Countermeasures of Social Groups Segregation of Cities in China Today, *Urban Problems*, Vol. 6, 2003.

# Chapter 7
# The Identity and Community Life of Urban New Immigrants

Although they are both citizens, they have the distinction between "urban new immigrants" and "locals". This segregation reflects the fact that there is still a "boundary" between outsiders and locals. Among the many reasons for this segregation, the first to be criticised is the household registration system. Zhu Yu believes that "the non-permanent migration of migrant population is an abnormal phenomenon caused by the household registration system. Once the household registration system is abolished, this non-permanent migration will be replaced by permanent migration."[1] Guo Xinghua's series of studies show that there are three types of community segregation, namely, location segregation, voluntary segregation and involuntary segregation, which affects the integration of the immigrant population into cities. Among them, the "voluntary segregation" particularly keeps rural migrants from becoming urban residents.[2] In addition, more research shows that the barrier to the integration of immigrants into cities comes from "subjective resistance of the migrant population", such as insufficient economic strength, poor ability to live in cities, and low degree of psychological recognition of cities. At the same time, the barrier also comes from "objective factors in the social system", such as the household registration system, urban management policies, and employment systems.[3]

However, no matter what kind of barriers, it can't stop the trend of migrant population immigrating to cities and becoming urban residents. Researches by scholars such as Ma Jiujie and Zhu Yu show that 20% to 24% of China's 140 million migrants are willing to immigrate permanently to cities. In other words, there are nearly 30

[1] Zhu Yu, Households Registration System Reform and the Dwelling Willing of Floating Population in the Areas They Are Dwelling and Its Restrictive Mechanism, *South China Population*, Vol. 3, 2004.

[2] Guo Xinghua, Chu Huijuan, From Rural Areas to Urban Areas: Integration and Segregation, *Jianghai Academic Journal*, Vol. 3, 2004.

[3] Wang Chunguang, Social Identity of the New Generation of Rural Migrants and Merger of Urban and Rural, *Sociological Studies*, Vol. 3, 2001.



D. Zhou, *Urban Migrants in China*,
https://doi.org/10.1007/978-981-99-3114-9_7

million migrant populations in China who are willing to become urban residents.[4] So, how do this migrant population achieve this transformation of social identity? Cai He and Wang Jin studied the migrant population from the perspective of rural migrants' choice of rural household registration, and believed that rural migrants' willingness to immigrate to cities depends on whether their choices are individual migration motivation based on "economic rational choice" or institutional legitimacy pressure based on "social rational choice".[5] Cai He and Wang Jin believe that the former may allow rural migrants to live and work permanently in cities, but this does not mean that rural migrants will give up their rural household registration and choose urban household registration. While the latter may cause rural migrants to choose systematic permanent migration, that is, to give up rural household registration and choose urban household registration. What is particularly inspiring is that their research raises the issue of "social status and permanent migration" and concludes that "rural migrants who are discriminated against in cities and have greater institutional legitimacy pressure actually have stronger desires to obtain urban household registration and achieve institutional permanent migration".[6]

The above mentioned "reverse selection" phenomenon of rural migrants currently living in cities remind us to pay attention to the issue of community identity and community acceptance of urban immigrants living in cities, including rural migrants. However, it is a pity that most of the current research related to community governance, community construction, community development, and community identification in cities are aimed at household registration residents in the community, while the majority of studies on migrant populations are from a macro perspective such as urban identification. There are very few studies that treat urban immigrants as specific and independent research objects to investigate their community lifestyle and their relationships with urban residents. If Cai He and Wang Jin's research first made a breakthrough in this regard, then Tong Xing and Ma Xiheng's research on the Y community in Shanghai can be regarded as an experimental move to explore this issue. Tong Xing and Ma Xiheng believe that the integration of urban immigrants into the city will go through three stages, firstly from the "binary community" to "the others in the harmonious community", and finally to "homogeneous identity". In other words, whether urban immigrants can finally integrate into urban society depends on whether they can combine with urban residents to form a "community city". Besides, Tong Xing and Ma Xiheng also put forward specific countermeasures such as "community overall development" and "community participation" that rely on communities to promote the integration of urban immigrants into cities.[7]

[4] Zhu Yu, Households Registration System Reform and the Dwelling Willing of Floating Population in the Areas They Are Dwelling and Its Restrictive Mechanism, *South China Population*, Vol. 3, 2004.

[5] Cai He, Wang Jin, Study on the Willingness of Permanent Migration of Rural Migrants, *Sociological Studies*, Vol. 6, 2007.

[6] Cai He, ed., *Rural Migrant in the Process of Urbanization: A Study from the Pearl River Delta*, Beijing: Social Sciences Academic Press (China), 2009.

[7] Tong Xing, Ma Xiheng, "The Others in Harmonious Community" and "Reconciliation": Community Integration of Urban Immigrants, *Social Science Research*, Vol. 1, 2008.

Although Cai He, Wang Jin, Tong Xing, and Ma Xiheng's research put forward the issue that the urban community environment affects whether rural migrants choose permanent migration and ultimately whether they can integrate into urban communities, their research does not involve specific issues such as the identity and community life of urban immigrants. And this chapter will make up for this deficiency through quantitative and qualitative data.

## 7.1 The Identity of Urban Immigrants

The survey shows that about 50% of urban immigrants consider themselves to be outsiders in cities. Among them, the proportion of intellectual immigrants is relatively low (see Table 7.1). And 16.87% of intellectual immigrants consider themselves to be locals, which is significantly higher than other types of urban immigrants. And fewer labour immigrants consider themselves to be locals, on the contrary, most labour immigrants consider themselves to be people who live in cities temporarily or outsiders in cities. Moreover, the willingness of labour immigrants to settle in cities is significantly lower than that of intellectual immigrants and entrepreneurial immigrants. However, the desire of labour immigrants to let their children develop in cities is stronger than that of other types of urban immigrants. As we can see from Table 7.2, in terms of whether they want their children to develop in cities in the future, the willingness of young labour immigrants is higher than that of intellectual immigrants, and the willingness of senior labour immigrants is also higher than that of entrepreneurial immigrants. However, with regard to the question of "whether the city is better than hometown", young labour immigrants have the opposite attitude to senior labour immigrants. Young labour immigrants don't think that the city is better than their hometown, while senior labour immigrants think that the city is better than their hometown.

In terms of the standard of distinction, labour immigrants tend to regard income, work, and status as their standard of distinction from locals. Moreover, labour immigrants believe that status difference is one of the biggest obstacles to their interaction with locals. Therefore, it is difficult for labour immigrants to identify themselves

**Table 7.1** The identity of urban new immigrants (Unit: %)

| Items | Intellectual immigrants | Young labour immigrants | Senior labour immigrants | Entrepreneurial immigrants |
|---|---|---|---|---|
| Outsiders | 47.53 | 53.27 | 54.47 | 54.75 |
| Locals | 16.87 | 7.34 | 7.05 | 9.95 |
| People who live in cities temporarily | 32.42 | 38.56 | 37.77 | 31.94 |
| Uncertain | 7.18 | 2.83 | 2.71 | 3.36 |
| Total (n) | 984 | 918 | 369 | 864 |

**Table 7.2** The identification of urban new immigrants to the city (Unit: %)

| Items | | Intellectual immigrants | Young labour immigrants | Senior labour immigrants | Entrepreneurial immigrants |
|---|---|---|---|---|---|
| Whether the city is better than hometown | Yes | 53.56 | 49.13 | 67.75 | 59.38 |
| | No | 46.44 | 49.89 | 32.25 | 39.81 |
| | Uncertain | – | 0.98 | – | 0.81 |
| Being equal to locals | Yes | 41.46 | 37.04 | 32.25 | 44.44 |
| | No | 57.52 | 61.33 | 67.31 | 53.71 |
| | Uncertain | 1.02 | 1.63 | 2.44 | 1.85 |
| Willing to settle in the city | Yes | 63.81 | 53.45 | 56.98 | 66.63 |
| | No | 36.19 | 46.55 | 43.02 | 33.37 |
| | Uncertain | – | – | – | – |
| Expecting my children to develop in the city in the future | Yes | 43.19 | 51.20 | 71.54 | 67.39 |
| | No | 46.34 | 39.11 | 22.49 | 26.62 |
| | Uncertain | 10.47 | 9.69 | 7.96 | 7.99 |
| Total (n) | | 984 | 918 | 369 | 864 |

as "locals" (see Figs. 7.1 and 7.2). Intellectual immigrants believe that "life-style" and "social roles" are the main distinctions between them and locals. Besides, they encounter relatively few obstacles in interacting with locals. And 37.06% of intellectual immigrants believe that they have no obstacles in interacting with locals, so a considerable proportion of intellectual immigrants identify themselves as locals. In studying the identification of urban immigrants to the city in Shanghai, Zhang Wenhong found that status identity, cultural identity and group identity have a tendency of differential identity, while professional identity, group identity and status identity have a tendency of consistent identity, and the latter is consistent with the status of labour immigrants' identity.

In the following, the authors will continue to analyse the identity of urban immigrants through cases.

Case: Mr Han/24 years old/Bachelor's degree/IT engineer/Been in Hangzhou for two years/ Now sharing an apartment with four college classmates

Mr Han said, "In fact, identity is related to personal attitude. As long as I identify myself as locals, then I am locals. The difference between me and locals is that locals have a deeper understanding of Hangzhou and have a larger social network in Hangzhou. They will completely put their future plans in Hangzhou and will not consider other cities. I now have a collective household registration in Hangzhou, so I don't think I am an outsider in Hangzhou."

Case: Mr Cheng/24 years old/ Bachelor's degree/Software R&D staff/Been in Hangzhou for five years/Now renting an apartment alone

Mr Cheng said, "I feel that I am not much different from locals. Except that locals already have houses in Hangzhou, and outsiders have to buy houses in Hangzhou. And locals can eat

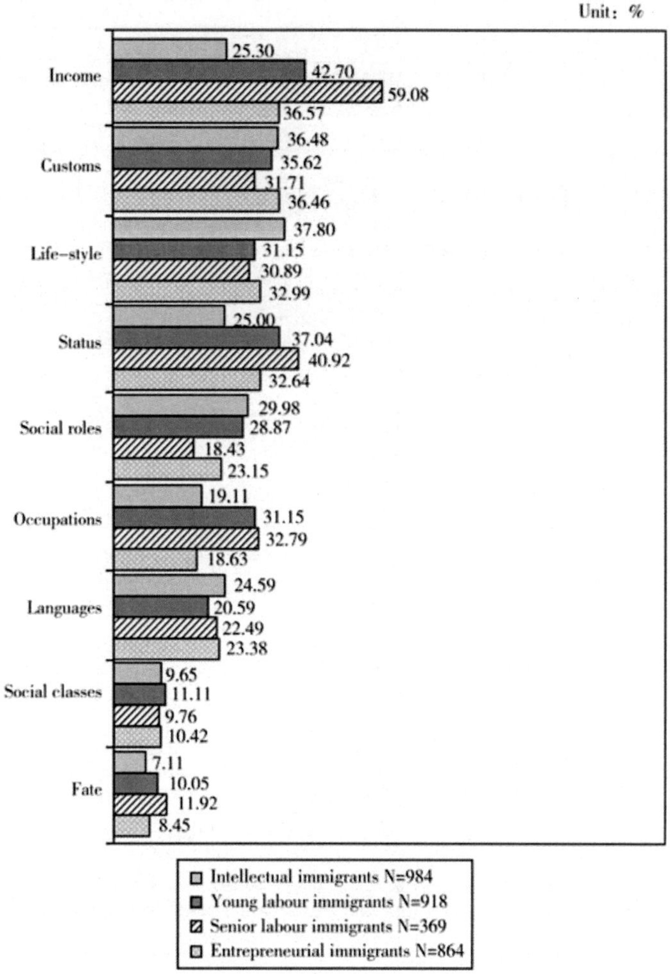


**Fig. 7.1** The distinction between urban new immigrants and locals

with their families every day, while outsiders can only eat alone. Besides, locals have many relatives in Hangzhou, while few outsiders have relatives in Hangzhou. I haven't interacted with many local people, so I don't know much about them. And I don't think I am an outsider in Hangzhou. After all, I'm surrounded by people from all over China, and everyone speaks Mandarin."

Case: Ms Zhang/20 years old/High school education degree/Seller/Been in Hangzhou for less than four years/Living in a company dormitory

Ms Zhang came to Hangzhou with her parents in 2005. Her daily social circle is very small, which makes her feel a little regret. Her place of work is at school, and she has to work four to five days a week, so she rarely has time to meet other people, and she mainly interacts with her colleagues and customers. She believes that some locals still exclude outsiders. She told the authors that locals usually take the initiative to ask where others come

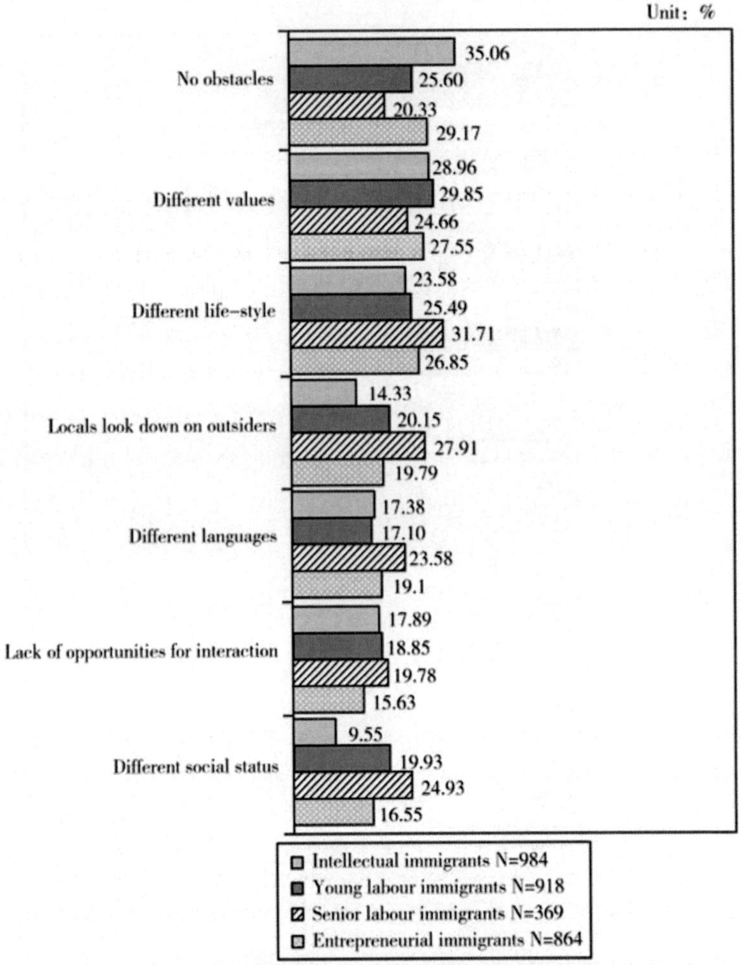


**Fig. 7.2** Obstacles to interaction between urban new immigrants and locals

from. And after knowing that other people are outsiders, the locals usually have nothing to talk with outsiders. For Ms Zhang, interacting with outsiders in Hangzhou makes her feel more respected, and she also resonates more with outsiders in Hangzhou. So compared with locals, Ms Zhang has a closer relationship with outsiders in Hangzhou.

The perceptions of urban immigrants mentioned above were confirmed in the authors' research of locals.

Case: Ms Zhou/Locals in Hangzhou/67 years old/Retired employee/Own house

Ms Zhou said, "There are many outsiders in the building where I live, and I have little interaction with them, but I still greet them when I meet them. They may be busy at work, so we only encounter them when they are off work. Besides, most of them only interact with their acquaintances, such as fellow villagers and colleagues. They are not familiar with locals. What's more, they are highly mobile, and we may just get acquainted with them before they move away. Tenants who live here are all coming and going. It is difficult for us to establish a stable relationship with them, so we cannot get close to them."

Case: Mr Chen/From Wenzhou City, Zhejiang Province/28 years old/Been in Hangzhou for eight years/Master of Medicine/Have been working in a hospital in Hangzhou for more than a year/Has purchased a house

Mr Chen said, "Our lifestyle is changed in a subtle process. For example, my current lifestyle has changed a lot, which is completely different from my original lifestyle in my hometown. But if I usually contact and interact with my fellow villagers, the situation may be different. This may be more easily felt as a student, because when we stay in the "ivory tower", we are less sensitive to everything outside. What I said is just an analogy, but it seems not so appropriate (laughter)."

Case: Mr Yao/23 years old/Ice cream shop clerk/Junior college education degree/Been in Hangzhou for three years/Currently renting in urban commercial housing

Mr Yao said, "In the past few years of living in Hangzhou, I mainly contacted and interacted with my colleagues and friends in the industry. In my daily life, I mainly interact with outsides in Hangzhou, but I have relatively few contacts with locals in Hangzhou. Perhaps because I am also an outsider, I feel that it is easier to get along with outsiders in Hangzhou, and it is easier to be friends with them and to exchange our inner thoughts and difficulties encountered in work. And when I get along with locals, I feel more tired because of the difference between North and South and the difference in personality, so I will unconsciously avoid contacts with locals. I think there is still a gap between me and locals, such as the difference in language, lifestyle and working hours. There is very little communication between me and locals, locals and I just greeted each other out of courtesy when we met in an elevator or somewhere in the community. I personally think that locals have a sense of superiority. They feel that we will affect their lives and other aspects in a bad way."

Whether form the perception of urban immigrants or perception of locals, the low degree of recognition and acceptance of urban immigrants by locals, which has created a large gap between the community where urban immigrants actually live and the community they expect, directly hindering their integration into the city. Nevertheless, the acquisition and enjoyment of urban lifestyle (see Table 7.3) by urban immigrants will not be affected. More than half of urban immigrants believe that their lifestyle is no different from that of urban residents, which is the result of a combination of options of "consistent in many ways" and "exactly consistent". More than 60% of intellectual immigrants believe their lifestyle is consistent with urban residents. However, nearly half of labour immigrants and entrepreneurial immigrants believe that their lifestyle is not consistent with those of urban residents. The reason why intellectual immigrants, labour immigrants, and entrepreneurial immigrants differ in their perceptions of the similarities and differences between their lifestyle and the lifestyle of urban residents lies in their difference in psychological identity. On the one hand, the identity of urban immigrants is diverse and complex. On the other hand, the acquisition of urban lifestyle is closely related to the education degree, occupation, age, and social experience of urban immigrants. However, it is obvious that labour immigrants and entrepreneurial immigrants have stability and relatively low plasticity in these aspects.

**Table 7.3** Differences between the current lifestyle of urban immigrants and locals

| Items | | N | % |
|---|---|---|---|
| Urban immigrants' lifestyle vs. locals' lifestyle | Exactly inconsistent | 332 | 10.48 |
| | Inconsistent in many ways | 968 | 30.56 |
| | Consistent in many ways | 1,398 | 44.13 |
| | Exactly consistent | 350 | 11.05 |
| | Uncertain | 120 | 3.79 |
| | Total | 3,168 | 100.00 |

## 7.2 The Community Life of Urban Immigrants

As mentioned earlier, there are many subjective and objective factors that affect the integration of urban immigrants into cities, such as the identification of urban immigrants to the city, professional stability, urban management system, social security system, etc. However, specifically for individuals, these factors may be concentrated on their direct feelings about their basic living space, that is, the community space in which their daily life occurs. In other words, the community life of urban immigrants directly affects their integration into the city. So, what constitutes urban immigrants' community life? The authors identified three main factors: population structure, living style, and housing type.

From Tables 7.4 and 7.5, we can see that more than 30% of labour immigrants and entrepreneurial immigrants and nearly 50% of intellectual immigrants choose the living style of living with another person. At the same time, three or four people living together are also common among urban immigrants. Nearly 80% of labour immigrants and entrepreneurial immigrants live with their wives and children, showing a trend of family immigration. This phenomenon shows that the main factor contributing to the formation of urban immigrants today is the emergence of the form of immigration in which families work and live in the same city, which encourages urban immigrants to choose a relatively stable lifestyle, such as choosing a stable career, living with their family members and neighbours who can help each other, so as to integrate into the city as soon as possible. Young labour immigrants have certain similarities with intellectual immigrants. Most of them choose to rent an apartment with "friends or colleagues". In addition, more than 20% of intellectual immigrants are stilling "living alone". These reflect that the majority of young labour immigrants and intellectual immigrants are still unmarried, and their economic condition is not well enough to "support a family".

At the same time, we should also see that there are a considerable number of senior labour immigrants and entrepreneurial immigrants who are stilling working hard with their spouses in the city, and their children are left in the countryside, which is also the biggest dilemma they currently face in the city. On the one hand, they want to take their children to the city to live with them, so that their children can receive the care and companionship of their parents, enjoy modern urban life

**Table 7.4** Life partners of three types of urban immigrants

| Items | | Types | | | | | | | |
|---|---|---|---|---|---|---|---|---|---|
| | | Total | | Intellectual immigrants | | Labour immigrants | | Entrepreneurial immigrants | |
| | | N | % | N | % | N | % | N | % |
| Life partners | Families | 1,330 | 41.98 | 226 | 22.97 | 484 | 36.67 | 620 | 71.76 |
| | Lovers | 291 | 9.19 | 133 | 13.52 | 102 | 7.73 | 56 | 6.48 |
| | Friends or colleagues | 963 | 30.40 | 366 | 37.20 | 506 | 38.33 | 91 | 10.53 |
| | Relatives | 125 | 3.95 | 29 | 2.95 | 68 | 7.15 | 28 | 3.24 |
| | Living alone | 568 | 17.93 | 248 | 27.20 | 215 | 16.29 | 105 | 12.15 |
| | Classmates | 9 | 0.28 | 5 | 0.51 | 4 | 0.30 | – | – |
| | Roommates | 14 | 0.44 | 7 | 0.71 | 5 | 0.38 | 2 | 0.23 |
| | Others | 7 | 0.22 | 1 | 0.10 | 6 | 0.45 | – | – |
| | Total | 3,168 | 100.00 | 984 | 100.00 | 1,320 | 100.00 | 864 | 100.00 |

**Table 7.5** Number of people living with three types of urban immigrants in the city

| Items | | Types | | | | | | | |
|---|---|---|---|---|---|---|---|---|---|
| | | Total | | Intellectual immigrants | | Labour immigrants | | Entrepreneurial immigrants | |
| | | N | % | N | % | N | % | N | % |
| B2_1. Total number of people | 2 | 1,017 | 39.16 | 353 | 47.83 | 386 | 37.00 | 278 | 36.77 |
| | 3 | 640 | 24.64 | 185 | 27.07 | 238 | 21.58 | 217 | 28.70 |
| | 4 | 454 | 17.48 | 124 | 16.80 | 183 | 16.59 | 147 | 19.44 |
| | 5 | 181 | 6.97 | 36 | 4.88 | 85 | 7.71 | 60 | 7.94 |
| | 6 | 138 | 7.31 | 25 | 3.39 | 82 | 7.43 | 31 | 4.10 |
| | 7 and above | 167 | 6.43 | 15 | 2.03 | 129 | 11.70 | 23 | 3.04 |
| | Total | 2,597 | 100.00 | 738 | 100.00 | 1,103 | 100.00 | 756 | 100.00 |

and its high-quality education resources. On the other hand, due to their low income and lack of housing in the city, they can only choose to leave their children at home and be taken care of by their older parents. However, the authors found through investigation that more and more urban immigrants are willing and working hard to create conditions to bring their children to the city to live together. This phenomenon of "family immigration" is increasing, which will accelerate the process of urban immigrants integrating into the city.

Interviewee: Mr Zheng/55 years old/From Jiangsu Province/High school education degree/ Owner of the "Tianjin steamed stuffed-buns"

Interview time: 23 February, 2009

It has been five years since Mr Zheng came to Hangzhou to open a steamed stuffed-buns shop. Before that, he worked in many cities in China, mainly engaged in clothing sales, bun shops, and so on. Now he runs his own bun shop, his wife and two daughters help him run the bun shop together. The family of four rented together in a 50 square metre rental house, and they have rented it for five years because they get along well with the landlord.

Q: Since you have been in Hangzhou for so many years and the business are also not bad, why not consider buying a house in Hangzhou?

A: Impossible! The housing prices in Hangzhou are too high, not to mention we have to raise three children.

Q: So, what kind of people do you think are more likely to become urban residents in Hangzhou?

A: If the children have great achievements and they are all married, we may consider buying a house in Hangzhou and settling down. But the key is that our children have no achievements and are not married, so we will not consider this matter. In any case, our children still have to stay and develop in Hangzhou, and it is impossible for them to return to their hometowns.

It can be seen that Mr Zheng, who brought his family to work in Hangzhou and struggle to achieve establishment in Hangzhou, is for the future of his children, not for himself. The essence of this type of family immigration is not the "family immigration" as we understand it, but it may be the foundation of the "second generation immigration". In other words, this type of family immigration is based on the hope that the second generation can successfully settle in the city. However, the housing problem is still the primary issue that they need to consider and solve before they settle in the city.

Survey statistics show that on average 11.43% of urban immigrants have self-purchased houses, nearly two-thirds of urban immigrants rent private houses, and nearly 20% of urban immigrants live in a collective dormitory provided by the company. However, the proportion of senior intellectual immigrants and entrepreneurial immigrants owning self-purchased houses is relatively high, exceeding 20%. For labour immigrants, in addition to renting private houses, some companies with good welfare will also provide them with dormitory, and only charge a relatively low rent. On the one hand, this reflects the company's welfare policy, on the other hand, it also facilitates the company's daily management of employees. As we all know, the place where urban immigrants live directly affects the quality of community management and the quality of life.

Nearly two-thirds of urban immigrants are renting houses, and 53.85% of urban immigrants are renting houses in "villages within cities" or small towns, and 34.72% of urban immigrants renting commercial housing in communities, which is especially true for senior intellectual immigrants, with up to 57.87% of intellectual immigrants living in commercial housing in communities. Moreover, cases show that a considerable part of the commercial housing in communities where senior intellectual immigrants live is their self-purchased houses. Besides, two-thirds of labour immigrants choose to rent in "villages within cities". In addition, urban immigrants also live in company dormitories, government low-rent housing, immigrant apartments, and some even live in their workplace. Although these forms of residence do not account for a large proportion of urban immigrants, it also reflects the general status of urban immigrants in the city "lack of fixed residence". Obviously, this situation has caused great difficulties for the identification of urban immigrants to the community and the current community management in cities.

Case: Mr. Du/26 years old/Bachelor's degree/Immigrated from Tongjiang County, Sichuan Province to Chengdu City in 2006/Now runs a consulting and training company/Known as "Boss Du"

"Boss Du" said, "Don't underestimate me because I am young, I have very rich experience. In 2003, I was admitted to a junior college in Jiangxi Province. In 2005, when I was a junior in college, I began to try various internships and jobs. And I had a short stay in Jiujiang City, Jiangxi Province and Guangzhou. In early 2006, I came to Chengdu for my girlfriend and worked as a senior manager in a well-known training management company. After graduation, I returned to my hometown of Tongjiang County and made the first real estate transaction, earning the first pot of gold in my life. In the spring of 2007, after marrying my girlfriend who had been in love with me for many years, I brought some funds and my accumulated experience and social networks to Chengdu to start my own business, and now I am also self-employed. The environment of the community I live in is very good, not only neat and clean, but also green. The two-bedroom house I live in is rented. Although the decoration of the house is rough, I will not spend money on it. After all, this house is rented. Moreover, I have bought my own house in Wenjiang, and I paid the house in one lump sum without paying the mortgage. Therefore, there is no need to waste too much money on the house I rent. And my career is in its infancy, so I still need to save money."

Obviously, "Boss Du" belongs to the category of "entrepreneurial immigrants" that this book focuses on. And the successful transition from renting a house to self-purchasing a house is not only a goal he has struggled for many years but also a sign that he has become a real urban immigrant. However, in reality, not all urban immigrants are in such a good situation as "Boss Du".

Case: Mr. Li/31 years old/From Xuancheng City, Anhui Province/Married/ Dropped out of school before finishing sixth grade/Has been a security guard at a maintenance station for five years/Initially worked in cities such as Dalian and Jinan and worked as a cement worker, welder, etc./Came to work in Yuhang District, Hangzhou City through the introduction of his fellow villagers in 2006

Mr. Li said, "The house where my wife and I live in is a self-built house by farmers here. Although it is far away from the company I work for and the transportation is not convenient, the rent is very cheap. The farmer's self-built house has three floors, and the landlord lives downstairs. The landlord is very nice, my wife and the landlord's wife often eat melon seeds and chat together. The people who live in the home are all outsiders in Hangzhou, and my working hours are inconsistent with them, so I rarely meet them and do not know their names. I think the place where I live is okay. Although it is relatively remote and dirty, it is cheaper and better than living in the city. However, the windows should be closed at night, because there are so many thieves, and people's things are often stolen at night, and sometimes even during the day. So, I put my mobile phone and wallet under my pillow when I sleep at night.

There is no entertainment here, and I rarely go out to play, and my wife does housework at home when she is not working. As you can see, we live far away from the urban area, so it is not convenient for us to go out to play, plus we need to work every day, so we rarely have leisure time, and we usually go home and watch TV after work. We only go out to play cards with our fellow villagers during the non-working holidays such as National Day and Labour Day, and we usually do not participate in any recreational activities. The fellow villager said that consumption in bars and tea houses is relatively high, so we never go there. Besides, we rarely encounter community officials, and we always ask the landlord for help if we have any problems. Once a community official came to me to check my temporary residence permit, and his service attitude was pretty good. However, we rarely met them in daily life."

Mr Li's living environment is the current status of most labour immigrants. They live in rental houses in "villages within cities" and live a semi-rural and semi-urban lifestyle. At the same time, the mixed living of urban immigrants and local villagers also brought new challenges to community management. The current community management of "villages within cities" is facing a major transformation, that is, the program of "changing village to residential community" promoted by the government, which transforms the original rural village committee system into an urban community neighbourhood committee. However, if the project is only completed at the structural level, not only can it not solve the problem of urbanisation of peasants, but it will also face the new problem of resettlement of urban immigrants living in "villages within cities".

Let's take a look at the community life of urban immigrants. When it comes to their "closely related content of community management", the top three choices for urban immigrants are "community security", "community environment" and "community service". However, they pay less attention to aspects such as "government construction of grassroots community", "community guarantee", "community employment", "community birth control", "community culture and education", "management of rental houses", "management of immigrants", etc. In addition, they have the lowest recognition of the author's hypothetical option of "withdrawal" of community management. In their view, community management is a "government matter" and has nothing to do with them (see Fig. 7.3).

In fact, in addition to the above reasons, the indifference of urban immigrants to community management is also closely related to our current community-building policies and measures that are oriented to urban residents, and regard urban immigrants as mobile, temporary community management objects. The lack of community building for urban immigrants, on the one hand, has caused urban immigrants to be indifferent to community management, on the other hand, it has also intangibly constructed and exacerbated the "boundary" between urban residents and urban immigrants living in the same community, resulting in de facto "dual community" and the invisible "bottom class of the community". Therefore, understanding and solving the current problems in community management and community construction must be put on the agenda.

Consistent with the content of community management concerned by urban immigrants, they also eagerly hope for a significant improvement in community security, community environment and community service. However, different types of urban immigrants have different perceptions and requirements of the community. More

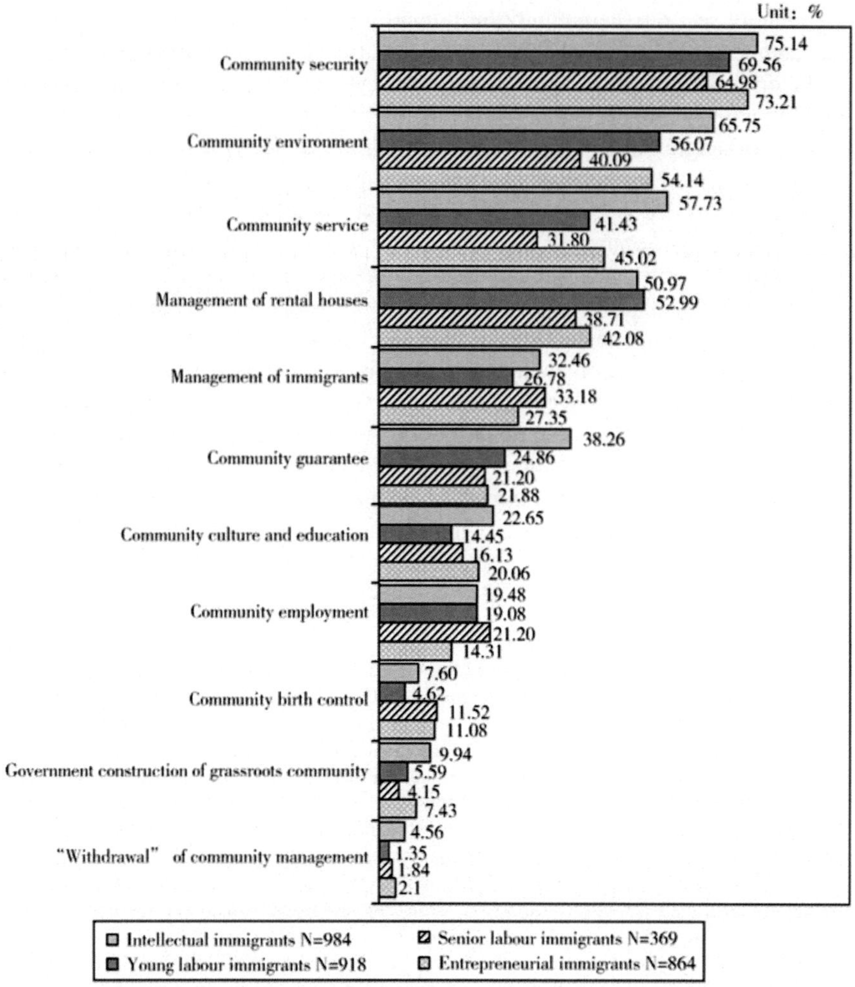


**Fig. 7.3** Contents of community management concerned by urban new immigrants

than 30% of senior labour immigrants and more than 20% of entrepreneurial immigrants are "indifferent" to community management (see the option "no"). However, compared with labour immigrants and entrepreneurial immigrants, intellectual immigrants pay special attention to all aspects of the content of community management involved. The authors divided all the options of community management content into four categories for specific analysis. The first category is the community organisation, including government construction of grassroots community, "withdrawal" of community management. The second category is the community service, including community service, community guarantee, community employment, and community culture and education. The third category is community management, including community security, community birth control, management of rental houses, and management of immigrants. The fourth category is community environment.

In terms of community construction, intellectual immigrants and entrepreneurial immigrants have higher requirements than labour immigrants. However, the three types of urban immigrants all showed indifference to the suggestions of community construction and "withdrawal" of community management. Besides, their desire for community service is also relatively low. This situation is caused by the current community service mainly designed for urban residents. Under the current institutional arrangements, even if urban immigrants have strong demands for community services, such demands cannot be realised in practice. Therefore, urban immigrants show a strong pragmatism close to reality. Their appeal for community management is obviously higher than that of community service. This is because the current community management is closely related to their community life, and only in this respect will they deal with community officials. Interestingly, the requirements of urban immigrants for the community environment are more apparent. In their view, the so-called community environment is the space they can imagine and perceive in reality, which is an urban space different from the rural or urban space where they originally lived.

Case: Mr. Wu/24 years old/Master's degree/Software R&D engineer/Have been in Hangzhou for two years/Renting a house with four people

Mr. Wu said, "The community does not organise any activities for residents to participate in, so we rarely deal with the neighbourhood committee. The only time we deal with the neighbourhood committee is to buy rat poison. I also did not feel the existence of the services or management provided by the community, but the community is relatively safe. I am currently renting a house in this community, so I don't have many expectations for this community, but I have more expectations for the management of the community where I will buy the house in the future. My basic requirements for the community are safety and convenience. In addition, I hope that the community can keep the environment clean and green, and can organise public activities and recreational activities, such as organising recreational activities or sports activities for the elderly. Moreover, the community should be equipped with public sports facilities for residents to do exercises. The community should also organise communication activities between neighbours to provide opportunities for interaction between neighbours so that community residents can have a harmonious neighbourhood relationship."

Case: Mr Zhu/28 years old/unmarried/Master's degree/Software company employee/Have been in Hangzhou for 8 years/Now renting an apartment of two bedrooms and a living room with other people

Mr Zhu said, "I rarely deal with community officials because we basically don't need to ask them for help. If there is a chance, I don't mind participating in the election of the neighbourhood committee. I think it's beneficial to participate in some of these activities, and it can also be counted as my own valuable experience. I am always busy with work, and I feel that participating in these activities occasionally will make life more enjoyable, so I will not refuse such an opportunity. I think the community guarantee, government construction of grassroots community, and community security need to be further strengthened so that we can truly feel that living in the community is like living with a big family. Now everyone is busy with work and has no special feelings for the community. For us, the community is just a collection of houses, where everyone lives. And many people don't even know their neighbours, they are like passers-by who do not interact with each other. Therefore, I think it is better to further refine the community service so that everyone can feel its existence and have a sense of belonging to the community."

So, what are the attitudes and considerations of community officials?

Interviewee: Mr. Li/Director of the community, Interview time: 9 April, 2009

Mr. Li said, "I have been a community director for two years. I was selected through the examination organised by the street office. In addition, this position was confirmed only after the election of community residents. The affairs that our community needs to manage are extremely cumbersome, such as low-security issues for households with financial difficulties, illegal buildings in the community, floor leakage, disorderly parking of vehicles, stolen bicycles, etc. Every day there are various problems that we need to deal with, but we have a shortage of staff, with only seven public officials, so we are very busy. The community is located in the old town. And the main residents here are locals, with a registered population of 5,565 and registered immigrants of about 2,300. These immigrants are mainly construction workers and small business owners engaged in service industries. We can't even manage all the affairs of community residents, let alone strengthen the management of immigrants. Besides, we also do not have a department dedicated to immigration management. Immigrants may sometimes turn to us for things like children's education and temporary residence permits. However, they basically solve the rest of their own affairs or ask their landlord for help."

It can be seen that although household registration residents and urban immigrants are the objects of community management, community neighbourhood committees have different management attitudes and measures towards the two. However, this does not mean that there is a clear boundary between household registration residents and urban immigrants. On the contrary, the two in the same community are actually in a closely connected but independent distribution. Moreover, urban immigrants lack the ability to self-organise, so they have not yet become a force of "community isolation".

As can be seen from Table 7.6, in the existing social organisation structure in China, the participation rate of urban immigrants in official and private social organisations is very low. Nearly half of urban immigrants have not participated in any organisation, which is especially true for labour immigrants and entrepreneurial immigrants, whose lives are almost confined to their own small social circles. In contrast, intellectual immigrants participate in a wide range of social organisations, such as the Communist Youth League of China, volunteer organisations, fellow villagers' clubs, the Communist Party of China, Labour Unions, public entertainment organisations, and some professional and technical associations. However, the survey found that even though urban immigrants participate in certain organisations, their actual role in the organisation is not very meaningful to their real life, and their actual participation rate is also very low. Most senior intellectual immigrants only act as members of some organisations, such as the Communist Youth League of China, the Communist Party of China, Labour Unions, other political parties and other political official organisations, but they rarely participate in the organisation's specific activities. This is true even for senior intellectual immigrants, not to mention other types of urban immigrants. On the contrary, some informal and private organisations, such as fellow villagers' clubs, volunteer organisations, and some community-oriented entertainment organisations, usually become the only and rare organisations for urban immigrants. Therefore, urban immigrants can be said to be "groups without formal and official organisations".

However, the lack of formal and official organisations of urban immigrants does not mean that there is no discipline. As the authors mentioned earlier, the reason why urban immigrants living in urban communities rarely interact with locals in the same community is that they are busy with work and have no time to communicate with locals. However, another main reason for this phenomenon is that most urban immigrants live in their own small social circles. That is to say, urban immigrants have basically maintained their original social network even in cities. At the same time, in order to develop in the city, urban immigrants have to build new social relationships, which does not mean breaking their original social networks and establishing a brand new social network, but to expand and build a new social network based on their original social network formed by kinship and geographical relationships. As a result, urban immigrants often form what we call informal social organisations. The survey found that labour immigrants spontaneously formed organisations such as the "worker's home". However, under the management system of associations and organisations, it is difficult for such organisations to be legally recognised. Therefore, these organisations are usually in a corporate or "underground" state, and are in a legal grey zone. Besides, intellectual immigrants, especially senior intellectual immigrants living in urban communities, have formed "virtual community organisations" like the QQ[8] group, providing immigrants with a platform and space to manage and discuss the affairs of their communities. However, at present, all these so-called self-organisation of urban immigrants are not "substantial", but are only loose aggregates in the community.

So, in the eyes of urban immigrants, how should community-building be carried out?

Interviewee: Mr. Liu/47 years old/Master's degree/Have been in Hangzhou for five years/A middle-level administrator of a company/Has purchased a house

Mr. Liu said, "The community is literally a management committee in a small area. Therefore, it should first be familiar with the area and the residents living in the area, and it should also try to integrate into the community. If it cannot be integrated into the community or dissociated from the community, then it will not be able to manage the residents in the community well. So, what role should it play in the community? I think it should first be integrated into the community, find out the problems in the community, and be familiar with the residents in the community. It can obtain the public opinion of the community through investigation and research and then propose and implement some related management measures. I think it should have a clear role positioning, that is, it should be a community servant rather than a community manager. If it wrongly plays the role of the community manager, then it can never manage the community well. Only when it plays the role of a community servant, that is, provides corresponding services when the community residents encounter difficulties and tries its best to meet the needs of the community residents within the scope of the law, can it exert its maximum effectiveness.

[8] QQ (Tencent QQ) is an internet-based instant messaging software developed by Tencent with reference to ICQ. It was launched in February 1999 and is a social software giant in China (by translator).

**Table 7.6** Organisations and activities that urban immigrants participated

| Items | | Types | | | | | | | |
|---|---|---|---|---|---|---|---|---|---|
| | | Total | | Intellectual immigrants | | Labour immigrants | | Entrepreneurial immigrants | |
| | | N | % | N | % | N | % | N | % |
| Organisations and activities | Labour Unions | 291 | 9.19 | 163 | 16.57 | 98 | 7.42 | 30 | 3.48 |
| | Communist Youth League of China | 728 | 22.99 | 380 | 38.62 | 248 | 18.79 | 100 | 11.59 |
| | Communist Party of China | 262 | 8.27 | 178 | 18.09 | 42 | 3.18 | 42 | 4.87 |
| | Other political parties | 7 | 0.22 | 3 | 0.30 | 2 | 0.15 | 2 | 0.23 |
| | Fellow villagers clubs | 625 | 19.73 | 225 | 22.87 | 233 | 17.65 | 167 | 19.35 |
| | Chamber of commerce | 157 | 4.96 | 20 | 2.03 | 25 | 1.89 | 112 | 12.98 |
| | Volunteer organisations | 499 | 17.76 | 252 | 27.61 | 171 | 12.95 | 76 | 8.81 |
| | Public entertainment organisations | 461 | 14.56 | 179 | 18.19 | 201 | 17.23 | 81 | 9.39 |
| | Professional and technical organisations | 268 | 8.46 | 126 | 12.80 | 70 | 7.30 | 72 | 8.34 |
| | Religious organisations | 89 | 2.81 | 36 | 3.66 | 31 | 2.35 | 22 | 2.55 |
| | Company activities | 8 | 0.25 | 5 | 0.51 | 2 | 0.15 | 1 | 0.12 |
| | Others | 20 | 0.63 | 9 | 0.91 | 8 | 0.61 | 3 | 0.35 |
| | N/a | 1,338 | 42.25 | 242 | 24.59 | 650 | 49.24 | 446 | 51.68 |
| | Total | 3,167 | 100.00 | 984 | 100.00 | 1,320 | 100.00 | 863 | 100.00 |

Q: Could you give me an example to talk about how the community should be managed?

A: For example, I am a resident in the community, and I may encounter some problems or difficulties in real life, such as pension problems, medical security issues, education of children and grandchildren, etc. These problems can hardly be solved by me alone, and I also can't enjoy the unit's management and services. At this time, community management and services are needed. I think the community can provide services on issues of particular concern to the residents of the community, such as medical care, social security, pensions, and the care of the elderly living alone. Moreover, the community neighbourhood committee can fully understand the concerns of the community residents through research, and then arrange and solve these problems according to priority, and provide corresponding services. In this way, the community neighbourhood committee will definitely win the recognition of the residents of the community.

Q: As you mentioned earlier, many issues can also be resolved through the unit's management and services, so is the management and services provided by the community still necessary in this case? Besides, comparing the form of management and service provided by the unit to the form of management and service provided by the community, which one do you think is better?

A: From the perspective of social development, the reform from "unit person" to "societal member"[9] is an inevitable trend. The unit only provides management and services for individual careers, but I think people should integrate into their communities after retirement, and personal medical care, pensions and other issues should be managed by communities where they live.

Q: Do you think it is necessary for the community to manage locals and immigrants in Hangzhou separately?

A: I don't think this is necessary. Although there are differences between locals and immigrants, as long as immigrants are in a certain community, then they are part of the community. And if the immigrant leaves that community and enters another community, then the immigrant becomes part of another community. Therefore, there is no need to manage locals and immigrants separately. Moreover, people are legally equal after birth, so I think it is inhuman and illegal to treat locals and immigrants differently.

Mr. Liu's idea about the community actually gave us a good revelation, that is, in the current process of community-building, the important factor of urban immigrants needs to be taken into consideration. On this basis, we will explore the feasibility and desirability of constructing a new urban "community mobilizing".

The practice of community-building in China began in 1999. So far, the urban community model with street offices and community neighbourhood committees as the main structure has basically been formed. The primary feature of this model is that it is not strictly a self-governing organisation of citizens, but the embodiment of state power building in urban communities. For example, Yang Min believes that the

[9] "Unit person" or "societal member" mainly refers to the relationship between the individual and the unit, the individual and the society and the related consciousness formed by it. Under China's planned economic system, every employee is objectively confined to the narrow space of "units" and has a fixed and comprehensive relationship with units. Essentially, the employee is actually in a passive position of being an "unit person". However, with the transition from China's planned economic system to China's market economic system, the basis of the original relationship between units and employees began to change. Both the unit and the employees have obtained a right of two-way choice. Employees who once belonged to the unit have their own choices and become "societal member".

building of urban communities in China is actually a top-down "national governance unit" to solve problems of social integration and social control in cities after the disintegration of the "unit system", rather than a regional social community that can promote the formation of the "public domain" or the development of civil society. Under this framework, the Chinese government advocates mobilising residents to participate in the building of community sanitation, community security, community services, and community culture through community neighbourhood committees, but discourages residents from expressing their own interests through channels other than community neighbourhood committees.[10]

Another important feature of urban community model in China is that its community-building is based solely on the registered residents of urban communities. Little attention has been paid to the connection between community-building and urban immigrants, let alone the inclusion of urban immigrants into actual practices of community-building. However, we must know the fact that in the current structure of the community population, urban immigrants have become part of the community members, and sometimes even become the main component of the community population. Therefore, it is very necessary to carry out institutional innovation in community-building to adapt to changes in community structure and changes in the community population. However, discussing "political topics" such as "community autonomy" or "withdrawal of community management" within the framework of community-building under China's state governance is meaningless for community innovation. In this case, how to create a new type of "community mobilizing" under the structure of China's state governance?

The survey shows that the identity and community life of urban immigrants have directly affected their integration into the city and current community-building. Specifically, urban immigrants are scattered in cities, most of them choose to rent private houses, and few urban immigrants own self-purchased houses. It can be said that urban immigrants are at the bottom class of society. Besides, the community life of urban immigrants is in an "unorganised but disciplined" state. Even if urban immigrants form self-organisations, these organisations are still in a legal grey zone, and some are even "invisible organisations". Moreover, urban immigrants tend to identify themselves as "outsiders" in terms of identity, and the degree of self-identification and recognition by others is relatively low. Furthermore, urban immigrants show a clear indifference to community management, and their participation rate in community activities is very low. Last but not least, the lifestyle of urban immigrants is basically the same as that of urban residents. However, in the process of community-building led by governments at all levels, urban immigrants are often overlooked.

In the eyes of labour immigrants, the "acquaintance society" of rural communities is composed of blood-related family ethics, vertical social structure, and shared

[10] Yang Min, Community as State Governance Unit: A Case Study on Residents' Community Participation and Cognition in the Process of Community Building Campaign, *Sociological Studies*, Vol. 4, 2007.

culture, beliefs, and values of the community. In the expectation of intellectual immigrants and entrepreneurial immigrants (including registered residents of the community), the community should have community identity, such as community spirit, community awareness, community culture, etc., and should not pay attention to differences in household registration, wealth, and social status of community residents. During Fei Xiaotong's research on community-building in Shanghai in the mid-1990s, the cadre of the local neighbourhood committee proposed to build the urban community into a new type of "acquaintance society". However, Fei Xiaotong believes that community-building is not just an "economic" or "administrative" issue, "but also has anthropological significance, which is to say that it is also a human problem. And the solution to this problem not only requires pertinence and standardisation but also requires flexibility and innovation".[11] So, how to reflect the "flexibility and innovation" in community-building? Fei Xiaotong outlined a "blueprint" for community-building in China. Firstly, we must enable those citizens released from the "unit system" of the planned economic system after the reform and opening up to inherit the traditional spirit of solidarity and mutual assistance that has existed in ancient China. Secondly, it is necessary to develop social relationships of cooperation and mutual assistance among community neighbourhoods to adapt them to the new pattern of cities. And make them gradually change from passive psychology that depends on the arrangement of superiors to a spirit of community independence that everyone cares about each other and can solve problems by themselves. Lastly, it is very important to further develop grassroots self-governing organisations from community self-care and form a "grassroots democracy" of socialism with Chinese characteristics.

Obviously, the realisation of the above "blueprint" requires a long process of social construction, which includes both "reform of personal private life" and "reform of public life", such as the construction of civil society. Du Weiming mentioned in his discussion of modern China that a family-centric society is difficult to transform into civil society.[12] If this "family-centric society" is narrowly understood as a modern rural society in China, is it still possible to create a civil society in an urban community centred on state governance? Ma Guoqing believes that this is not only possible but also feasible. However, this so-called civil society is not the civil society referred to in the context of Western society, but a Chinese-style "family-oriented civil society". Ma Guoqing believes that Chinese society has a natural foundation for building a "family-oriented civil society", that is, family ethics, a continued vertical social structure, and cultural resources such as "strata" and "relationship" in Chinese society.[13] As for the specific measures and strategies for building urban communities including urban immigrants, it is beyond the scope of this article.

---

[11] Fei Xiaotong, Chinese Modernisation: Rethinking the Rebuilding of Urban Communities, *Jiangsu Social Sciences*, Vol. 1, 2001.

[12] Du Weiming, *Modern Spirit and Confucian Tradition*, Beijing: SDX Joint Publishing Company, 1997.

[13] Ma Guoqing, The Foundation of a Family-Oriented Civil Society: Family Theory and Continued Vertical Society, *Academic Research*, Vol. 8, 2007.

# Chapter 8
# The Adaptation and Integration of Urban New Immigrants

## 8.1 Introduction

Over the past 40 years of reform and opening up, China has made remarkable achievements in the process of industrialisation and urbanisation. Accompanied by which, China's population migration has also entered the most active period. According to statistics, China's migrant population reached 211 million in 2009, with an average age of about 27.3 years old. Among them, 78.7% have rural household registration and are mainly young and middle-aged people. These people are mainly employed in manufacturing, wholesale and retail, and social service industries, and they are mostly engaged in low-paid or high-risk industries.[1]

In order to survive and have a better life, a large number of urban immigrants migrate between different regions, supporting the rapid development of China's economy. They migrate from the countryside to the city, from the simplicity to the prosperity, thus closely linking the countryside with the city. On the one hand, they have provided cheap labour for cities and created huge wealth. On the other hand, they have brought income to the countryside and promoted the economic development of the countryside.

A survey conducted by the National Bureau of Statistics shows that nearly 70% of rural migrants believe that they can adapt to urban life to a large extent, and more than half of rural migrants hope to stay in the city for further development.[2] This is especially true for the new generation of urban immigrants. They enter society (city) directly after graduating from school, so they lack agricultural experience and are more adaptable to urban life. However, they still face various difficulties in the process of integrating into the city they migrated to. From a macro perspective, first of

[1] Peng Xizhe, *The Changing Population of China*, Wiley-Blackwell, 2000, p. 180.

[2] Li Ke, Liu E, The Process of Urbanisation Urgently Needs Migrant Workers to Integrate into the City, *Theory Front*, Vol. 23, 2009.



D. Zhou, *Urban Migrants in China*,
https://doi.org/10.1007/978-981-99-3114-9_8

all, urban immigrants suffer from institutional exclusion (discrimination). "Unreasonable institutional arrangements allow certain social groups to obtain relatively more profit-making opportunities and welfare benefits, while other social groups are trapped in institutional poverty".[3] Secondly, differences in lifestyles, values, and social psychology have also increased the difficulty for urban immigrants to integrate into the city. From a micro perspective, the breakage in social networks, the lack of comprehensive capabilities, and the lack of relevant professional knowledge, skills and experience also make it difficult for urban immigrants to truly integrate into the city and settle in the city to complete the immigration process.

So far, many disciplines such as sociology, demography, economics and public policy have done a lot of research on population migration and the status quo, process and characteristics of migrants' integration into the place of immigration. However, indicators and analysis dimensions used by various scholars are different, so they usually have different understandings, definitions and measurement standards for same or similar concepts. Therefore, in order to better grasp the ideas for further research, the authors need to clarify the research scope of each scholar, clarify the definition of the concepts they have put forward, and briefly summarise their research conclusions on the degree, process, obstacles, and influencing factors of integrating into cities of the migrant population, such as rural migrants, domestic immigrants, etc.

From the existing research on the social integration of urban immigrants, we can find that the degree of social integration of urban immigrants is generally low and the difficulty of social integration is relatively high. Urban immigrants generally face problems such as low incomes, narrow social networks, low recognition of citizens and governments of cities they migrated to, and a weak sense of belonging. Although rural migrants can catch up with the life of urban citizens within eight hours of the working system, their lives outside of the eight-hour working system still seem incompatible with the lives of urban citizens.[4]

After migrating to the city, how can the migrant population adapt to the city and integrate into it? In the 1890s, the Chicago School, represented by American sociologist Parker, began to study how urban immigrants who came to the United States from Europe migrated to and adapted to the new environment. Western theoretical discussions on the relationship between immigrants and mainstream society can be sorted out into two major schools of "assimilation theory" and "multiculturalism" according to their basic orientation.[5] According to the "assimilation theory", cross-border immigrants generally have to go through three stages of settlement, adaptation and assimilation in the receiving country. For immigrants, only by learning, adapting to, and accepting the local lifestyle and cultural values, and discarding their original

[3] Zhong Dajun, *Examining the Inequality of National Treatment: China under the Dual Structure*, Beijing: China Workers Publishing House, 2002.

[4] Zhu Li, Adaptation to Urban Life: Taking the Spiritual and Cultural Life of Migrating Population for Example, *Journal of Hehai University*, Vol. 3, 2005.

[5] Li Minghuan, An Overview of the Western Theories of International Migration in the 20th Century, *Journal of Xiamen University*, Vol. 4, 2000.

social and cultural traditions and habits, can assimilation and integration be achieved. In contrast, according to the "multiculturalism", immigrants will reshape their place of life through their different cultural backgrounds, social experiences and values, which will also help construct a diversified social and economic order.

From the perspective of urban immigrants' identification and sense of belonging to the city, rural migrants have strong geographical identification and weak group identification with the city.[6] Rural migrants have a high degree of recognition of urban life. The data shows that 49.17% of rural migrants like urban life very much, 27.19% of rural migrants don't like urban life very much, only 3.16% of rural migrants clearly stated that they don't like urban life, and rural migrants with an unclear attitude towards urban life accounted for 18.18%. However, taking Beijing as an example, and comparing the attitudes of rural migrants towards Beijing, their residential communities, and locals in Beijing, it can be found that the degree of rural migrants' identification with these three is gradually reduced. In daily life, rural migrants can directly feel the attitudes of locals towards them through specific life experiences, thus forming their own intuitive evaluations and feelings.[7]

By examining whether the actor is willing to interact with other people or other social groups, we can measure the perception of social members of the distance between them or between them and other social group members.[8] From the perspective of the social network of urban immigrants, the interpersonal interaction of urban immigrants in the city is still limited to kinship and geographical relationships, such as relatives, fellow villagers, and colleagues. However, the interaction between urban immigrants and locals is mainly limited to production, work, etc., and there are few emotional exchanges. What's more, some urban immigrants have no interaction with locals at all. On the whole, urban immigrants do not have a strong sense of belonging to urban citizens.

The integration of urban immigrants and cities is actually a power game between the two parties. The power of cities is mainly manifested in institutional arrangements and social attitudes, while the power of urban immigrants is mainly manifested in the capital they possess for integration. According to Bourdieu's theory, capital can be divided into different forms such as economic capital, cultural capital, and social capital. The lack of capital for the integration of urban immigrants, as well as the existence of urban institutional discrimination and social attitudes, are main obstacles to the integration of urban immigrants into local society.

The household registration system and the various urban systems built on it constitute an institutional barrier for urban immigrants to integrate into cities. A series of basic public services such as social security, employment, education, medical health based on the household registration system have not yet broken through the dual division of urban and rural areas, which has become the primary obstacle to urbanisation

[6] Mi Qingcheng, Analysis on the Sense of Belonging of Rural Migrants, *Youth Studies*, Vol. 3, 2004.

[7] Li Weidong, The Social Fusion of Rural Migrants in View of Social Distance, *Social Sciences of Beijing*, Vol. 6, 2007.

[8] Guo Xinghua, Chu Huijuan, From Rural to Urban: Integration and Isolation, *Jianghai Academic Journal*, Vol. 3, 2004.

and the urbanisation of rural residents and has widened the gap between rural residents and urban residents in all aspects. "Unreasonable institutional arrangements allow certain social groups to obtain relatively more profit-making opportunities and welfare benefits, while other social groups are trapped in institutional poverty."[9] Becker and others put forward the theory of human capital, and believe that the economic success of immigrants in the immigration country depends on their human capital level. However, the disunity of China's urban labour market makes this theory not applicable here. There are multiple levels of labour markets in China, such as the talent market, the labour market for urban workers, and the labour market for urban migrant workers. What's more, due to various tangible and intangible restrictions on the employment of urban immigrants by urban management departments, most urban immigrants are isolated to the relatively low-end secondary labour market and become the bottom class of urban society.[10]

In terms of social attitudes, urban residents and rural migrants still have a clear sense of division between the inside and outside of the group to a certain extent. Urban residents and rural migrants stand in their respective positions and regard each other as a completely different group from their own. Some urban residents regard rural migrants as "outsiders", believing that they have no right to enjoy the superior environment of the city, and treat them psychologically as "aliens". Some urban residents not only have a prejudice against rural migrants but also express their discrimination against rural migrants through behaviour. Although this type of urban citizen is a minority, the negative impact of such discrimination cannot be underestimated. Because it is easy for rural migrants to misunderstand this minority attitude as the general attitude of the general public, which in turn will arouse the resentment of rural migrants and lead to group friction and conflict.[11]

As we all know, the economy is the foundation. Only on the basis of economic integration can urban immigrants have the capital for social, psychological or cultural integration. In terms of the income of urban immigrants, on the one hand, urban immigrants face the problems of different pay for equal work and no access to engage in local high-paying industries under China's long-standing household registration system. On the other hand, two-thirds of urban immigrants are below the junior high school education degree.[12] The lack of human capital is also one of the reasons for the low income of urban immigrants.

Cultural capital includes competence-based cultural capital in the form of behaviour, language style, taste, lifestyle, and values, and institutionalised cultural capital in the form of diplomas and professional certificates. The lack of cultural capital of rural migrants directly hinders their integration into urban mainstream

---

[9] Zhong Dajun, *Examining the Inequality of National Treatment: China Under the Dual Structure*, Beijing: China Workers Publishing House, 2002.

[10] Li Qiang, Labour Market of Urban Migrant Workers in China, *Journal of Dalian Minzu University*, Vol. 7, 2000.

[11] Zhu Li, Group Prejudice and Discrimination, *Jianghai Academic Journal*, Vol. 6, 2001.

[12] Wang Guangling, Yang Chen, Analysis on the Difficulties of Social Integration of Rural Migrants Under the Background of Financial Crisis, *Theory Journal*, Vol. 3, 2010.

society. In addition, the lack of cultural capital also negatively affects the accumulation of human capital and social capital of rural migrants, which leads to the cultural exclusion of urban citizens towards rural migrants and further hinders their integration into cities.[13] According to a survey conducted by the National Bureau of Statistics of the Agricultural Research Team, about 85% of rural migrants are below the junior high school education degree, including junior high school education degree, elementary school education degree and illiteracy.[14] Whats' more, nearly 24% of rural migrants have received various skills training, that is to say, more than 76% of rural migrants have not received any skills training.[15]

The concept of "social capital" originally evolved from the concept of "capital" in economics. Social capital is obtained from resources embedded in social networks, and it is rooted in social networks and social relationships.[16] The social capital and social network of immigrants have an important influence on every aspect of their immigration process. Zhao Yandong pointed out that social capital is extremely important in the process of obtaining the economic status of rural migrants, which may be more significant than human capital and other factors. What's more, the human capital possessed by rural migrants may also depend on their social capital to fully function.[17] It can be said that economic behaviour is not passively "embedded" in social relationships, but rural migrants rely on social relationships to actively "carry out" their economic behaviour.

Qu Jingdong believes that the social network of rural migrants is structured around the homogeneous relationship of kinship, geographical relationship and industry relationship, and influences the process of constructing the life world of rural migrants.[18] On one hand, the economic and spiritual support provided by the social network based on primary groups can enable rural migrants who have just migrated to the city to quickly adapt to the environment, and prevent them from becoming losers of urbanisation to a certain extent. However, on the other hand, this kind of social network has strengthened the sub-social ecological environment in which rural migrants survive,

---

[13] Liu Huiwu, Cultural Capital and Integration into Cities of Rural Migrants, *Rural Economy*, Vol. 7, 2007.

[14] A survey conducted by the National Bureau of Statistics shows that 110 million rural migrants in China went out to work in 2003. Retrieved at 23 May, 2020, https://news.sina.com.cn/c/2004-05-16/09322541943s.shtml.

[15] Research Group of the State Council Research Office, *Research Report on Chinese Rural Migrants*, Beijing: China Yan Shi Press, 2006.

[16] Ren Yuan, Wu Minle, Social Integration of Floating Population in Urban China: A Literature Review, *Population Research*, Vol. 3, 2006.

[17] Zhao Yandong, Wang Fenyu, The Acquisition and Determinants of Economic Status of the Migrant Population in Urban and Rural Areas, *Chinese Journal of Population Science*, Vol. 4, 2002.

[18] Qu Jingdong, The Strength of Relations in the Life World, the Life Trajectory of Rural Migrants, in Ke, Lanjun & Li, Hanlin, *Villagers in the City, Migrant Population in Big Cities in China*, Beijing: Central Compilation & Translation Press, 2001.

and protected the traditional ideas and petty-farmer consciousness of rural migrants, hindering their sense of identity and belonging to the city.[19]

The social integration of urban immigrants into the city is affected by the comprehensive effects of institutional arrangements, social networks, cultural education, skill training, and social values. As the number of urban immigrants continues to increase and urban integration issues become increasingly prominent, systematic institutional reforms and conceptual changes are urgently needed to promote urban immigrants to better adapt to cities where they migrate. As we all know, the realisation of social integration is of great significance to the construction of a diversified socioeconomic order.

To sum up, there are different opinions on the social integration of urban immigrants, but they all focus on several basic issues. The first issue is how to measure and what indicator system is used to measure the level and degree of social integration of urban immigrants? The second issue is what is the current status of the level and degree of social integration of urban immigrants in China? The third issue is what are the factors that affect the social integration of urban immigrants? The fourth issue is how to promote the process of social integration of urban immigrants? In order to respond to these issues, in the following, the authors will use the survey data of urban immigrants in 6 cities in China to analyse.

## 8.2 Evaluation Indicator System for Urban Adaptation and Integration of Urban Immigrants

There are a lot of academic studies on population migration, and the angle of analysis is also different. After the mid-1990s, scholars paid more and more attention to the issue of social integration.

The concept of "social integration" originated from the field of immigration research. According to the theory of international immigration, due to the cultural differences between the destination of immigration and the original place of residence, immigrant groups often appear "non-integrated", that is, immigrants generally exhibit three different living conditions after immigrating into the destination of immigration, namely "mosaic-like group division", "multiculturalism" and "being away from the mainstream society".[20] After a group enters a new environment, they will face obstacles in various aspects in the process of becoming a member of this new environment, such as system construction, resource allocation, culture and customs, values and so on.

Early studies have shown that although rural migrants have changed their social life in the process of migrating from rural to urban areas, their social network based on

[19] Zhu Li, Debate on the Urban Adaptability of the Rural Migrants Strata, *Jianghai Academic Journal*, Vol. 6, 2002.

[20] Wu Xinhui, Paying Attention to the Social Integration of Children of Migrants, *Chinese Journal of Society*, Vol. 9, 2004.

blood relationships has not fundamentally changed. Therefore, they can "migrate" to the city relatively smoothly with this social network, but it is still difficult for them to truly "integrate" into the city. Some scholars pointed out that to become a real urban new citizen is not only a matter of regional change, but also a process of the gradual transformation of ideas, behaviours, lifestyles, and forms of social organisations to the urban paradigm. Abroad studies further show that the integration of immigrants in the receiving country involves three stages of settlement, adaptation and complete assimilation.[21] At the same time, Zhu Kaojin and Liu Ruiqing also put forward some different viewpoints, and they believe that the key to judge the integration of immigrants lies in the identity of the actors. In other words, as long as immigrants believe that they have achieved integration, even if there is a large gap between them and urban residents in many aspects of urban life, they can still be considered to have achieved urban integration. On the contrary, if immigrants believe that they have not achieved urban integration, even if their living standards and other objective aspects catch up or even surpass that of urban residents, they cannot be considered to have achieved complete urban integration.[22]

How migrant population integrate into the city has become a hot and classic topic in the field of urban research, and scholars also have their own ideas on the definition of "social integration". Many scholars believe that social integration refers to the cohesiveness of different individuals or groups with a certain group, which represents the degree of participation and identity of individuals in a certain group and the degree of interdependence between group members.

From the perspective of the interaction between the two parties of integration, Ren Yuan and Wu Minle pointed out that social integration is a process of mutual coordination and adaptation between individuals, between different groups or between different cultures, intending to build a harmonious society.[23] Liu Jian'e further proposed that social integration is the realisation of a harmonious society based on the development of social capital in the dynamic process of structural adjustment and self-adaptation of individual, group and community in a specific society, which is specifically manifested in promoting the equal distribution of social resources, promoting the effective delivery of social services, realising extensive social rights, achieving equal participation in the creation and sharing of economic wealth and social civilisation, and improving the quality of social life of disadvantaged groups.[24]

Taking rural migrant as the analysis group, Zhu Li pointed out that the urban integration of rural migrant refers to the process in which rural migrant reach the urban living standard, adapt to the urban lifestyle, and find a sense of belonging in the city. Zhu further divide the urban integration of rural migrant into three levels: the

---

[21] Huang Xiaoyan, The Social Integration Dimensions and Methods of New Urban Citizens, *Social Scientist*, Vol. 3, 2010.

[22] Zhu Kaojin, Liu Ruiqing, Research on the Social Support Network and Urban Integration of Young Rural Migrants, *Youth Studies*, Vol. 8, 2007.

[23] Ren Yuan, Wu Minle, Social Integration of Floating Population in Urban China: A Literature Review, *Population Research*, Vol. 3, 2006.

[24] Liu Jian'e, A Study on Rural–Urban Migrants' Practical Strategies for Social Inclusion: Towards a Community Inclusion Approach, *Chinese Journal of Sociology*, Vol. 1, 2006.

economic level, the social level and the psychological level; and believe that these three different aspects are progressive: economic reach is the foundation; the social reach is the further requirements reflect the breadth of urban integration; psychological cognition or belongingness reflects the depth of participation in urban life. Rural migrant is fully integrated into urban society only when the social and psychological adaptions are completed.[25]

In recent years, many scholars have conducted in-depth research on the social integration of the migrant population from the disciplines of economics, sociology, demography, and political science, and have achieved certain achievements. However, at present, there is no unified view of the structural dimensions of social integration, and there is also no unified standard for the measurement and indicators of integration.

As early as the 1930s, Fei Xiaotong had already conducted relevant research. In the book *Peasant Life in China*, Fei Xiaotong pointed out that in the long-term agricultural society, outsiders must truly integrate into the local community through three methods: land-purchase, property-purchase and intermarriage.[26] Subsequent studies have increasingly refined the indicators of social integration. Zhu Li believes that there are three levels of social integration, namely economic level, social level and psychological level or cultural level, and the three levels of social integration are progressive. Firstly, economic adaptation is the basis for immigrants to establish themselves in cities. Secondly, the social level is a further requirement for immigrants to integrate into urban life, reflecting the breadth of their integration into urban life. Last but not least, psychological adaptation belongs to the spiritual category, reflecting the depth of immigrants' participation in urban life, and only when the migrant population adapts at the psychological and cultural levels can they fully integrate into urban society.[27]

Based on the research on the social integration of rural migrants in the Three Gorges area, Feng Xiaotian proposed that social integration should be analysed from five dimensions: family economy, daily life, relationship with locals, production and labour, and community identity.[28] In addition, the difference between the dialects of the original place of residence and the destination of immigration, the changes in local customs and values, and the gap between the immigrants' demand for housing and the actual purchase rate, all directly or indirectly affect the degree of integration. Li Ke and Liu E pointed out that rural migrants' income, housing conditions, protection of basic rights and interests, degree of satisfaction with employment and life in the city, relationship with urban residents, recognition of their current identity and so

[25] Zhu Li, Debate on the Urban Adaptability of Rural Migrants Strata, Jianghai Academic Journal, Vol. 6, 2002.

[26] Fei, Xiaotong, *Peasant Life in China*, Beijing: The Commercial Press, 2001.

[27] Zhu Li, Debate On the Urban Adaptability of the Rural Migrants Strata, *Jianghai Academic Journal*, Vol. 6, 2002.

[28] Feng Xiaotian, Taking Roots: Social Adaptation of Rural Migrants in the Three Gorges Area, *Sociological Studies*, Vol. 5, 2004.

on should be used as the main evaluation indicators to judge whether rural migrants have successfully integrated into the city.[29]

Through multidimensional comparison, Zhang Wenhong and Lei Kaichun selected 14 indicators of social integration to quantitatively analyse the degree of social integration of urban immigrants. These indicators are career stability, mastery of dialect, familiarity with customs, acceptance of cultural values, number of relatives who can accompany, degree of identity, the scope of social interaction, number of frequent contacts, willingness to purchase real estate, social-psychological distance, household registration status, satisfaction with society, satisfaction with career and satisfaction with housing. Among them, "mastery of dialect", "familiarity with customs", "acceptance of cultural values", "career stability", "number of relatives who can accompany", "household registration system" and other indicators mainly refer to relevant dimensions in the study of international immigration, while indicators such as "degree of identity", "scope of social interaction", "social-psychological distance", "number of frequent contacts", "satisfaction with society", "satisfaction with career", "satisfaction with housing" and other indicators are the result of further refinement of the relevant indicators of social integration of domestic immigrants. Besides, the "willingness to purchase real estate" is a new indicator designed in consideration of continuously rising housing prices in major cities in China. In addition, indicators such as "number of relatives who can accompany", "career stability", "household registration status" and "number of frequent contacts" are objective indicators, while others are subjective evaluation indicators. From the perspective of the results of factor analysis, the three indicators of "mastery of dialect", "familiarity with customs" and "acceptance of cultural values" are named as "cultural integration" factors, "satisfaction with society", "satisfaction with career" and "satisfaction with housing" are named as "psychological integration" factors, "career stability", "degree of identity" and "household registration status" are named as factors of "integration of identity", "number of relatives who can accompany" and "willingness to purchase real estate" are named as "economic integration" factors.[30]

The authors believe that any immigrant's adaptation and integration to the new environment cannot be a static and sudden result, but a continuous and dynamic process that is constantly advancing. W. R. Bohning once proposed the "four-stage theory" of immigration and settlement. He believes that from migration and settlement to survival and development, immigrants will go through the following four stages. In the first stage, new immigrants, mainly young guest workers, arrived at the destination of immigration. At this stage, immigrants have two main characteristics, namely, remitting money to their hometown and expecting to return home after the work period is over. In the second stage, some immigrants stayed behind and built a social network based on blood or geographical relationships that would help them survive in the new environment. In the third stage, family reunion, awareness of

[29] Li Ke, Liu E, Urbanisation Urgently Needs Rural Migrants to Achieve Urban Integration, *Theory Front*, Vol. 23, 2009.

[30] Zhang Wenhong, Lei Kaichun, The Urban New Immigrants' Social Inclusion: Internal Structure, Present Situation and Influential Factors, *Sociological Studies*, Vol. 5, 2008.

long-term settlement, yearning for the immigration country, and the formation and development of the same ethnic group and community make immigrants increasingly inclined to settle permanently in the immigration country. In the fourth stage, immigrants strive for citizenship and other social rights and legal status in order to survive better and equally. At this time, the government of the immigration country played an important role. Whether immigrants finally obtain equal citizenship status or are excluded is closely related to the country's policies and its political, economic, social, and cultural conditions.[31] The authors believe that W. R. Bohning has well summarised the dynamic process of immigrants adapting and integrating into the new environment, and provides a good frame of reference for the authors to establish an indicator system for evaluating the status quo of adaptation and integration of urban immigrants, and the authors' deduction of this indicator system is shown in Table 8.1.

## 8.3 The Evaluation of the Status Quo of Adaptation and Integration of Urban Immigrants

Based on the above-mentioned indicator system, the authors conducted a questionnaire-design related to the status quo of adaptation and integration of urban immigrants, and the results of the survey are reported below.

### *8.3.1 Focus of Life*

Whether to remit money to their hometown and the amount of remittance is an important indicator for measuring the integration of urban immigrants into the local society, which directly reflects whether the focus of consumption of urban immigrants is placed in the immigration city. And as we all know, the focus of consumption is a reflection of the focus of life. Remittances from urban immigrants to their hometown indicate that they have economic ties to the place of emigration. Moreover, the higher the proportion of remittances to their hometown in urban immigrants' income and expenditure, the more their focus of life is toward their hometown, that is, the lower their integration into the place of migration.

To this end, the authors used questionnaires to obtain information on the number of remittances of the respondents to their hometown last month, their other expenses, such as food, housing, communications, transportation, entertainment and leisure, interpersonal interaction, etc., and monthly disposable income. Based on these survey data, the authors can estimate whether urban immigrants remit money to their hometown and the amount of remittances.

[31] Zhou Yu'e, Ruan Zhengyu, The Status Quo and Trend of Theoretical Research on Contemporary International Immigration, *Jinan Journal* (Philosophy & Social Sciences), Vol. 2, 2003.

**Table 8.1** Evaluation indicator system of the status quo of adaptation and integration of urban immigrants

| Stages of immigration | Theme of integration | Degree of integration | Characteristics of behavior | Evaluation indicators |
|---|---|---|---|---|
| The first stage | The shift of focus of life | Lowest | Remitting money to their hometown | Whether to remit money to their hometown |
| | | | | The proportion of remittance to hometown in personal income |
| | | | | The proportion of remittance to hometown in personal expenditure |
| The second stage | The formation of social support networks | Lower | Building a social network based on blood or geographical relationships that would help them survive in the new environment | Whether there is a best friend in the immigration city |
| The third stage | The willingness to long-term settlement | Higher | Family reunion | Whether to live with their families in the immigration city |
| | | | Awareness of long-term settlement and yearning for the immigration country | The willingness to long-term settlement |
| The fourth stage | The fight for equal rights | Highest | The fight for citizenship and other social rights and legal status | The willingness to participate in the election of the community or neighbourhood committee |

The statistical results show that the proportion of urban immigrants who remit money to their hometown reaches 32.6%, the amount of remittances to their hometown accounts for an average of 32.6% of their income, and the amount of remittances to their hometown accounts for an average of 38.2% of their household living expenses. In other words, about two-thirds of urban immigrants have shifted their focus of life to the place of migration, and the degree of integration into the place of migration is relatively high. Based on various indicators, among the three types of urban immigrants, labour immigrants tend to focus more on their hometowns and have the lowest degree of integration into the place of migration, followed by entrepreneurial immigrants, and intellectual immigrants have the highest degree of integration into the place of migration. In terms of different regions, urban immigrants in Guangzhou and Dongguan in Guangdong Province have the lowest degree of integration into the city. Besides, the proportion of immigrants who remit money to their hometown, the proportion of remittance to their hometown in their income, and the proportion of remittance to their hometown in their expenditure in Zhengzhou are the lowest among several cities, indicating that urban immigrants in Zhengzhou have the highest degree of integration into the city. And the indicator data of other cities have little difference, indicating that the degree of integration of urban immigrants in these cities is moderate (see Table 8.2).

### 8.3.2 The Social Support Network

Whether urban immigrants have entered or established a social network in the place of migration is another important indicator of their integration into the local society. On the one hand, entering or establishing a social network in the place of migration is itself an important form of social integration. On the other hand, social networks provide social support for urban immigrants and become the basis and advantage for their further integration into local society.

The authors obtained information about the respondent's three best friends through the questionnaire, including whether their three best friends were in the city, their hometown or other places. If none of their three best friends is in the city, the authors will mark their social network integration as 1, which represents the lowest degree of integration. If one of their three best friends is in the city, the authors will mark their social network integration as 2, which means a low degree of integration. If two of their three best friends are in the city, the authors will mark their social network integration as 3, which represents a high degree of integration. And if their three best friends are all in the city, the authors will mark their social network integration as 4, which represents the highest degree of integration.

Statistics show that 39.0% of urban immigrants' social networks have all shifted to the place of migration, indicating a high degree of integration. 25.7% of urban immigrants' social networks have tilted towards the place of migration, indicating a relatively high degree of integration. And about one-third of urban immigrants have a low degree of social network integration. From the perspective of the types of urban immigrants, there is no obvious difference between intellectual immigrants, labour immigrants and entrepreneurial immigrants. From a regional perspective, urban immigrants in several coastal cities such as Guangzhou, Dongguan, and Hangzhou have a low degree of social network integration, while those in relatively inland or less developed cities such as Zhengzhou, Chengdu, and Shenyang have a higher degree of social network integration (see Table 8.3).

**Table 8.2** Indicator data for urban new immigrants remitting money to their hometown (Unit: %)

| Indicator | Intellectual immigrants | Labour immigrants | Entrepreneurial immigrants | Guangzhou | Dongguan | Shenyang | Chengdu | Hangzhou | Zhengzhou | Total |
|---|---|---|---|---|---|---|---|---|---|---|
| The proportion of urban immigrants who remit money to their hometown | 25.6 | 37.0 | 34.3 | 39.3 | 43.2 | 30.2 | 27.9 | 33.4 | 21.5 | 32.6 |
| The proportion of remittance to urban immigrants' hometown in their income (taking urban immigrants who remit money to their hometown as the sample base) | 28.7 | 41.0 | 22.4 | 38.4 | 36.5 | 30.9 | 28.7 | 26.8 | 28.7 | 32.6 |
| The proportion of remittance to urban immigrants' hometown in their expenditure (taking urban immigrants who remit money to their hometown as the sample base) | 34.9 | 44.4 | 31.1 | 43.1 | 43.2 | 36.7 | 37.7 | 30.4 | 32.3 | 38.2 |
| Sample size | 952 | 1239 | 832 | 516 | 526 | 513 | 519 | 446 | 503 | 3023 |

**Table 8.3** The degree of social network integration of urban new immigrants

| How many of your three best friends are in your city? | Intellectual immigrants | Labour immigrants | Entrepreneurial immigrants | Guangzhou | Dong guan | Shen yang | Chengdu | Hang zhou | Zhengzhou | Total |
|---|---|---|---|---|---|---|---|---|---|---|
| Zero (1 point) (%) | 13.4 | 15.3 | 18.2 | 18.2 | 19.7 | 11.4 | 13.6 | 21.2 | 9.3 | 15.5 |
| One (2 points) (%) | 18.8 | 20.0 | 20.7 | 23.4 | 26.9 | 21.2 | 17.6 | 12.4 | 15.6 | 19.8 |
| Two (3 points) (%) | 28.3 | 25.2 | 23.7 | 22.5 | 26.7 | 27.7 | 23.4 | 24.3 | 29.7 | 25.7 |
| Three (4 points) (%) | 39.5 | 39.5 | 37.4 | 35.9 | 26.7 | 39.8 | 45.5 | 42.0 | 45.4 | 39.0 |
| Total (%) | 100 | 100 | 100 | 100 | 100 | 100 | 100 | 100 | 100 | 100 |
| Average (points) | 2.9 | 2.9 | 2.8 | 2.8 | 2.6 | 3.0 | 3.0 | 2.9 | 3.1 | 2.9 |
| Standard deviation | 1.1 | 1.1 | 1.1 | 1.1 | 1.1 | 1.0 | 1.1 | 1.2 | 1.0 | 1.1 |
| Sample size | 984 | 1320 | 864 | 538 | 580 | 528 | 552 | 452 | 518 | 3168 |

### 8.3.3 The Willingness to Long-Term Settlement

The willingness to long-term settlement is also the consciousness to take root, which is a stage that immigrants must pass through to complete the immigration process and fully integrate into the place of migration, and is also the main manifestation of the immigrant's higher degree of integration into the place of migration. The authors used the interviewee's degree of acceptance of the sentence "I am willing to settle in this city" to measure the willingness of urban immigrants to settle in the city, which are divided into "strongly agree", "agree", "not sure", "disagree" and "strongly disagree", and are marked as 5 points, 4 points, 3 points, 2 points and 1 point respectively.

The statistical results show that the willingness of urban immigrants to settle in the cities where they migrated is generally at a relatively high level, and 60.7% of urban immigrants have a clear willingness to settle in the city. From the perspective of the types of urban immigrants, labour immigrants have weaker willingness to settle in cities, while entrepreneurial immigrants and intellectual immigrants have a stronger willingness to settle in cities. From a regional perspective, urban immigrants in Guangzhou and Dongguan in Guangdong Province have weaker willingness to settle in cities, while urban immigrants in Shenyang have stronger willingness to settle in cities, and urban immigrants in the remaining cities have relatively moderate willingness to settle in cities (see Table 8.4).

The willingness to long-term settlement is a manifestation of the awareness and attitude of urban immigrants to take root in the place of migration, while family reunion is the direct and actual manifestation of urban immigrants taking root in the place of migration. Family reunion directly reflects the overall shift in the focus of the life of urban immigrants and reflects the separation of urban immigrants from their original life bonds and the main social connections with their hometowns. In addition, family reunion provides the strongest social support network for urban immigrants to integrate into the place of migration, which is conducive to the accelerated integration of urban immigrants into the new social system. In the questionnaire, the authors asked the respondent whether they lived with their families, so as to understand the status quo of the family reunion of urban immigrants.

The statistical results show that the proportion of urban immigrants achieving family reunion in the place of migration is very high, 87.2% of urban immigrants are married and live with their families. From the perspective of the types of urban immigrants, the proportion of labour immigrants achieving family reunion is significantly low, while the proportion of entrepreneurial immigrants and intellectual immigrants achieving family reunion reaches 100%. From a regional perspective, the proportion of urban immigrants in Guangzhou and Dongguan in Guangdong Province that achieve family reunion is obviously low, while the proportion of urban immigrants in Hangzhou that achieve family reunion is higher, and the proportion of urban immigrants in other cities that achieve family reunion is in the middle (see Table 8.5).

**Table 8.4** The willingness of urban new immigrants to settle in the city

| Degree of acceptance of the sentence "I am willing to settle in this city" | Intellectual immigrants | Labour immigrants | Entrepreneurial immigrants | Guangzhou | Dongguan | Shenyang | Cheng du | Hangzhou | Zhengzhou | Total |
|---|---|---|---|---|---|---|---|---|---|---|
| Strongly agree (%) | 11.2 | 8.7 | 9.2 | 8.1 | 4.1 | 20.8 | 7.2 | 13.4 | 5.1 | 9.6 |
| Agree (%) | 52.6 | 45.7 | 57.4 | 45.0 | 37.2 | 51.0 | 64.0 | 51.4 | 59.2 | 51.1 |
| Not sure (%) | 23.3 | 20.9 | 15.0 | 21.9 | 22.6 | 19.4 | 11.9 | 22.6 | 22.0 | 20.0 |
| Disagree (%) | 10.0 | 20.9 | 15.7 | 21.5 | 30.6 | 6.3 | 13.8 | 10.4 | 11.5 | 16.1 |
| Strongly disagree (%) | 2.9 | 3.8 | 2.8 | 3.5 | 5.5 | 2.5 | 3.0 | 2.3 | 2.2 | 3.2 |
| Total (%) | 100 | 100 | 100 | 100 | 100 | 100 | 100 | 100 | 100 | 100 |
| Average (points) | 3.6 | 3.3 | 3.5 | 3.3 | 3.0 | 3.8 | 3.6 | 3.6 | 3.5 | 3.5 |
| Standard deviation | 0.92 | 1.02 | 0.95 | 1.01 | 1.03 | 0.92 | 0.92 | 0.92 | 0.84 | 0.98 |
| Sample size | 967 | 1266 | 836 | 520 | 562 | 520 | 528 | 434 | 505 | 3069 |

**Table 8.5** The status quo of family reunion of urban new immigrants

| Married, who do you live with now? | Intellectual immigrants | Labour immigrants | Entrepreneurial immigrants | Guangzhou | Dongguan | Shen yang | Chengdu | Hangzhou | Zhengzhou | Total |
|---|---|---|---|---|---|---|---|---|---|---|
| Families (%) | 100.0 | 70.4 | 100.0 | 83.8 | 86.1 | 88.4 | 87.9 | 90.8 | 87.8 | 87.2 |
| Sample size | 141 | 514 | 531 | 229 | 267 | 138 | 248 | 173 | 131 | 1186 |

### 8.3.4 The Fight for Equal Rights

The fight for citizenship and other social rights and legal status is the highest degree of integration of urban immigrants into the place of migration, that is, complete and thorough integration.

The authors used the respondents' degree of acceptance of the sentence "I am equal to the locals" to measure the awareness of equality of urban immigrants, which are divided into "strongly agree", "agree", "not sure", "disagree" and "strongly disagree". From the perspective of the types of urban immigrants, labour immigrants have the weakest awareness of equality, and their approval rate for the aforementioned sentence is only 36.2%, while entrepreneurial immigrants have the strongest awareness of equality, with an approval rate of 45.3%, and intellectual immigrants have a slightly stronger awareness of equality than average. From a regional perspective, urban immigrants in developed cities such as Guangzhou, Dongguan, and Hangzhou have a weak awareness of equality, especially Dongguan, the most typical contemporary urban immigrant city in China. In contrast, urban immigrants in the two sub-developed cities of Shenyang and Chengdu have the strongest awareness of equality (see Table 8.6).

In order to make a clearer measurement of the degree of social integration of urban immigrants, the authors further standardised the above research results. Firstly, comparisons between different types of urban immigrants and different regions are based on the average level, and those above the average level are described as "high", those below the average level are described as "low", and those near the average level are described as "medium". Secondly, the overall evaluation is based on proportions, and those more than two-thirds are described as "high", those less than half are described as "low", and those between half and two-thirds are described as "medium". Lastly, the scoring method for the level of social integration of urban immigrants is 3 points for a "high", 2 points for a "medium", and 1 point for a "low". And the highest score is 21 points and the lowest score is 7 points. The evaluation results are shown in Table 8.7. In addition, the evaluation results show that:

1. The overall level of social integration of urban immigrants is above the middle level;
2. In general, the higher the immigration stage, the lower the level of social integration of urban immigrants;
3. The level of social integration of labour immigrants is significantly lower than that of entrepreneurial immigrants and intellectual immigrants;
4. Guangzhou and Dongguan are at a low level in almost all measurement indicators;
5. The relatively late-developing cities of Zhengzhou, Chengdu, and Shenyang have the highest level of social integration of urban immigrants.

**Table 8.6** The awareness of equality of urban new immigrants

| Degree of acceptance of the sentence "I am equal to the locals" | Intellectual immigrants | Labour immigrants | Entrepreneurial immigrants | Guangzhou | Dongguan | Shen yang | Chengdu | Hang zhou | Zhengzhou | Total |
|---|---|---|---|---|---|---|---|---|---|---|
| Strongly disagree (%) | 3.0 | 5.1 | 4.4 | 4.9 | 7.3 | 3.2 | 1.8 | 5.7 | 2.5 | 4.3 |
| Disagree (%) | 20.7 | 28.3 | 23.8 | 30.6 | 39.0 | 15.9 | 20.0 | 20.7 | 20.2 | 24.7 |
| Not sure (%) | 33.7 | 30.3 | 26.5 | 29.1 | 26.8 | 24.7 | 27.7 | 36.0 | 39.7 | 30.4 |
| Agree (%) | 34.8 | 32.3 | 38.4 | 30.8 | 25.0 | 45.2 | 42.8 | 30.1 | 34.4 | 34.7 |
| Strongly agree (%) | 6.7 | 3.9 | 6.8 | 4.5 | 1.9 | 11.0 | 7.0 | 6.4 | 3.1 | 5.6 |
| Approval rate (%) | 41.5 | 36.2 | 45.3 | 35.3 | 27.0 | 56.2 | 49.7 | 36.4 | 37.6 | 40.3 |
| Sample size | 984 | 1295 | 848 | 530 | 575 | 527 | 545 | 439 | 511 | 3127 |

**Table 8.7** The overall evaluation of level of social integration of urban new immigrants

| Stages of immigration | Theme of integration | Evaluation indicators | Intellectual immigrants | Labour immigrants | Entrepreneurial immigrants | Guangzhou | Dongguan | Shenyang | Chengdu | Hangzhou | Zhengzhou | Total |
|---|---|---|---|---|---|---|---|---|---|---|---|---|
| The first stage | The shift of focus of life | Whether to remit money to their hometown | High | Low | Medium | Low | Low | Medium | High | Medium | High | High |
| | | The proportion of remittance to hometown in personal income | Medium | Low | High | Low | Low | Medium | High | High | High | High |
| | | The proportion of remittance to hometown in personal expenditure | Medium | Low | High | Low | Low | Medium | Medium | High | High | Medium |
| The second stage | The formation of social support networks | Whether there is a best friend in the immigration city | Medium | Medium | Low | Low | Low | Medium | Medium | Low | High | High |
| The third stage | The willingness to long-term settlement | Whether to live with their families in the immigration city | High | Low | High | Low | Low | Medium | Medium | High | Medium | High |

(continued)

**Table 8.7** (continued)

| Stages of immigration | Theme of integration | Evaluation indicators | Intellectual immigrants | Labour immigrants | Entrepreneurial immigrants | Guangzhou | Dongguan | Shenyang | Chengdu | Hangzhou | Zhengzhou | Total |
|---|---|---|---|---|---|---|---|---|---|---|---|---|
| | | The willingness to long-term settlement | High | Low | High | Low | Low | High | Medium | Medium | Medium | Medium |
| The fourth stage | The fight for equal rights | The willingness to participate in the election of the community or neighbourhood committee | Medium | Low | High | Low | Low | High | High | Low | Medium | Low |
| The score of social integration level | | | 17 | 8 | 18 | 7 | 7 | 16 | 17 | 15 | 18 | 17 |

## 8.4 Influencing Factors of Urban Immigrants' Adaptation and Integration

### 8.4.1 Overview of Relevant Literature and Theory

There are many studies on the factors influencing the social integration of urban immigrants. Cai He and Wang Jin concluded that the main factors affecting rural migrants' willingness to give up land and choose behavioural permanent migration are individual factors such as human capital and identification to urban lifestyle, while the factors that affect the willingness of rural migrants to transfer their household registration to immigration cities to choose institutional permanent migration are mainly regional factors and social factors such as institutional legal pressure.[32] In addition, the "new generation of rural migrants" have almost no agricultural experience, and their identification with the city exceeds their identification with the countryside.[33] Ren Yuan found that the migrant population is in the process of gradual settlement in the city. And the longer they stay in the city, the higher the probability that they will continue to stay in the city for a long time.[34] Zhang Wenhong and Lei Kaichun found that there is a positive correlation between the settlement plan of urban immigrants and their adoption of local festivals and their expectations of their children.[35] Li Ruojian pointed out that the two most important factors affecting the residence of the migrant population are occupation and housing. White-collar workers have relatively good working conditions and income. Not only are their jobs relatively stable, but they also have a certain social status, so they are more likely to live in the place of migration for a long time. In addition, those who have their own houses are basically those who plan to settle for a long time.[36] Zhang Zheng believes that annual income, expected working years and housing prices are the main factors affecting rural migrants and their families to settle in cities.[37] Wang Yijie found that age, education degree, and living time in cities in the variables of "socioeconomic status", and the "emotional relationship" and "instrumental relationship" at the two ends of the social network continuum significantly affect the willingness

---

[32] Cai He, Wang Jin, Study on the Willingness of Permanent Migration of Rural Migrants, *Sociological Studies*, Vol.6, 2007.

[33] Lin Peng, Research on Young Rural Migrants Marching towards the "Post-survival Era", China Rural Studies Website, retrieved at 3 November, 2003, http://www.cucc.org.cn/html/74-3/3087.htm.

[34] Ren Yuan, "Gradually Settlement" and "Residence Determines Residence": Analysis of the Residential Mode of the Immigrants in Shanghai, *Chinese Journal of Population Science*, Vol. 3, 2006.

[35] Zhang Wenhong, Lei Kaichun, Structural Model of Social Identity of Urban Immigrants, *Sociological Studies*, Vol. 4, 2009.

[36] Li Ruojian, A Preliminary Analysis on the Settlement and Mobility of the Migrant Population in Guangdong Province, *Population Research*, Vol. 6, 2007.

[37] Zhang Zheng, Settling in the City or Returning to the Hometown for Development? Analysis of the Life Cycle of Rural Migrants' Decision on Migration, *Chinese Rural Economy*, Vol. 7, 2006.

of rural migrants to stay in cities.[38] Zhang Jijiao believes that social networks are social capital for the migrant population to work in cities. Besides, social networks are not single and fixed but are combined and variable. People at different stages of migration and employment often use different social relationship networks.[39] Zhu Kaojin and Liu Ruiqing pointed out that there is a positive correlation between the different types of social support networks that young rural migrants mainly rely on and their different degree of urban integration, that is, the higher the degree of urban integration, the greater the role of the formal social support network they rely on. Conversely, the lower the degree of urban integration, the more they rely on informal social support networks.[40] Wang Chunlan and Ding Jinhong believe that income has become the most significant factor influencing the willingness to settle in cities under the current background that earning a living is the main purpose of the migrant population going out. Besides, the income also affects other factors, such as the attractiveness of the city, including employment opportunities brought about by urban development, the degree of social integration into the place of migration, and household registration system, the status of marriage and family, including whether spouse and children move along and the children's schooling status, and personal characteristics, including age, gender, length of time in the place of migration, all of which have a significant impact on the migrant population's willingness to settle in the city.[41] Chen Wenzhe and Zhu Yu studied the dynamic changes and internal differences of the migrant population's willingness to settle in the city and found that there are obvious differences in the migrant population's willingness to settle in the city and its changes. Firstly, the proportion of women and unmarried people who choose to settle in the place of migration has risen much more than men and married people. Secondly, the willingness of the migrant population from outside the province to return to their hometowns has been significantly weakened. Lastly, the willingness to settle in the place of migration increases with the improvement of the education degree of the migrant population, that is to say, the positive effect of education becomes more prominent.[42] Huang Qian found that the willingness of rural migrants to settle in cities is generally weak. In addition, rural migrants' age, employment status, monthly income, land in their hometown, housing, pension, and the degree of social integration have a significant impact on their willingness to settle in the city, while gender, marriage, education degree, occupation and social

---

[38] Wang Yijie, Analysis of Influencing Factors Affecting Rural Migrants' Willingness to Settle in Cities, *Jiangsu Social Sciences*, Vol. 5, 2005.

[39] Zhang Jijiao, The "Differential Mode of Association" from Rural to Urban Version: With Examples of the Migrant Population's Employment in Cities, *Ethno-National Studies*, Vol. 6, 2004.

[40] Zhu Kaojin, Liu Ruiqing, Research on the Social Support Network and Urban Integration of Young Rural Migrants, *Youth Studies*, Vol. 8, 2007.

[41] Wang Chunlan, Ding Jinhong, Analysis of Influencing Factors of the Migrant Population's Willingness to Settle in the City, *South China Population*, Vol. 1, 2007.

[42] Chen Wenzhe, Zhu Yu, Changes and Internal Differentiation of the Migrant Population's Willingness to Settlement: Evidence from Four Cities in Fujian Province, *South China Population*, Vol. 2, 2008.

capital have no significant influence on their willingness to settle in the city.[43] Yao Jun analyses from four aspects and believes that in terms of institutional factors, whether to engage in formal employment has a positive and significant impact on the willingness of rural migrants to settle in the city. In terms of social life action factors, the ability to purchase houses and the current living conditions have no significant impact on whether rural migrants settle in cities. In terms of social psychological factors, the length of time in the city has a positive effect on the willingness of rural migrants to settle in the city, but the hypothesis of the positive correlation between the social support network and the willingness to settle in the city does not hold. In terms of personal characteristics, the assumption of a positive correlation between income and the willingness to settle in the city is proved. Besides, gender also has a significant influence on the willingness of rural migrants to settle in the city, while the assumption of a positive correlation between marriage, education degree, age and the willingness to settle in the city does not hold.[44] Xiong Bo and Shi Renbing found that occupation, income and housing have a significant impact on the willingness of rural migrants to settle in the city, while gender, age and education degree have no significant impact on the willingness of rural migrants to settle in the city.[45] Wu Xinglu divided rural migrants into two groups: those who are willing to settle in the city and those who are unwilling to settle in the city, and found that the average education degree, working time in cities, frequency of job changes, and income of the former were significantly higher than those of the latter.[46] Ren Yuan and Dai Xingji also found that family migration is more prone to long-term settlement in cities than individual migration. Besides, the stronger the migrant population's tendency to save money, the lower their tendency to settle in cities for a long time.[47] A survey of rural migrants in Chengdu by Zeng Xuhui and Qin Wei in 2000 showed that rural migrants' identification to urban communities and their high trust in the government would make them more inclined to stay in the city. And those who choose to stay in the city show alienation from the land both in their behaviour and attitude. In addition, the current situation of children and parents does not show statistical significance in the influence of the sample's tendency to stay in cities, while whether the sample has a spouse and whether the spouse is in the city shows statistical significance in the influence of the sample's tendency to stay in cities, and those whose spouse is in

---

[43] Huang Qian, An Empirical Analysis on the Factors Affecting the Rural Migrants' Willingness to Settle in Cities: Based on the Survey of Five Cities, *Journal of Shanxi University of Finance and Economics*, Vol. 4, 2008.

[44] Yao Jun, Survey on the Willingness to Settle in Cities of Rural Migrants, An Empirical Analysis in Three Cities of Jiangsu Province, *Urban Problems*, Vol. 9, 2009.

[45] Xiong Bo, Shi Renbing, An Analysis on the Factors Affecting the Rural Migrants' Willingness to Settle Permanently in the City, *South China Population*, Vol. 2, 2007.

[46] Wu Xinglu, An Empirical Study on the Influencing Factors of Rural Migrants' Decision-making on Settlement-Oriented Migration, *Population & Economics*, Vol. 1, 2005.

[47] Ren Yuan, Dai Xingyi, Model Analysis of the Tendency of the Migrant Population to Settle in Cities for a Long Time, *South China Population*, Vol. 4, 2003.

the city are most inclined to stay in the city.[48] Based on research on rural migrants in Jiangxi, He Pichan and Guo Jinfeng found that rural migrants' high income, high education degree, and long years of going out cause them to have a more obvious tendency to immigration, while whether they enjoy social security benefits has no effect on their tendency to immigration.[49] Li Yaqin and Li Kairong believe that the main reason why rural migrants who have already gone out to work still have not achieved complete immigration is the result of a combination of many factors, such as the duration of the labour contract and the education of children in the city, the rural land system, the human capital of rural migrants, and the government's low degree of organisation of the employment of rural migrants, public management services, employment and living environment, social security, etc.[50] Zhu Yu believes that the migrant population's willingness to stay in the city is closely related to a series of non-household-registration factors such as their viability and family strategies in the place of migration, fluctuations in market demand, and corporate employment strategies.[51] The push–pull model on the factor of transfer decision-making for rural migrants' settlement in cities constructed by Xiong Caiyun shows that individual differences in rural migrants' gender, education degree, marital status, length of work, and income have a direct impact on their settlement in cities. Among the pull factors from the rural areas, the lack of professional skills of rural migrants and concerns about the care of their elderly parents are the most critical, while rural income, land transfer, and love for the homeland do not pose too many obstacles to rural migrants' decisions on settling in cities. Among the push factors from cities, housing, children's education, social security, equal employment, etc. are the most critical. Besides, social demographic variables have a significant impact on the social integration of urban immigrants.[52] Zhang Wenhong and Lei Kaichun also explained in different dimensions, pointing out that the degree of the economic integration of female urban immigrants is significantly higher than that of males. Besides, the degree of economic integration and identity integration of married urban immigrants is significantly higher than that of unmarried urban immigrants, while the degree of cultural integration of the former is significantly lower than that of the latter. The education degree has a significant positive impact on the identity integration of urban immigrants, income has a significant positive impact on the psychological integration of urban immigrants, and length of residence has a significant positive

[48] Zeng Xuhui, Qin Wei, Analysis on the Influencing Factors of Rural Migrants' Tendency to Stay in the City, *Population & Economics*, Vol. 3, 2003.

[49] He Pichan, Guo Jinfeng, An Analysis of Rural Migrants' Tendency to Immigration Under Institutional Constraints, *Chinese Rural Economy*, Vol. 10, 2004.

[50] Li Yaqin, Li Kairong, Research on the Effectiveness of Settlement-Oriented Migration of Rural Migrants, *Rural Economy*, Vol. 12, 2007.

[51] Zhu Yu, Household Registration System Reform and the Dwelling Willing of the Floating Population in the Areas They Are Dwelling and Its Restrictive Mechanism, *South China Population*, Vol. 3, 2004.

[52] Xiong Caiyun, Push–Pull Model on the Factor of Transfer Decision-making for Rural Migrants' Settling Down in the City and Empirical Analysis, *Issues in Agricultural Economy*, Vol. 3, 2007.

impact on economic integration, identity integration and cultural integration of urban immigrants.[53]

To sum up, most of the above-mentioned studies on the influencing factors of social integration of urban immigrants focus on individual factors, such as immigrants' demographic characteristics, human capital, length of stay in cities, income, behaviour and habits, family structure, etc., while existing studies pay relatively little attention to social environmental factors. At the same time, there are many studies conducted with a simple sample of rural migrants, but there is a lack of research conducted from the perspective of immigrants in a general sense. Furthermore, there are many studies focusing on the factors that affect the willingness of urban immigrants to settle in the city, while there are few studies on other factors that affect the social integration of urban immigrants. Therefore, the authors attempt to advance the study of factors affecting the social integration of immigrants from three aspects based on the existing research. Firstly, the authors will replace simple samples of rural migrants with samples of "urban immigrants". Secondly, the authors will pay more attention to the impact of the living environment of rural migrants on their willingness to settle in cities. Lastly, the authors will study the influencing factors of the degree of social integration of urban immigrants at all four levels according to the ideas in Sect. 8.2, and establish a set of composite regression models.

### 8.4.2 Research Methods

The dependent variables of this study include 4 groups of 7 variables extracted in Sect. 8.2, and the authors will first use Table 8.8 to introduce the measurement methods of dependent variables.

Based on the conclusions of previous studies above, the authors choose the following independent variables of personal identity or human capital: (1) using gender as a dummy variable, and the male takes the value 1, while the female takes the value 0; (2) using specific age as the interval variable; (3) taking ordinal variables of education degree as interval variables, primary school education degree and below takes the value 1, junior high school education degree takes the value 2, and high school education degree and above takes the value 3; (4) taking ordinal variables of monthly disposable income as interval variables and dividing it into 10 levels. RMB1000 and below takes the value 1, while RMB5001 and above takes the value 10, and the middle part is separated by 1 level every RMB500; (5) the authors used the interviewee's degree of acceptance of the sentence "I hope my children will also develop in this city" to measure their consideration of the future of their children, which are divided into "strongly agree", "agree", "not sure", "disagree" and "strongly

[53] Zhang Wenhong, Lei Kaichun, Structural Model of Social Identity of Urban Immigrants, *Sociological Studies*, Vol. 4, 2009.

**Table 8.8** The measurement method of the dependent variable of social integration

| Grouping of variables | Variables | Measurement methods | Levels of measurement |
|---|---|---|---|
| The first group: variables of the shift of focus of life | Whether to remit money to their hometown | Yes = 1, No = 0 | Dummy variables |
| | The proportion of remittance to hometown in personal income | Actual proportion | Ratio variables |
| | The proportion of remittance to hometown in personal expenditure | Actual proportion | Ratio variables |
| The second group: variables of social support network | Whether there is a best friend in the immigration city | Zero = 1, One = 2, Two = 3, Three = 4 | Interval variables |
| The third group: variables of the willingness to long-term settlement | Whether to live with their families in the immigration city | Yes = 1, No = 0 | Dummy variables |
| | The willingness to long-term settlement | Strongly agree = 5, Agree = 4, Not sure = 3, Disagree = 2, Strongly disagree = 1 | Interval variables |
| The fourth group: variables of the fight for equal rights | Degree of acceptance of the sentence "I am equal to the locals" | Strongly agree = 5, Agree = 4, Not sure = 3, Disagree = 2, Strongly disagree = 1 | Interval variables |

disagree", and are marked as 5 points, 4 points, 3 points, 2 points and 1 point respectively; (6) using housing types as dummy variables; (7) using the specific years of residence in the current city as interval variables.

"Living environment" is an important independent variable in this study, including the following indicators. (1) Satisfaction with public welfare. The authors measured the respondents' degree of satisfaction with public welfare by comparing the differences between the respondents and the locals, which are divided into four levels, namely "being completely consistent with locals", "being a little different from locals but enjoyed most of the welfare", "being very different from locals and only enjoyed a small part of the welfare" and "being completely inconsistent with locals", and are marked as 4 points, 3 points, 2 points and 1 point respectively. (2) Satisfaction with housing prices. Respondents' degree of satisfaction with housing prices is divided into five levels, namely "very satisfied", "satisfied", "moderate", "dissatisfied" and "very dissatisfied", and are marked as 5 points, 4 points, 3 points, 2 points and 1 point respectively. (3) Satisfaction with the price of commodities. Respondents' degree of satisfaction with price of commodities is divided into five levels, namely "very satisfied", "satisfied", "moderate", "dissatisfied" and "very dissatisfied", and are marked as 5 points, 4 points, 3 points, 2 points and 1 point respectively. (4) Satisfaction with living conditions. Respondents' degree of satisfaction with living conditions is divided into five levels, namely "very satisfied", "satisfied", "moderate", "dissatisfied" and "very dissatisfied", and are marked as 5 points, 4 points, 3 points, 2 points and 1 point respectively. (5) Interpersonal environment. The authors used the interviewee's degree of acceptance of the sentence "There is much of the milk of human kindness in the city where I am now" to measure the city's interpersonal environment, which is divided into five levels, namely "strongly agree", "agree", "not sure", "disagree" and "strongly disagree", and are marked as 5 points, 4 points, 3 points, 2 points and 1 point respectively. (6) Social exclusion. The authors asked the interviewees "After several years of living in the current city, what do you think of the attitudes of the locals towards the migrant population", and the answers include "being always better than before", "being better than before", "being worse than before" and "nothing changed", which are marked as 4 points, 3 points, 2 points, and 1 point respectively. (7) Political exclusion. The authors asked the interviewees "What do you think of the attitude of the community officials in your community towards you", and the answers include "being considerate", "being inconsiderate" and "being discriminate against the migrant population", which are marked as 3 points, 2 points, and 1 point respectively.

When the dependent variable is a dichotomous dummy variable, the value is 0 or 1, which is suitable for establishing a logistic regression model, and the statistical model is as follows: $\mathrm{Y} = \ln(p_1/p_0) = \alpha + \sum \beta_i \mathrm{X}_i (i = 1, 2, 3, \ldots, n)$; when the dependent variable is a continuous interval variable, it is suitable to establish a multiple linear regression model, and the statistical model is $\mathrm{Y} = \beta_0 + \beta_1 \mathrm{X}_{i1} + \beta_2 \mathrm{X}_{i2} + \beta_3 \mathrm{X}_{i3} + \cdots + \beta_k \mathrm{X}_{ik}$.

### 8.4.3 Descriptive Statistical Results of Variables

The descriptive statistical results of the dependent variables have been reported in Sect. 8.2, and the descriptive statistical results of the independent variables are shown in Table 8.9.

The statistical results show that the proportions of intellectual immigrants, labour immigrants and entrepreneurial immigrants among all respondents of the survey are 31.1%, 41.7% and 27.2% respectively. And among cities where respondents of survey live, Guangzhou accounted for 17.0%, Dongguan accounted for 18.3%, Shenyang accounted for 16.7%, Chengdu accounted for 17.4%, Hangzhou accounted for 14.3%, and Zhengzhou accounted for 16.4%.

There are 60.4% of urban migrants clearly expressed their willingness to settle in their current city, only 19.2% of urban migrants clearly expressed their unwillingness to settle in their current city, and another 0.5% of urban migrants were temporarily unable to make a clear statement.

The sample of urban immigrants studied by the authors is roughly equally divided between men and women, with an average age of 28.14. Their education degree is

**Table 8.9** Descriptive statistical results of variables

| Variables | Options | Sample size (number of people) | Proportions (%) | Cumulative down (%) | Average | Standard deviation |
|---|---|---|---|---|---|---|
| I'm willing to settle in the current city | Strongly disagree | 99 | 3.2 | 3.2 | | |
| | Disagree | 493 | 16.1 | 19.3 | | |
| | Not sure | 614 | 20.0 | 39.3 | | |
| | Agree | 1568 | 51.1 | 90.4 | | |
| | Strongly agree | 295 | 9.6 | 100.0 | | |
| | Total | 3069 | 100.0 | | 3.48 | 0.98 |
| Gender | Male | 1586 | 50.1 | 50.1 | | |
| | Female | 1582 | 49.9 | 100.0 | | |
| | Total | 3168 | 100.0 | | | |
| Education degree | Elementary school education degree and below | 167 | 5.3 | 5.3 | | |
| | Junior high school education degree | 831 | 26.2 | 31.5 | | |
| | High school education degree (including technical secondary school) | 967 | 30.5 | 62.0 | | |
| | Junior college education degree | 620 | 19.6 | 81.6 | | |
| | Bachelor's degree and above | 583 | 18.4 | 100.0 | | |
| Education degree | Total | 3168 | 100.0 | | | |
| Monthly disposable income | Less than RMB1000 | 454 | 14.4 | 14.4 | | |
| | RMB1001–1500 | 930 | 29.5 | 43.8 | | |
| | RMB1501–2000 | 575 | 18.2 | 62.1 | | |
| | RMB2001–2500 | 339 | 10.7 | 72.8 | | |
| | RMB2501–3000 | 242 | 7.7 | 80.5 | | |
| | RMB3001–3500 | 177 | 5.6 | 86.1 | | |

(continued)

**Table 8.9** (continued)

| Variables | Options | Sample size (number of people) | Proportions (%) | Cumulative down (%) | Average | Standard deviation |
|---|---|---|---|---|---|---|
| | RMB3501–4000 | 104 | 3.3 | 89.4 | | |
| | RMB4001–4500 | 66 | 2.1 | 91.4 | | |
| | RMB4501–5000 | 71 | 2.2 | 93.7 | | |
| | More than RMB5001 | 199 | 6.3 | 100.0 | | |
| | Total | 3157 | 100.0 | | | |
| I hope my children can develop in this city in the future | Strongly disagree | 97 | 3.4 | 3.4 | | |
| | Disagree | 428 | 14.9 | 18.2 | | |
| | Not sure | 614 | 21.3 | 39.5 | | |
| | Agree | 1364 | 47.3 | 86.8 | | |
| | Strongly agree | 379 | 13.2 | 100.0 | | |
| | Total | 2882 | 100.0 | | 3.52 | 1.01 |
| Whether to own self-purchased houses | No | 2806 | 88.6 | 88.6 | | |
| | Yes | 362 | 11.4 | 100.0 | | |
| | Total | 3168 | 100.0 | | | |
| Length of residence in the current city | 2 years | 895 | 28.3 | 28.3 | | |
| | 3 years | 855 | 27.0 | 55.2 | | |
| | 4 years | 606 | 19.1 | 74.4 | | |
| | 5 years | 812 | 25.6 | 100.0 | | |
| | Total | 3168 | 100.0 | | 3.42 | 1.15 |
| Satisfaction with public welfare in the current city | Did not enjoy any public welfare at all | 613 | 20.7 | 20.7 | | |
| | There are huge differences in the public welfare between migrant population and locals, and migrant population only enjoyed a small part of the public welfare | 1113 | 37.5 | 58.2 | | |

(continued)

**Table 8.9** (continued)

| Variables | Options | Sample size (number of people) | Proportions (%) | Cumulative down (%) | Average | Standard deviation |
|---|---|---|---|---|---|---|
| Satisfaction with public welfare in the current city | There are a little differences in the public welfare between migrant population and locals, and migrant population enjoyed most of the public welfare | 1053 | 35.5 | 93.7 | | |
| | Being completely consistent with locals in terms of public welfare | 187 | 6.3 | 100.0 | | |
| | Total | 2966 | 100.0 | | 2.27 | 0.86 |
| Satisfaction with housing prices in the current city | Strongly dissatisfied | 684 | 21.6 | 21.6 | | |
| | Dissatisfied | 1069 | 33.7 | 55.3 | | |
| | Moderate | 1207 | 38.1 | 93.4 | | |
| | Satisfied | 180 | 5.7 | 99.1 | | |
| | Strongly satisfied | 28 | 0.9 | 100.0 | | |
| | Total | 3168 | 100.0 | | 2.31 | 0.90 |
| Satisfaction with commodity prices in the current city | Strongly dissatisfied | 78 | 2.5 | 2.5 | | |
| | Dissatisfied | 514 | 16.2 | 18.7 | | |
| | Moderate | 1828 | 57.7 | 76.4 | | |
| | Satisfied | 692 | 21.8 | 98.2 | | |
| | Strongly satisfied | 56 | 1.8 | 100.0 | | |
| | Total | 3168 | 100.0 | | 3.04 | 0.74 |
| Satisfaction with life in the current city | Strongly dissatisfied | 22 | 0.7 | 0.7 | | |

(continued)

**Table 8.9** (continued)

| Variables | Options | Sample size (number of people) | Proportions (%) | Cumulative down (%) | Average | Standard deviation |
|---|---|---|---|---|---|---|
| | Dissatisfied | 115 | 3.7 | 4.4 | | |
| | Moderate | 1826 | 58.5 | 62.9 | | |
| | Satisfied | 1008 | 32.3 | 95.2 | | |
| | Strongly satisfied | 149 | 4.8 | 100.0 | | |
| | Total | 3120 | 100.0 | | 3.37 | 0.67 |
| There is much of the milk of human kindness in the city where I am now | Strongly disagree | 121 | 3.9 | 3.9 | | |
| | Disagree | 495 | 15.9 | 19.8 | | |
| | Not sure | 1421 | 45.7 | | | |
| | Agree | 950 | 30.6 | | | |
| | Strongly agree | 122 | 3.9 | | | |
| | Total | 3109 | 100.0 | | 3.15 | 0.87 |
| Social exclusion (after several years of living in the current city, what do you think of the attitude of locals towards migrant population?) | Noting changed | 1142 | 36.0 | 36.0 | | |
| | Being worse than before | 73 | 2.3 | 38.4 | | |
| | Being better than before | 1322 | 41.7 | 80.1 | | |
| | Always good | 631 | 19.9 | 100.0 | | |
| | Total | 3168 | 100.0 | | 2.46 | 1.17 |
| Political exclusion (what do you think of the attitude of the community officials in your community towards you?) | Being discriminate against the migrant population | 149 | 4.8 | 4.8 | | |
| | Being inconsiderate | 1914 | 61.3 | 66.1 | | |
| | Being considerate | 1059 | 33.9 | 100.0 | | |
| | Total | 3122 | 100.0 | | 2.29 | 0.55 |

mainly junior and senior high school, and their average length of residence in their current city is 2.42 years.

The monthly disposable income of most urban immigrants is less than RMB2500, and 47.5% of urban immigrants have monthly disposable income between RMB1001 and RMB2000. It can be seen that the income of urban immigrants is generally not high. In addition, only 11.4% of urban immigrants live in self-purchased houses, and 42.0% of urban immigrants live with their families.

There are 60.5% of urban immigrants hope that their children can live and develop in their current cities in the future, and only 18.3% of urban immigrants do not want their children to live and develop in their current cities in the future.

Urban immigrants' satisfaction with housing prices in their current cities is very low. Only 6.6% of urban immigrants are satisfied with housing prices in their current cities, and 55.3% of urban immigrants express dissatisfaction with housing prices in their current cities, while the remaining urban immigrants are moderately satisfied with the housing prices in their current cities.

The enjoyment of public welfare by urban immigrants is also not optimistic. Only 6.3% of urban immigrants said that they enjoyed the same public welfare as locals, 20.7% of urban immigrants said that they did not enjoy any public welfare at all, 37.5% of urban immigrants said that they only enjoyed a small part of public welfare, and 35.5% of urban immigrants said that there are some differences in the public welfare between them and the locals, but they still enjoyed most of the public welfare.

Urban immigrants' evaluation of commodity prices in their current cities is relatively moderate, and 57.7% of urban immigrants are moderately satisfied with commodity prices in their current cities. 23.6% of urban immigrants expressed satisfaction with the prices of commodities in their current cities, while 18.7% of urban immigrants expressed dissatisfaction with the prices of commodities in their current cities.

Urban immigrants' satisfaction with life is generally at a medium and upper-medium level. 58.5% of urban immigrants said that they were moderately satisfied with life, while 37.1% of urban immigrants said they were satisfied with life, and 4.4% of urban immigrants said they were not satisfied with life. It can be seen that the satisfaction rate is significantly higher than the dissatisfaction rate.

There are 61.6% of urban immigrants believe that locals have a better attitude towards them than before, only 2.3% of urban immigrants believe that the attitude of locals towards them is worse than before, and 36.0% of urban immigrants believe that the attitude of locals towards them has not changed.

There are 66.1% of urban immigrants believe that community officials do not care about them, and even feel that they are discriminated against by community officials. Only 33.9% of urban immigrants believe that community officials care about them, while the rest of the urban immigrants do not comment on this.

Urban immigrants have a good overall evaluation of the interpersonal environment of their current cities. Only 19.8% of urban immigrants are dissatisfied with the interpersonal environment of their current cities, 34.5% of urban immigrants are satisfied with the interpersonal environment of their current cities and think that there is much of the milk of human kindness in their current cities, and the remaining 45.7% of urban immigrants believe that the interpersonal environment of their current cities is at a medium level.

### 8.4.4 The Regression Model

The authors performed regression analysis on dependent variables and independent variables and performed 7 analyses based on 4 types of 7 dependent variables. Among them, the authors will use logistic regression model for 2 dummy variables, namely "whether to remit money to the hometown" and "whether to live with the family in the immigration city". Besides, the authors will use multiple linear regression model for 5 ratio variables and interval variables, namely "the proportion of remittance to the hometown in personal income", "the proportion of remittance to the hometown in personal expenditure", "whether there is a best friend in the immigration city", "whether to live with the family in the immigration city", "the willingness to long-term settlement" and "degree of acceptance of the sentence 'I am equal to locals'".

**Research Results and Findings**

(1) Whether to remit money to the hometown

The statistical results show (see Table 8.10) that the various variables of the regression model have limited interpretation of the impact of whether urban immigrants remit money to their hometowns, with an $R^2$ value of 0.064. And the main variables that affect whether urban immigrants remit money to their hometown are as follows:

1. Age. The older the age of urban immigrants ($p < 0.01$), the stronger their willingness to remit money to their hometown. Under the circumstance that other factors remain unchanged, for every year of increase in the age of urban immigrants, their willingness to remit money to their hometown will increase by 0.04 times.
2. Education degree. The higher the education degree of urban immigrants ($p < 0.01$), the weaker their willingness to remit money to their hometown. Under the circumstance that other factors remain the same, the willingness of urban immigrants to remit money to their hometown will be reduced by 0.19 times for each level of improvement in their education degree.
3. Self-purchased houses. When urban immigrants have self-purchased houses ($p < 0.05$), their willingness to remit money to their hometown will be reduced. Under the circumstance that other factors remain unchanged, the willingness of urban immigrants who own self-purchased houses to remit money to their hometown is 0.28 times lower than that of urban immigrants who do not own self-purchased houses.
4. Interpersonal environment (there is much of the milk of human kindness in current cities). The higher the satisfaction of urban immigrants with the interpersonal environment ($p < 0.01$) of their current city, the weaker their willingness to remit money to their hometown. Under the circumstance that other factors remain unchanged, every degree of increase in urban immigrants' satisfaction with the interpersonal environment of their current city will reduce their willingness to remit money to their hometown by 0.09 times.

**Table 8.10** Logistic regression model of factors affecting whether urban immigrants remit money to their hometown

| Independent variables | B | S.E | Wald | df | Sig | Exp (B) |
|---|---|---|---|---|---|---|
| Constant | −1.51 | 0.41 | 13.68 | 1 | 0.000*** | 0.22 |
| Gender | 0.03 | 0.08 | 0.17 | 1 | 0.679 | 1.04 |
| Age | 0.04 | 0.01 | 35.90 | 1 | 0.000*** | 1.04 |
| Education degree | −0.21 | 0.04 | 27.01 | 1 | 0.000*** | 0.81 |
| Monthly disposable income | 0.01 | 0.01 | 0.45 | 1 | 0.504 | 1.01 |
| I hope my children can develop in this city in the future | 0.03 | 0.04 | 0.44 | 1 | 0.509 | 1.03 |
| Whether to own self-purchased houses | −0.33 | 0.14 | 5.61 | 1 | 0.018** | 0.72 |
| Length of residence in the current city | 0.04 | 0.04 | 1.23 | 1 | 0.268 | 1.04 |
| Satisfaction with public welfare in the current city | 0.02 | 0.05 | 0.19 | 1 | 0.661 | 1.02 |
| Satisfaction with housing prices in the current city | 0.08 | 0.05 | 2.51 | 1 | 0.113 | 1.08 |
| Satisfaction with commodity prices in the current city | −0.04 | 0.06 | 0.48 | 1 | 0.486 | 0.96 |
| Satisfaction with life in the current city | −0.01 | 0.07 | 0.04 | 1 | 0.845 | 0.99 |
| There is much of the milk of human kindness in the city where i am now | −0.09 | 0.05 | 3.09 | 1 | 0.079* | 0.91 |
| Social exclusion (after several years of living in the current city, what do you think of the attitude of locals towards migrant population?) | 0.09 | 0.04 | 5.47 | 1 | 0.019** | 1.09 |
| Political exclusion (what do you think of the attitude of the community officials in your community towards you?) | 0.15 | 0.08 | 3.84 | 1 | 0.05* | 1.17 |

*Note* (1) 2 Log likelihood = 3309.685, Nagelkerke $R^2 = 0.064$; (2) the sample size is 2620; (3) meaning of prompt symbols: * represents $p < 0.1$, ** represents $p < 0.05$, *** represents $p < 0.01$

5. Social exclusion (after several years of living in the current city, what do you think of the attitudes of locals towards the migrant population?). The weaker the sense of social exclusion ($p < 0.05$) felt by urban immigrants, the stronger their willingness to remit money to their hometown. Under the circumstance that other factors remain the same, for every degree of decrease in the sense of social exclusion felt by urban immigrants, their willingness to remit money to their hometown will increase by 0.09 times.
6. Political exclusion (what do you think of the attitude of the community officials in your community towards you?). The weaker the sense of political exclusion ($p < 0.1$) felt by urban immigrants, the stronger their willingness to remit money to their hometown. Under the circumstance that other factors remain the same, for every degree of decrease in the sense of political exclusion felt by urban

immigrants, their willingness to remit money to their hometown will increase by 0.17 times.

(2) The proportion of remittance to the hometown in personal income

The statistical results show (see Table 8.11) that the various variables of the regression model have limited interpretation of the impact of the proportion of remittance to the hometown in urban immigrants' personal income, with an $R^2$ value of 0.031. And the main variables that affect the proportion of remittance to the hometown in urban immigrants' personal income are as follows:

1. Age. The older the age of urban immigrants ($p < 0.01$), the greater the proportion of remittance to the hometown in urban immigrants' personal income. Under the circumstance that other factors remain unchanged, for every year of increase in the age of urban immigrants, the proportion of remittance to the hometown in urban immigrants' personal income will increase by 0.2%.
2. Education degree. The higher the education degree of urban immigrants ($p < 0.01$), the smaller the proportion of remittance to the hometown in urban immigrants' personal income. Under the circumstance that other factors remain the same, the proportion of remittance to the hometown in urban immigrants' personal income will decrease by 2% for each level of improvement in their education degree.
3. Self-purchased houses. When urban immigrants have self-purchased houses ($p < 0.01$), the proportion of remittance to the hometown in urban immigrants' personal income will decrease. Under the circumstance that other factors remain unchanged, the proportion of remittance to the hometown in personal income of urban immigrants who own self-purchased houses is 4% lower than that of urban immigrants who do not own self-purchased houses.
4. Satisfaction with housing prices in current city. The higher the satisfaction of urban immigrants with housing prices ($p < 0.05$) in their current city, the greater the proportion of remittance to the hometown in urban immigrants' personal income. Under the circumstance that other factors remain the same, every degree of increase in urban immigrants' satisfaction with housing prices in their current city will increase the proportion of remittance to the hometown in urban immigrants' personal income by 1%.
5. Social exclusion (after several years of living in the current city, what do you think of the attitudes of locals towards the migrant population?). The weaker the sense of social exclusion ($p < 0.05$) felt by urban immigrants, the greater the proportion of remittance to the hometown in urban immigrants' personal income. Under the circumstance that other factors remain unchanged, for every degree of decrease in the sense of social exclusion felt by urban immigrants, the proportion of remittance to the hometown in urban immigrants' personal income will increase by 1%.
6. Political exclusion (what do you think of the attitude of the community officials in your community towards you?). The weaker the sense of political exclusion ($p < 0.01$) felt by urban immigrants, the greater the proportion of remittance to

**Table 8.11** Multiple linear regression model of factors affecting the proportion of remittance to the hometown in urban immigrants' personal income

| Independent variables | Non-standard regression coefficient | | Standard regression coefficient (Beta) | T | Sig | Collinearity analysis | |
|---|---|---|---|---|---|---|---|
| | B | Standard deviation | | | | Tolerance | Vif |
| Constant | 0.06 | 0.04 | | 1.47 | 0.142 | | |
| Gender | 0.01 | 0.01 | 0.02 | 0.90 | 0.37 | 0.94 | 1.07 |
| Age | 0.002 | 0.00 | 0.06 | 2.93 | 0.003*** | 0.82 | 1.22 |
| Education degree | −0.02 | 0.00 | −0.11 | −5.01 | 0.000*** | 0.75 | 1.34 |
| Monthly disposable income | 0.00 | 0.00 | −0.02 | −0.95 | 0.34 | 0.79 | 1.26 |
| I hope my children can develop in this city in the future | 0.00 | 0.00 | −0.02 | −0.87 | 0.386 | 0.87 | 1.15 |
| Whether to own self-purchased houses | −0.04 | 0.01 | −0.05 | −2.63 | 0.009*** | 0.89 | 1.12 |
| Length of residence in the current city | 0.00 | 0.00 | 0.02 | 0.76 | 0.449 | 0.88 | 1.14 |
| Satisfaction with public welfare in the current city | 0.00 | 0.01 | 0.02 | 0.81 | 0.415 | 0.84 | 1.19 |
| Satisfaction with housing prices in the current city | 0.01 | 0.01 | 0.05 | 2.19 | 0.029** | 0.84 | 1.19 |
| Satisfaction with commodity prices in the current city | 0.00 | 0.01 | −0.01 | −0.69 | 0.488 | 0.85 | 1.18 |
| Satisfaction with life in the current city | 0.01 | 0.01 | 0.02 | 0.75 | 0.456 | 0.81 | 1.24 |

(continued)

**Table 8.11** (continued)

| Independent variables | Non-standard regression coefficient | | Standard regression coefficient (Beta) | T | Sig | Collinearity analysis | |
|---|---|---|---|---|---|---|---|
| | B | Standard deviation | | | | Tolerance | Vif |
| There is much of the milk of human kindness in the city where I am now | −0.01 | 0.01 | −0.03 | −1.48 | 0.139 | 0.79 | 1.27 |
| Social exclusion (after several years of living in the current city, what do you think of the attitude of locals towards migrant population?) | 0.01 | 0.00 | 0.05 | 2.21 | 0.027** | 0.90 | 1.11 |
| Political exclusion (what do you think of the attitude of the community officials in your community towards you?) | 0.02 | 0.01 | 0.05 | 2.60 | 0.009*** | 0.90 | 1.11 |

*Note* (1) $R^2 = 0.036$, Adjusted $R^2 = 0.031$, F = 6.647; (2) the sample size is 2499; (3) meaning of prompt symbols: * represents $p < 0.1$, ** represents $p < 0.05$, *** represents $p < 0.01$

the hometown in urban immigrants' personal income. Under the circumstance that other factors remain unchanged, for every degree of decrease in the sense of political exclusion felt by urban immigrants, the proportion of remittance to the hometown in urban immigrants' personal income will increase by 2%.

(3) The proportion of remittance to the hometown in personal expenditure

The statistical results show (see Table 8.12) that the various variables of the regression model have limited interpretation of the impact of the proportion of remittance to the hometown in urban immigrants' personal expenditure, with an $R^2$ value of 0.048. And the main variables that affect the proportion of remittance to the hometown in urban immigrants' personal expenditure are as follows:

**Table 8.12** Multiple linear regression model of factors affecting the proportion of remittance to the hometown in urban immigrants' personal expenditure

| Independent variables | Non-standard regression coefficient | | Standard regression coefficient (Beta) | T | Sig | Collinearity analysis | |
|---|---|---|---|---|---|---|---|
| | B | Standard deviation | | | | Tolerance | Vif |
| Constant | 0.17 | 0.04 | | 4.30 | 0.000*** | | |
| Gender | 0.00 | 0.01 | 0.00 | 0.08 | 0.937 | 0.97 | 1.03 |
| Age | 0.002 | 0.00 | 0.05 | 2.62 | 0.009*** | 0.84 | 1.19 |
| Education degree | −0.03 | 0.00 | −0.18 | −8.57 | 0.000*** | 0.81 | 1.23 |
| Monthly disposable income | 0.00 | 0.00 | −0.02 | −1.23 | 0.218 | 0.94 | 1.06 |
| I hope my children can develop in this city in the future | −0.01 | 0.00 | −0.04 | −1.80 | 0.073* | 0.87 | 1.15 |
| Whether to own self-purchased houses | −0.05 | 0.01 | −0.07 | −3.53 | 0.000*** | 0.90 | 1.11 |
| Length of residence in the current city | 0.00 | 0.00 | −0.01 | −0.26 | 0.799 | 0.89 | 1.12 |
| Satisfaction with public welfare in the current city | 0.01 | 0.01 | 0.03 | 1.65 | 0.098* | 0.84 | 1.19 |
| Satisfaction with housing prices in the current city | 0.00 | 0.00 | 0.01 | 0.69 | 0.488 | 0.85 | 1.18 |
| Satisfaction with commodity prices in the current city | 0.00 | 0.01 | 0.02 | 0.79 | 0.432 | 0.86 | 1.17 |
| Satisfaction with life in the current city | 0.00 | 0.01 | −0.01 | −0.25 | 0.801 | 0.81 | 1.23 |

(continued)

**Table 8.12** (continued)

| Independent variables | Non-standard regression coefficient | | Standard regression coefficient (Beta) | T | Sig | Collinearity analysis | |
|---|---|---|---|---|---|---|---|
| | B | Standard deviation | | | | Tolerance | Vif |
| There is much of the milk of human kindness in the city where i am now | −0.01 | 0.01 | −0.04 | −2.00 | 0.045** | 0.79 | 1.26 |
| Social exclusion (after several years of living in the current city, what do you think of the attitude of locals towards migrant population?) | 0.01 | 0.00 | 0.06 | 2.83 | 0.005*** | 0.91 | 1.10 |
| Political exclusion (what do you think of the attitude of the community officials in your community towards you?) | 0.01 | 0.01 | 0.03 | 1.48 | 0.14 | 0.90 | 1.11 |

*Note* (1) $R^2 = 0.053$, Adjusted $R^2 = 0.048$, $F = 10.429$; (2) the sample size is 2618; (3) meaning of prompt symbols: * represents $p < 0.1$, ** represents $p < 0.05$, *** represents $p < 0.01$

1. Age. The older the age of urban immigrants ($p < 0.01$), the greater the proportion of remittance to the hometown in urban immigrants' personal expenditure. Under the circumstance that other factors remain unchanged, for every year of increase in the age of urban immigrants, the proportion of remittance to the hometown in urban immigrants' personal expenditure will increase by 0.2%.
2. Education degree. The higher the education degree of urban immigrants ($p < 0.01$), the smaller the proportion of remittance to the hometown in urban immigrants' personal expenditure. Under the circumstance that other factors remain the same, the proportion of remittance to the hometown in urban immigrants' personal expenditure will decrease by 3% for each level of improvement in their education degree.

3. Hoping my children can develop in this city in the future too. The greater the willingness of urban immigrants to let their children develop in this city in the future ($p < 0.1$), the lower the proportion of their remittances to their hometown in their personal expenditure. Under the circumstance that other factors remain unchanged, the proportion of remittance to the hometown in urban immigrants' personal expenditure will decrease by 1% for each level of increase in their willingness to let their children develop in this city in the future.
4. Self-purchased houses. When urban immigrants have self-purchased houses ($p < 0.01$), the proportion of their remittances to their hometown in their personal expenditure will decrease. Under the circumstance that other factors remain the same, the proportion of remittance to the hometown in the personal expenditure of urban immigrants who own self-purchased houses is 5% lower than that of urban immigrants who do not own self-purchased houses.
5. Satisfaction with public welfare. The higher the satisfaction of urban immigrants with public welfare ($p < 0.1$), the greater the proportion of remittance to the hometown in urban immigrants' personal expenditure. Under the circumstance that other factors remain the same, every degree of increase in urban immigrants' satisfaction with public welfare will increase the proportion of remittance to the hometown in urban immigrants' personal expenditure by 1%.
6. Interpersonal environment (there is much of the milk of human kindness in current cities). The higher the satisfaction of urban immigrants with the interpersonal environment ($p < 0.05$) of their current city, the smaller the proportion of remittance to the hometown in urban immigrants' personal expenditure. Under the circumstance that other factors remain unchanged, every degree of increase in urban immigrants' satisfaction with the interpersonal environment of their current city will reduce their the proportion of remittance to the hometown in urban immigrants' personal expenditure by 1%.
7. Social exclusion (after several years of living in the current city, what do you think of the attitudes of locals towards the migrant population?). The weaker the sense of social exclusion ($p < 0.01$) felt by urban immigrants, the greater the proportion of remittance to the hometown in urban immigrants' personal expenditure. Under the circumstance that other factors remain unchanged, for every degree of decrease in the sense of social exclusion felt by urban immigrants, the proportion of remittance to the hometown in urban immigrants' personal expenditure will increase by 1%.

(4) Whether urban immigrants have a best friend in the immigration city

The statistical results show (see Table 8.13) that the various variables of the regression model have limited interpretation of the impact of whether there is a best friend in the immigration city, with an $R^2$ value of 0.038. And the main variables that affect whether urban immigrants have a best friend in the immigration city are as follows:

1. Education degree. The higher the education degree of urban immigrants ($p < 0.01$), the greater the number of best friends they have in the immigration city.

**Table 8.13** Multiple linear regression model of factors affecting whether urban immigrants have a best friend in the immigration city

| Independent variables | Non-standard regression coefficient | | Standard regression coefficient (Beta) | T | Sig | Collinearity analysis | |
|---|---|---|---|---|---|---|---|
| | B | Standard deviation | | | | Tolerance | Vif |
| Constant | 1.62 | 0.21 | | 7.83 | 0.000*** | | |
| Gender | 0.04 | 0.04 | 0.02 | 1.04 | 0.3 | 0.97 | 1.03 |
| Age | 0.00 | 0.00 | −0.03 | −1.28 | 0.201 | 0.84 | 1.19 |
| Education degree | 0.08 | 0.02 | 0.09 | 4.05 | 0.000*** | 0.81 | 1.23 |
| Monthly disposable income | 0.00 | 0.00 | −0.02 | −0.88 | 0.377 | 0.94 | 1.06 |
| I hope my children can develop in this city in the future | 0.11 | 0.02 | 0.10 | 4.89 | 0.000*** | 0.87 | 1.15 |
| Whether to own self-purchased houses | 0.15 | 0.07 | 0.04 | 2.22 | 0.027** | 0.90 | 1.11 |
| Length of residence in the current city | 0.12 | 0.02 | 0.12 | 6.00 | 0.000*** | 0.89 | 1.12 |
| Satisfaction with public welfare in the current city | −0.04 | 0.03 | −0.03 | −1.38 | 0.168 | 0.84 | 1.19 |
| Satisfaction with housing prices in the current city | −0.01 | 0.03 | −0.01 | −0.25 | 0.804 | 0.85 | 1.18 |
| Satisfaction with commodity prices in the current city | −0.03 | 0.03 | −0.02 | −1.00 | 0.316 | 0.85 | 1.17 |
| Satisfaction with life in the current city | 0.05 | 0.03 | 0.03 | 1.34 | 0.181 | 0.81 | 1.23 |

(continued)

**Table 8.13** (continued)

| Independent variables | Non-standard regression coefficient | | Standard regression coefficient (Beta) | T | Sig | Collinearity analysis | |
|---|---|---|---|---|---|---|---|
| | B | Standard deviation | | | | Tolerance | Vif |
| There is much of the milk of human kindness in the city where I am now | 0.06 | 0.03 | 0.05 | 2.16 | 0.031** | 0.79 | 1.26 |
| Social exclusion (after several years of living in the current city, what do you think of the attitude of locals towards migrant population?) | −0.01 | 0.02 | −0.01 | −0.48 | 0.633 | 0.90 | 1.11 |
| Political exclusion (what do you think of the attitude of the community officials in your community towards you?) | 0.11 | 0.04 | 0.06 | 2.77 | 0.006*** | 0.90 | 1.11 |

*Note* (1) $R^2 = 0.043$, Adjusted $R^2 = 0.038$, F = 8.384; (2) the sample size is 2620; (3) meaning of prompt symbols: * represents $p < 0.1$, ** represents $p < 0.05$, *** represents $p < 0.01$

Under the circumstance that other factors remain the same, the number of best friends that urban immigrants have in the immigration city will increase by 0.08 for each level of improvement in their education degree.

2. Hoping my children can develop in this city in the future too. The greater the willingness of urban immigrants to let their children develop in this city in the future ($p < 0.1$), the greater the number of best friends they have in the immigration city. Under the circumstance that other factors remain unchanged, the number of best friends that urban immigrants have in the immigration city will increase by 0.11 for each level of increase in their willingness to let their children develop in this city in the future.
3. Self-purchased Houses. When urban immigrants have self-purchased houses ($p < 0.05$), the number of best friends they have in the immigration city will increase. Under the circumstance that other factors remain the same, the number of best friends in the immigration city of urban immigrants who own self-purchased houses is 0.15 more than that of urban immigrants who do not own self-purchased houses.

4. Length of residence in the current city. The longer the time that urban immigrants live in the current city ($p < 0.01$), the greater the number of best friends they have in the immigration city. Under the circumstance that other factors remain the same, the number of best friends that urban immigrants have in the immigration city will increase by 0.12 for every additional year they live in the current city.
5. Interpersonal environment (there is much of the milk of human kindness in current cities). The higher the satisfaction of urban immigrants with the interpersonal environment ($p < 0.05$) of their current city, the greater the number of best friends they have in the immigration city. Under the circumstance that other factors remain unchanged, the number of best friends that urban immigrants have in the immigration city will increase by 0.06 for every degree of increase in their satisfaction with the interpersonal environment of their current city.
6. Social exclusion (after several years of living in the current city, what do you think of the attitudes of locals towards the migrant population?). The weaker the sense of social exclusion ($p < 0.01$) felt by urban immigrants, the greater the number of best friends they have in the immigration city. Under the circumstance that other factors remain unchanged, the number of best friends that urban immigrants have in the immigration city will increase by 0.11 for every degree of decrease in the sense of social exclusion felt by urban immigrants.

(5) Whether to live with the family in the immigration city

The statistical results show (see Table 8.14) that the various variables of the regression model have a certain degree of explanation for the impact of whether urban immigrants live with the family in the immigration city, with an $R^2$ value of 0.116. And the main variables that affect whether urban immigrants live with the family in the immigration city are as follows:

1. Gender. The proportion of male urban immigrants ($p < 0.01$) living with their families in immigration cities is significantly higher than that of female urban immigrants. Under the circumstance that other factors remain unchanged, the number of male urban immigrants living with their families in the immigration city is 1.02 times higher than that of female urban immigrants.
2. Education degree. The higher the education degree of urban immigrants ($p < 0.01$), the less likely they are to live with their families in the immigration city. Under the circumstance that other factors remain the same, the probability of urban immigrants living with their families in the immigration city will decrease by 0.3 times for each level of improvement in their education degree.
3. Monthly disposable income. The higher the monthly disposable income of urban immigrants ($p < 0.01$), the more likely they are to live with their families in the immigration city. Under the circumstance that other factors remain unchanged, every time the monthly disposable income of urban immigrants increases by RMB500, the probability of urban immigrants living with their families in the immigration city increases by 0.13 times.
4. Self-purchased houses. When urban immigrants have self-purchased houses ($p < 0.01$), they are more likely to live with their families in the immigration city.

**Table 8.14** Logistic regression model of factors affecting whether urban immigrants live with the family in the immigration city

| Independent variables | B | S.E | Wald | df | Sig | Exp (B) |
|---|---|---|---|---|---|---|
| Constant | −0.12 | 0.80 | 0.02 | 1 | 0.876 | 0.88 |
| Gender | 0.70 | 0.17 | 17.63 | 1 | 0.000*** | 2.02 |
| Age | 0.00 | 0.01 | 0.00 | 1 | 0.944 | 1.00 |
| Education degree | −0.36 | 0.09 | 17.48 | 1 | 0.000*** | 0.70 |
| Monthly disposable income | 0.12 | 0.03 | 11.90 | 1 | 0.001*** | 1.13 |
| I hope my children can develop in this city in the future | 0.14 | 0.08 | 2.63 | 1 | 0.105 | 1.15 |
| Whether to own self-purchased houses | 1.26 | 0.28 | 19.75 | 1 | 0.000*** | 3.53 |
| Length of residence in the current city | 0.14 | 0.07 | 3.72 | 1 | 0.054* | 1.15 |
| Satisfaction with public welfare in the current city | −0.23 | 0.10 | 5.41 | 1 | 0.02** | 0.79 |
| Satisfaction with housing prices in the current city | 0.00 | 0.09 | 0.00 | 1 | 0.977 | 1.00 |
| Satisfaction with commodity prices in the current city | 0.02 | 0.12 | 0.02 | 1 | 0.888 | 1.02 |
| Satisfaction with life in the current city | 0.17 | 0.13 | 1.73 | 1 | 0.188 | 1.18 |
| There is much of the milk of human kindness in the city where I am now | −0.09 | 0.10 | 0.88 | 1 | 0.349 | 0.91 |
| Social exclusion (after several years of living in the current city, what do you think of the attitude of locals towards migrant population?) | −0.02 | 0.07 | 0.04 | 1 | 0.836 | 0.98 |
| Political exclusion (what do you think of the attitude of the community officials in your community towards you?) | 0.09 | 0.14 | 0.41 | 1 | 0.52 | 1.09 |

*Note* (1) −2 Log likelihood = 1022.465, Nagelkerke $R^2 = 0.116$; (2) the sample size is 1100; (3) meaning of prompt symbols: * represents $p < 0.1$, ** represents $p < 0.05$, *** represents $p < 0.01$

Under the circumstance that other factors remain the same, urban immigrants who have self-purchased houses are 2.53 times more likely to live with their families in the immigration city than that of urban immigrants who do not own self-purchased houses.

5. Length of residence in the current city. The longer the time that urban immigrants live in the current city ($p < 0.1$), the more likely they are to live with their families in the immigration city. Under the circumstance that other factors remain the same, the probability of urban immigrants living with their families in the immigration city will increase by 0.15 times for every additional year they live in the current city.
6. Satisfaction with public welfare. The higher the satisfaction of urban immigrants with public welfare ($p < 0.05$), the less likely they are to live with their families in the immigration city. Under the circumstance that other factors remain the same,

the probability of urban immigrants living with their families in the immigration city will decrease by 0.21 times for every degree of increase in their satisfaction with the public welfare.

(6) The willingness to long-term settlement

The statistical results show (see Table 8.15) that the various variables of the regression model have a certain degree of explanation for the impact of whether urban immigrants have the willingness to long-term settlement, with an $R^2$ value of 0.324. And the main variables that affect whether urban immigrants have the willingness to long-term settlement are as follows:

1. Gender. The willingness of male urban immigrants ($p < 0.01$) to settle in the immigration city is significantly higher than that of female urban immigrants. Under the circumstance that other factors remain unchanged, male urban immigrants' willingness to long-term settlement in immigration cities is 0.09 levels higher than that of female urban immigrants.
2. Age. The older the age of urban immigrants ($p < 0.05$), the lower their willingness to settle in the immigration city for a long time. Under the circumstance that other factors remain the same, for every year of increase in the age of urban immigrants, their willingness to settle in immigration cities for a long time will be reduced by 0.01 levels.
3. Education degree. The higher the education degree ($p < 0.01$) of urban immigrants, the higher their willingness to settle in the immigration city for a long time. Under the circumstance that other factors remain unchanged, urban immigrants' willingness to settle in the immigration city for a long time will increase by 0.12 levels for each level of improvement in their education degree.
4. Hoping my children can develop in this city in the future too. The greater the willingness of urban immigrants to let their children develop in this city in the future ($p < 0.01$), the higher their willingness to settle in the immigration city for a long time. Under the circumstance that other factors remain unchanged, urban immigrants' willingness to settle in the immigration city for a long time will increase by 0.46 levels for each level of increase in their willingness to let their children develop in this city in the future.
5. Self-purchased houses. When urban immigrants have self-purchased houses ($p < 0.01$), their willingness to settle in the immigration city for a long time is higher. Under the circumstance that other factors remain the same, the willingness of urban immigrants who own self-purchased houses to settle in the immigration city is 0.17 levels higher than that of urban immigrants who do not own self-purchased houses.
6. Length of residence in the current city. The longer the time that urban immigrants live in the current city ($p < 0.1$), the higher their willingness to settle in the immigration city for a long time. Under the circumstance that other factors remain the same, urban immigrants' willingness to settle in the immigration city for a long time will increase by 0.04 levels for every additional year they live in the current city.

**Table 8.15** Multiple linear regression model of factors affecting whether urban immigrants have the willingness to long-term settlement

| Independent variables | Non-standard regression coefficient | | Standard regression coefficient (Beta) | T | Sig | Collinearity analysis | |
|---|---|---|---|---|---|---|---|
| | B | Standard deviation | | | | Tolerance | Vif |
| Constant | 0.50 | 0.16 | | 3.19 | 0.001*** | | |
| Gender | 0.09 | 0.03 | 0.04 | 2.73 | 0.006*** | 0.97 | 1.03 |
| Age | −0.01 | 0.00 | −0.04 | −2.34 | 0.019** | 0.84 | 1.19 |
| Education degree | 0.12 | 0.02 | 0.15 | 8.29 | 0.000*** | 0.81 | 1.23 |
| Monthly disposable income | 0.01 | 0.00 | 0.03 | 2.10 | 0.036** | 0.95 | 1.06 |
| I hope my children can develop in this city in the future | 0.46 | 0.02 | 0.47 | 27.09 | 0.000*** | 0.87 | 1.15 |
| Whether to own self-purchased houses | 0.17 | 0.05 | 0.06 | 3.26 | 0.001*** | 0.90 | 1.11 |
| Length of residence in the current city | 0.04 | 0.01 | 0.05 | 2.90 | 0.004*** | 0.89 | 1.13 |
| Satisfaction with public welfare in the current city | 0.00 | 0.02 | 0.00 | −0.17 | 0.864 | 0.84 | 1.19 |
| Satisfaction with housing prices in the current city | −0.04 | 0.02 | −0.03 | −1.98 | 0.048** | 0.84 | 1.18 |
| Satisfaction with commodity prices in the current city | 0.05 | 0.02 | 0.04 | 2.01 | 0.044** | 0.85 | 1.17 |
| Satisfaction with life in the current city | 0.11 | 0.03 | 0.07 | 4.11 | 0.000*** | 0.81 | 1.23 |

(continued)

Table 8.15 (continued)

| Independent variables | Non-standard regression coefficient | | Standard regression coefficient (Beta) | T | Sig | Collinearity analysis | |
|---|---|---|---|---|---|---|---|
| | B | Standard deviation | | | | Tolerance | Vif |
| There is much of the milk of human kindness in the city where i am now | 0.14 | 0.02 | 0.13 | 7.07 | 0.000*** | 0.79 | 1.26 |
| Social exclusion (after several years of living in the current city, what do you think of the attitude of locals towards migrant population?) | 0.01 | 0.01 | 0.01 | 0.67 | 0.505 | 0.90 | 1.11 |
| Political exclusion (what do you think of the attitude of the community officials in your community towards you?) | −0.01 | 0.03 | −0.01 | −0.30 | 0.765 | 0.90 | 1.11 |

*Note* (1) $R^2 = 0.328$, Adjusted $R^2 = 0.324$, $F = 89.183$; (2) the sample size is 2573; (3) meaning of prompt symbols: * represents $p < 0.1$, ** represents $p < 0.05$, *** represents $p < 0.01$

7. Satisfaction with housing prices in current city. The higher the satisfaction of urban immigrants with housing prices ($p < 0.05$) in their current city, the lower their willingness to settle in the immigration city for a long time. Under the circumstance that other factors remain the same, urban immigrants' willingness to settle in the immigration city for a long time will decrease by 0.04 levels for every degree of increase in urban immigrants' satisfaction with housing prices in their current cities.
8. Satisfaction with commodity prices in current city. The higher the satisfaction of urban immigrants with commodity prices ($p < 0.05$) in their current city, the higher their willingness to settle in the immigration city for a long time. Under the circumstance that other factors remain the same, urban immigrants' willingness to settle in the immigration city for a long time will increase by 0.05 levels for every degree of increase in urban immigrants' satisfaction with commodity prices in their current cities.
9. Satisfaction with life in current city. The higher the satisfaction of urban immigrants with life ($p < 0.01$) in their current city, the higher their willingness

to settle in the immigration city for a long time. Under the circumstance that other factors remain the same, urban immigrants' willingness to settle in the immigration city for a long time will increase by 0.11 levels for every degree of increase in urban immigrants' satisfaction with life in their current cities.

10. Interpersonal environment (there is much of the milk of human kindness in current cities). The higher the satisfaction of urban immigrants with the interpersonal environment ($p < 0.01$) of their current city, the higher their willingness to settle in the immigration city for a long time. Under the circumstance that other factors remain unchanged, urban immigrants' willingness to settle in the immigration city for a long time will increase by 0.14 levels for every degree of increase in urban immigrants' satisfaction with the interpersonal environment in their current cities.

(7) Degree of acceptance of the sentence "I am equal to locals"

The statistical results show (see Table 8.16) that the various variables of the regression model have a certain degree of explanation for the impact of urban immigrants' degree of acceptance of the sentence "I am equal to locals", with an $R^2$ value of 0.324. And the main variables that affect urban immigrants' degree of acceptance of the sentence "I am equal to locals" are as follows:

1. Gender. Male urban immigrants' ($p < 0.05$) degree of acceptance of the sentence "I am equal to locals" is higher than that of female urban immigrants. Under the circumstance that other factors remain unchanged, male urban immigrants' degree of acceptance of the sentence "I am equal to locals" is 0.03 levels higher than that of female urban immigrants.
2. Age. The older the age of urban immigrants ($p < 0.01$), the lower their degree of acceptance of the sentence "I am equal to locals". Under the circumstance that other factors remain the same, for every year of increase in the age of urban immigrants, their degree of acceptance of the sentence "I am equal to locals" will be reduced by 0.01 levels.
3. Education degree. The higher the education degree ($p < 0.01$) of urban immigrants, the higher their degree of acceptance of the sentence "I am equal to locals". Under the circumstance that other factors remain unchanged, urban immigrants' degree of acceptance of the sentence "I am equal to locals" will increase by 0.1 levels for each level of improvement in their education degree.
4. Monthly disposable income. The higher the monthly disposable income of urban immigrants ($p < 0.01$), the higher their degree of acceptance of the sentence "I am equal to locals". Under the circumstance that other factors remain unchanged, every time the monthly disposable income of urban immigrants increases by RMB500, their degree of acceptance of the sentence "I am equal to locals" will increase by 0.01 levels.
5. Self-purchased houses. When urban immigrants have self-purchased houses ($p < 0.01$), their degree of acceptance of the sentence "I am equal to locals" is higher. Under the circumstance that other factors remain the same, the degree of acceptance of the sentence "I am equal to locals" of urban immigrants who

**Table 8.16** Multiple linear regression model of factors affecting urban immigrants' degree of acceptance of the sentence "I am equal to locals"

| Independent variables | Non-standard regression coefficient | | Standard regression coefficient (Beta) | T | Sig | Collinearity analysis | |
|---|---|---|---|---|---|---|---|
| | B | Standard deviation | | | | Tolerance | Vif |
| Constant | 0.33 | 0.17 | | 1.98 | 0.047** | | |
| Gender | 0.09 | 0.03 | 0.04 | 2.53 | 0.011** | 0.97 | 1.03 |
| Age | −0.01 | 0.00 | −0.07 | −3.59 | 0.000*** | 0.84 | 1.19 |
| Education degree | 0.10 | 0.02 | 0.11 | 5.85 | 0.000*** | 0.81 | 1.23 |
| Monthly disposable income | 0.01 | 0.00 | 0.04 | 2.28 | 0.023** | 0.94 | 1.06 |
| I hope my children can develop in this city in the future | 0.02 | 0.02 | 0.02 | 0.97 | 0.333 | 0.87 | 1.15 |
| Whether to own self-purchased houses | 0.23 | 0.06 | 0.08 | 4.25 | 0.000*** | 0.90 | 1.11 |
| Length of residence in the current city | 0.02 | 0.02 | 0.02 | 0.97 | 0.331 | 0.89 | 1.13 |
| Satisfaction with public welfare in the current city | 0.09 | 0.02 | 0.08 | 4.01 | 0.000*** | 0.84 | 1.19 |
| Satisfaction with housing prices in the current city | 0.05 | 0.02 | 0.05 | 2.56 | 0.011** | 0.84 | 1.18 |
| Satisfaction with commodity prices in the current city | 0.04 | 0.02 | 0.03 | 1.68 | 0.093* | 0.86 | 1.17 |
| Satisfaction with life in the current city | 0.23 | 0.03 | 0.15 | 8.03 | 0.000*** | 0.81 | 1.23 |

(continued)

**Table 8.16** (continued)

| Independent variables | Non-standard regression coefficient | | Standard regression coefficient (Beta) | T | Sig | Collinearity analysis | |
|---|---|---|---|---|---|---|---|
| | B | Standard deviation | | | | Tolerance | Vif |
| There is much of the milk of human kindness in the city where I am now | 0.35 | 0.02 | 0.31 | 15.81 | 0.000*** | 0.79 | 1.26 |
| Social exclusion (after several years of living in the current city, what do you think of the attitude of locals towards migrant population?) | 0.03 | 0.02 | 0.04 | 2.00 | 0.046** | 0.90 | 1.11 |
| Political exclusion (what do you think of the attitude of the community officials in your community towards you?) | 0.02 | 0.03 | 0.01 | 0.64 | 0.525 | 0.90 | 1.11 |

*Note* (1) $R^2 = 0.236$, Adjusted $R^2 = 0.232$, F = 57.057; (2) the sample size is 2598; (3) meaning of prompt symbols: * represents $p < 0.1$, ** represents $p < 0.05$, *** represents $p < 0.01$

own self-purchased houses is 0.23 levels higher than that of urban immigrants who do not own self-purchased houses.

6. Satisfaction with public welfare in current cities. The higher the satisfaction of urban immigrants with public welfare ($p < 0.05$) in their current city, the higher their degree of acceptance of the sentence "I am equal to locals". Under the circumstance that other factors remain the same, urban immigrants' degree of acceptance of the sentence "I am equal to locals" will increase by 0.09 levels for every degree of increase in urban immigrants' satisfaction with public welfare in their current cities.
7. Satisfaction with housing prices in current cities. The higher the satisfaction of urban immigrants with housing prices ($p < 0.05$) in their current city, the higher their degree of acceptance of the sentence "I am equal to locals". Under the circumstance that other factors remain the same, urban immigrants' degree of acceptance of the sentence "I am equal to locals" will increase 0.05 levels for every degree of increase in urban immigrants' satisfaction with housing prices in their current cities.

8. Satisfaction with commodity prices in current cities. The higher the satisfaction of urban immigrants with commodity prices ($p < 0.1$) in their current city, the higher their degree of acceptance of the sentence "I am equal to locals". Under the circumstance that other factors remain the same, urban immigrants' degree of acceptance of the sentence "I am equal to locals" will increase by 0.04 levels for every degree of increase in urban immigrants' satisfaction with commodity prices in their current cities.
9. Satisfaction with life in current cities. The higher the satisfaction of urban immigrants with life ($p < 0.01$) in their current city, the higher their degree of acceptance of the sentence "I am equal to locals". Under the circumstance that other factors remain the same, urban immigrants' degree of acceptance of the sentence "I am equal to locals" will increase by 0.23 levels for every degree of increase in urban immigrants' satisfaction with life in their current cities.
10. Interpersonal environment (there is much of the milk of human kindness in current cities). The higher the satisfaction of urban immigrants with the interpersonal environment ($p < 0.01$) of their current city, the higher their degree of acceptance of the sentence "I am equal to locals". Under the circumstance that other factors remain unchanged, urban immigrants' degree of acceptance of the sentence "I am equal to locals" will increase by 0.35 levels for every degree of increase in urban immigrants' satisfaction with the interpersonal environment in their current cities.
11. Social exclusion (after several years of living in the current city, what do you think of the attitudes of locals towards the migrant population?). The weaker the sense of social exclusion ($p < 0.05$) felt by urban immigrants, the higher their degree of acceptance of the sentence "I am equal to locals". Under the circumstance that other factors remain unchanged, urban immigrants' degree of acceptance of the sentence "I am equal to locals" will increase by 0.03 levels for every degree of decrease in the sense of social exclusion felt by urban immigrants.

In order to show the significance of each independent variable to each dependent variable more comprehensively, the authors summarised the above seven regression models into Table 8.17, from which we can see that:

(1) Gender has no significant effect on several variables such as the shift of focus of life and social support network, but it has a significant effect on several variables such as the willingness to long-term settlement and the fight for equal rights. The proportion of male urban immigrants living with their families in immigration cities is 1.02 times higher than that of female urban immigrants. Besides, male urban immigrants' willingness to long-term settlement and degree of acceptance of the sentence "I am equal to locals" is both 0.09 levels higher than that of female urban immigrants.

(2) On the whole, age is negatively correlated to the degree of social adaptation and integration. The older the urban immigrants, the lower their degree of social adaptation and integration. For every year of increase in the age of

**Table 8.17** A summary of regression models of factors affecting the adaptation and integration of urban immigrants

| Independent variables | The shift of focus of life | | | Social support networks | The willingness to long-term settlement | | The fight for equal rights |
|---|---|---|---|---|---|---|---|
| | Whether to remit money to the hometown | The proportion of remittance to the hometown in personal income | The proportion of remittance to the hometown in personal expenditure | Whether there is a best friend in the immigration city | Whether to live with families in the immigration city | The willingness to long-term settlement | The degree of acceptance of the sentence "I am equal to locals" |
| Constant | −1.51*** | 0.06 | 0.17*** | 1.62*** | −0.12 | 0.5*** | 0.33** |
| Gender | 0.03 | 0.01 | 0.00 | 0.04 | 0.7*** | 0.09*** | 0.09** |
| Age | 0.04*** | 0.002*** | 0.002*** | 0.00 | 0.00 | −0.01** | −0.01*** |
| Education degree | −0.21*** | −0.02*** | −0.03*** | 0.08*** | −0.36*** | 0.12*** | 0.1*** |
| Monthly disposable income | 0.01 | 0.00 | 0.00 | 0.00 | 0.12*** | 0.01** | 0.01** |
| I hope my children can develop in this city in the future | 0.03 | 0.00 | −0.01* | 0.11*** | 0.14 | 0.46*** | 0.02 |
| Whether to own self-purchased houses | −0.33** | −0.04*** | −0.05*** | 0.15** | 1.26*** | 0.17*** | 0.23*** |
| Length of residence in the current city | 0.04 | 0.00 | 0.00 | 0.12*** | 0.14* | 0.04*** | 0.02 |

(continued)

**Table 8.17** (continued)

| Independent variables | The shift of focus of life | | | Social support networks | The willingness to long-term settlement | | The fight for equal rights |
|---|---|---|---|---|---|---|---|
| | Whether to remit money to the hometown | The proportion of remittance to the hometown in personal income | The proportion of remittance to the hometown in personal expenditure | Whether there is a best friend in the immigration city | Whether to live with families in the immigration city | The willingness to long-term settlement | The degree of acceptance of the sentence "I am equal to locals" |
| Satisfaction with public welfare in the current city | 0.02 | 0.00 | 0.01* | −0.04 | −0.23** | 0.00 | 0.09*** |
| Satisfaction with housing prices in the current city | 0.08 | 0.01** | 0.00 | −0.01 | 0.00 | −0.04** | 0.05** |
| Satisfaction with commodity prices in the current city | −0.04 | 0.00 | 0.00 | −0.03 | 0.02 | 0.05** | 0.04* |
| Satisfaction with life in the current city | −0.01 | 0.01 | 0.00 | 0.05 | 0.17 | 0.11*** | 0.23*** |
| There is much of the milk of human kindness in the city where I am now | −0.09* | −0.01 | −0.01** | 0.06** | −0.09 | 0.14*** | 0.35*** |
| Social exclusion | 0.09** | 0.01** | 0.01*** | −0.01 | −0.02 | 0.01 | 0.03** |

(continued)

**Table 8.17** (continued)

| Independent variables | The shift of focus of life | | | Social support networks | The willingness to long-term settlement | | The fight for equal rights |
|---|---|---|---|---|---|---|---|
| | Whether to remit money to the hometown | The proportion of remittance to the hometown in personal income | The proportion of remittance to the hometown in personal expenditure | Whether there is a best friend in the immigration city | Whether to live with families in the immigration city | The willingness to long-term settlement | The degree of acceptance of the sentence "I am equal to locals" |
| Political exclusion | 0.15* | 0.02*** | 0.01 | 0.11*** | 0.09 | −0.01 | 0.02 |
| Models | Logistic regression | Multiple linear regression | Multiple linear regression | Multiple linear regression | Logistic regression | Multiple linear regression | Multiple linear regression |
| Sample size | 2620 | 2499 | 2618 | 2620 | 1100 | 2573 | 2598 |
| Statistics of models | −2LL = 3309.685 | $R^2 = 0.036$ | $R^2 = 0.053$ | $R^2 = 0.043$ | −2LL = 1022.465 | $R^2 = 0.328$ | $R^2 = 0.236$ |
| | Nagelkerke $R^2 =$ 0.064 | Adjusted $R^2 =$ 0.031 | Adjusted $R^2 =$ 0.048 | Adjusted $R^2 =$ 0.038 | Nagelkerke $R^2 =$ 0.116 | Adjusted $R^2 =$ 0.324 | Adjusted $R^2 =$ 0.232 |
| | | F = 6.647 | F = 10.429 | F = 8.384 | | F = 89.183 | F = 57.057 |

*Notes* Meaning of prompt symbols: * represents $p < 0.1$, ** represents $p < 0.05$, *** represents $p < 0.01$

urban immigrants, their probability of remitting money to their hometown will increase by 0.04 times, the proportion of remittance to their hometown in their personal income will increase by 0.2%, the proportion of remittance to their hometown in their personal expenditure will increase by 0.2%, their willingness to long-term settlement will decrease by 0.01 levels, and their degree of acceptance to the sentence "I am equal to locals" will also decrease by 0.01 levels.

(3) On the whole, the education degree is positively correlated to the degree of social adaptation and integration. The higher the education degree of urban immigrants, the higher their degree of social adaptation and integration. For each level of improvement in urban immigrants' education degree, their probability of remitting money to their hometown will decrease by 0.19 times, the proportion of remittance to their hometown in their personal income will decrease by 2%, the proportion of remittance to their hometown in their personal expenditure will decrease by 3%, the number of their best friends in the immigration city will increase by 0.08, their willingness to long-term settlement will increase by 0.12 levels, and their degree of acceptance to the sentence "I am equal to locals" will also increase by 0.1 levels. However, for each level of improvement in urban immigrants' education degree, their probability of living with their families in the immigration city will decrease by 0.3 times, which is an exception and shows that there are some intermediate variables and indirect factors between the education degree of urban immigrants and whether they live with their families in the immigration city.

(4) Monthly disposable income has no significant effect on several variables such as the shift of focus of life and social support network, but it has a significant effect on variables such as the willingness to long-term settlement and the fight for equal rights. The higher the monthly disposable income of urban immigrants, the stronger their willingness to long-term settlement and awareness of fighting for equal rights. Every time the monthly disposable income of urban immigrants increases by RMB500, their probability of living with their families in the immigration city will increase by 0.13 times, and their willingness to long-term settlement and degree of acceptance of the sentence "I am equal to locals" will both increase by 0.01 levels.

(5) The hope for children to also develop in the immigration city in the future has no significant effect on several variables such as the shift of focus of life and the fight for equal rights, but it has a significant effect on variables such as the social support network and the willingness to long-term settlement. For each level of improvement in urban immigrants' hope for their children to also develop in the immigration city in the future, the number of their best friends in the immigration city will increase by 0.11, and their willingness to long-term settlement will increase by 0.46 levels.

(6) The ownership of self-purchased houses is positively correlated to the degree of social adaptation and integration. Urban immigrants who own self-purchased houses are less willing to remit money to their hometown than those who do not own self-purchased houses. Thus, the former's proportion of remittances

to their hometown is naturally less than that of the latter. However, in terms of the number of best friends in the immigration city, the willingness to live with families, the willingness to long-term settlement and the awareness of fighting for equal rights, urban immigrants with self-purchased houses are significantly higher than those without self-purchased houses.

(7) The length of residence in the immigration city has no significant effect on several variables such as the shift of focus of life and the fight for equal rights, but it is positively correlated to the social support network and the willingness to long-term settlement. The longer the time that urban immigrants live in the current city, the greater the number of best friends they have in the immigration city, and the higher their willingness to settle in the immigration city. For every additional year urban immigrants live in the current city, the number of best friends they have in the immigration city will increase by 0.12, their probability of living with their families in the immigration city will increase by 0.15 times, and their willingness to long-term settlement will increase by 0.04 levels.

(8) Satisfaction with public welfare has no significant effect on several variables such as the shift of focus of life and social support network, but it has a significant effect on the fight for equal rights. For every degree of increase in urban immigrants' satisfaction with the public welfare, their degree of acceptance of the sentence "I am equal to locals" will increase by 0.09 levels.

(9) Satisfaction with housing prices in the current city has no significant effect on the social support network, but it has a significant effect on several variables such as the shift of focus of life, the willingness to long-term settlement and the fight for equal rights. For every degree of increase in urban immigrants' satisfaction with housing prices in their current city, their degree of acceptance of the sentence "I am equal to locals" will increase by 0.05 levels. In contrast, for every degree of increase in urban immigrants' satisfaction with housing prices in their current city, the proportion of remittance to their hometown in their personal income will increase by 1%, and their willingness to long-term settlement will decrease by 0.04 levels, which shows that there are some intermediate variables and indirect factors that need to be further studied.

(10) Satisfaction with commodity prices in the current city has no significant effect on several variables such as the shift of focus of life and social support network, but it has a significant effect on several variables such as the willingness to long-term settlement and the fight for equal rights. For every degree of increase in urban immigrants' satisfaction with commodity prices in their current city, their willingness to long-term settlement will increase by 0.05 levels, and their degree of acceptance of the sentence "I am equal to locals" will increase by 0.04 levels.

(11) Satisfaction with life in the current city has no significant effect on several variables such as the shift of focus of life and social support network, but it has a significant effect on several variables such as the willingness to long-term settlement and the fight for equal rights. For every degree of increase in urban

immigrants' satisfaction with life in their current city, their willingness to long-term settlement will increase by 0.11 levels, and their degree of acceptance of the sentence "I am equal to locals" will increase by 0.23 levels.

(12) On the whole, satisfaction with the interpersonal environment (there is much of the milk of human kindness in current cities) is positively correlated to the degree of social adaptation and integration. The higher the satisfaction of urban immigrants with the interpersonal environment of their current city, the higher their degree of social adaptation and integration. For every degree of increase in urban immigrants' satisfaction with the interpersonal environment, their probability of remitting money to their hometown will decrease by 0.09 times, the proportion of remittance to their hometown in their personal expenditure will decrease by 1%, the number of best friends they have in the immigration city will increase by 0.06, their willingness to long-term settlement will increase by 0.14 levels, and their degree of acceptance of the sentence "I am equal to locals" will increase by 0.35 levels.

(13) Social exclusion (after several years of living in the current city, what do you think of the attitudes of locals towards the migrant population?) has no significant effect on several variables such as the social support network and the willingness to long-term settlement but it has a significant effect on several variables such as the shift of focus of life and the fight for equal rights. For every degree of decrease in the sense of social exclusion felt by urban immigrants, their probability of remitting money to their hometown will increase by 0.09 times, the proportion of remittance to their hometown in their personal income and their personal expenditure will both increase by 1%. However, for every degree of decrease in the sense of social exclusion felt by urban immigrants, their degree of acceptance of the sentence "I am equal to locals" will increase by 0.35 levels. In view of the possible reasons, the authors analysed that the stronger the sense of social exclusion felt by urban immigrants, the stronger their need and desire for equality.

(14) Political exclusion (what do you think of the attitude of the community officials in your community towards you?) has no significant effect on several variables such as the willingness to long-term settlement and the fight for equal rights but it has a significant effect on several variables such as the shift of focus of life and the social support network. For every degree of decrease in the sense of political exclusion felt by urban immigrants, their probability of remitting money to their hometown will increase by 0.17 times, and the proportion of remittance to their hometown in their personal income will increase by 2%. However, for every degree of decrease in the sense of political exclusion felt by urban immigrants, the number of best friends they have in the immigration city will increase by 0.11. In view of the possible reasons, the authors analysed that the stronger the sense of political exclusion felt by urban immigrants, the more they need social support.

# Chapter 9
# The Social Mobility of Urban New Immigrants

In this Chapter, we will exam the mobility features for urban new immigrants, especially the labour immigrants, who showing a phenomenon of "return home" and "transference". We will elaborate the case of mobility for labour immigrant in a stationary factory, showing the sources, reasons, and the living strategy for immigrants.

## 9.1 Research on "Return" for Labour Immigrants

The researches on the "return home" for labour immigrants has always being the hot topic in academia. The author's paper The Eternal Pendulum: Study on the Mobility of Labour Immigrants,[1] has made a comprehensive investigation on the "return home" phenomenon. In the late 1990s, some studies on this phenomenon was started, but it is not until recent years that widespread attention has been aroused, and a series of research on labour immigrants' return or migrate appeared.

Academics made abundant discussion on the forming of "return home" mechanism. Some scholars apply the theory of "push and pull" on this subject, Ben Chenglong use "push and pull" theory to examine the return of labour immigrants in Nantong region, he considers the push factors from urban and the pull factors from rural is mechanism for labour immigrants returning.[2] Cheng Aihua based on his field study in Enshi Tujia and Miao Autonomous Prefectre of Hubei province, discussed the return of labour immigrants on the theory setting of "push and pull", she argued

[1] Zhou Daming, The Eternal Pendulum, Study on the Mobility of Labour Immigrants, Ke Lanjun, Li Hanlin, *Urban Villagers, The Floating Population in China's Big Cities*, Beijing: Central Compilation & Translation Press, 2001.

[2] Ben Chenglong, Going to the City or Going Home? Analysis of the Phenomenon of Migrant Workers and Farmers Returning to Cities under the New Situation, *Rural Economy*, Vol. 11, 2007.



D. Zhou, *Urban Migrants in China*,
https://doi.org/10.1007/978-981-99-3114-9_9

that the income gap between urban and rural is the basic reason for pulling the labour force from rural to urban.[3] Some other scholar such as Gao Qiang, Jia Haiming and Zhang Cheng, discuss the return of labour immigrants base on policy, subjective decision, push and pull and etc., they also examined the positives and negatives of the returning for immigrants.[4]

Some scholars consider the return of labour immigrants as a positive factor for the development of rural economy, for these people are the elites in the rural area. Yi Qingchuan, Yi Yuanhua and Tan Xiangjin consider the labour immigrants are elites from countryside, they have the greatest potential among the youth in rural area, they will play an important role in the rural development with their great energy. The authority shall seize this opportunity of returning young immigrants, make the most extend of use from these people, provide good environment for their development in the rural area, finally push forward the coordinated development of urban and rural areas and the construction of a harmonious society.[5] However, some scholars argued that the returning of labour immigrants is a regressive phenomenon, Liu Zheng argues, despite the return will bring some positive effects on the construction of rural economy, it is an regression from the perspective of urbanization, therefore, the treatment for labour immigrants in urban area shall be improved and their salary must raise to a reasonable degree, in order to protect the rights for labour immigrants, as well as to secure the urbanization progress.[6]

In author's view, among these fore mentioned researches, rarely see subjective studies that takes the labour immigrants as an initiative subject. Most researches see the labour immigrants as an object of the push and pull rural urban interactive mechanism, or the social system, it is an objective view on the return of labour immigrants, that did not see the initiative of subjects, but only see the return as an objective behavior under certain circumstances. Moreover, these researches did not differentiate the subdivisions of immigrants. Therefore, this chapter will exam the initiatives of return immigrants, their reason for return, and their true desire beyond the decision.

This book takes a stationary factory as a case, the stationary factory is located in Pearl River Delta. As a labour intensive factory, in pursuit of productive, cheap labour. Its original equipment, personnel and production technology have been kept relatively stable for more than ten years, and the senior employees of the factory

[3] Cheng Aihua, Research on the Dynamic Mechanism of Migrant Workers' Return in China, *Journal of Anhui Agricultural Sciences*, Vol. 2, 2007.

[4] Gao Qiang, Jia Haiming, Analysis on the Causes and Effects of Migrant Workers' Backflow, *Management of Agricultural Science and Technology*, Vol. 4, 2007. Zhang Cheng, Analysis of the Phenomenon of Migrant Workers' Return in China, *Journal of Nanjing Forestry University (Humanities and Social Sciences Edition)*, Vol. 3, 2007.

[5] Yi Qingchuan, Yi Yuanhua, Tan Yangjin, A Study on the Backflow of Migrant Workers' Elite Group to Construct New Countryside, *Rural Economy*, Vol. 9, 2007.

[6] Liu Zheng, The Paradox of Unlimited Supply of Labour Force: An Analysis on the Causes and Effects of Migrant Workers' Backflow, *Journal of Tsinghua University (Philosophy and Social Science)*, Vol. 3, 2006.

show the unique characteristics and flow track of labour immigrants who went out in the late 1980s, and have a continuous span in research time.

The key problem the stationery factory facing is "too frequent turnover of staff". The factory only has slightly more than 1000 employees, but more than 4000 people have left their jobs throughout the year. Flows fall into two broad categories: backflow and job-hopping. Employees of the personnel department of the stationery factory spend 2–3 h a day dealing with employees' dismission, and the ratio between the number of recruits and the number of dismission is almost 1:1. The loss of personnel makes the production line change every day, which is a fatal injury to labor-intensive enterprises, because this industry does not need high and new technologies, but in particular needs skilled workers. Moreover, the turnover of personnel makes the people in the production line adapt to the machine every day, seriously affects the production efficiency and wastes the training cost. As a result, the factory's bosses have urged lower-level managers to be kind to their employees and do everything possible to retain them, but the frequent personal turnover has not eased.

Returning is the most prominent behavioral strategy and personal choice in the practice of the older generation of labour immigrants. It is not simply the influence of age or life cycle, but also reflects the initiative differentiation and change occurring within the labour immigrants. We surveyed more than 20 individuals who migrated in the late 1980s and early 1990s, age between 35 and 45 years old, these labour immigrants have stay in cities for more than a decade, when asking their prospect for the future, more than 95% of the respondents answered "return home", but it's merely a will, only half of the respondents already taking actions, that is to say, half of the respondents was getting ready to return, and they will return in nearly one or two years, the other half of labour immigrants are still in wait-and-see state, which means their destiny is almost certain, but they still want to insist immigrating for some more years. Why do most labour immigrants make the personal choice of returning after working in cities for many years? Is this choice the end or the beginning of their working lives? What lies on their future path? What do they do when they return home?

### *9.1.1 Willingness and Reasons for Returning*

Since the emergence of massive domestic migration in early 1980s, China has changed greatly along time, with the enormous progress in economy, the labour immigrants themselves, who changed the history, also changed by the history. Those pioneers of labour immigrants who migrated in 1980s–1990s already in their middle-age and started family, some of them have improved skills and their self-estimate is raised; but most of them have lower their expectation for future, the choice of settle down in cities or return home is struggling, which causing psychological conflict and confusion. On the greater picture, the generation of labour immigrants is undergoing a renewal, new immigrants at the age of 18 or 19 is entering cities, they have better education compare to their predecessors, their identity to the rural community is weak, their capability to integrate in city lifestyle is better, moreover, their desire for future success is stronger. Aging for the senior labour immigrants, makes them inevitable to think about the path of future; the coming of new generation of labour immigrants, on the one hand, meaning the seniors have to face with the awkwardness of professional development, on the other hand, meaning their burden of life is easing in some way. Therefore, facing the fore mentioned circumstances, the retreating path for senior labour immigrants must be brought to consideration.

The questions are, who will return? Where shall they go? Why they return?

The author surveyed more than 20 respondents who immigrated in late 1980s and early 1990s, their age ranges from 35 to 45, they were in cities for more than ten years, when asked about the future plan, more than 95% of the respondents answered "return home", but it's merely a will, only half of the respondents already taking actions, that is to say, half of the respondents was getting ready to return, and they will return in nearly one or two years, the other half of labour immigrants are still in wait-and-see state, which means their destiny is almost certain, but they still want to insist in cities for some more years.

In fact, labour immigrants' return home decision, is closely related to whether they are willing to stay, and the possibility for them to stay in cities.

In the survey, the author found that high percentage of senior labour immigrants chose to return home, is caused by their awkward orientation for personal career in cities.

As for career development, those senior labour immigrants who worked in cities for more ten years share a common trait, which is insufficient education, before they migrated, 80% or more of them were ploughing farmland at home, so they generally do not have any skill for careers in cities. Therefore, at the beginning of employment, they worked as apprentices and learned skills and accumulated experience in practice. After more than a decade of working, they are basically set in their skills, or they have mastered the skills in a certain field, or they have risen to the position of foreman or manager. Technically, they are reluctant to switch to learning new skills, but they are also reluctant to do less technically demanding work. In terms of their status, they do not want to stay in the command position forever because of their own skills, experience and effort, but their status as labour immigrants make it difficult for them to have any room to rise. Therefore, many labour immigrants say that their future working in the city is very embarrassing, and they are basically satisfied with the current situation. A senior labour immigrant said: "I have been out to work for more than ten years. I have no education qualifications. It is not easy to get to where I am today. I think my working life is about to end. If I work hard, that's it. I can't want more, and it won't happen…".

From the perspective of life prospects, the labour immigrant is always a special group in the city. Although they live in the city, their household registration is in the countryside. Just as in the first case in this chapter, a considerable number of labour immigrants live in factories rather than live in cities. Their days seem to have nothing to do with the city they live in. They go to work, leave work, and move between dormitories and workplaces. Usually, there is no special place to go. They almost stay in their own rooms every day. Except for work, eating and sleeping, there are no special activities, and no special activities are needed. That is to say, they have been alienated from the city for more than ten years. They are the builders of the city, but they are excluded from the city service system. They are the producers in the city rather than the consumers. To a large extent in the city means to maintain this state throughout life, which is unacceptable for them. A senior labour immigrant said: "If it's not for making money, I don't think I will stay here. I don't belong here. It is impossible and unnecessary to settle down here. I have been here for so long and I rarely go out. This is not my home…".

From the perspective of life cycle, why "return" cannot form a boom like when farmers go out to work in large numbers. This is largely related to the life cycle and the situation in the critical period of life. In other words, although most people are willing to return home, but the time to return home is not the same. In fact, many people return home when they get married or give birth. The author has learned that many quitter (not job-hoppers) left their jobs because they returned to their hometowns to get married, give birth or have children. This happens to a higher proportion of female workers. For many female workers, marriage is a turning point in their working life. After marriage, they may rely on their husbands to make money, or they may take a break before going out to work. Childbirth is almost a major breaking point in their working life. During the childbirth period, wages are paid or medical expenses are reimbursed, so the phenomenon of resignation before marriage and before childbirth is more common. The 45-year-old is the "high incidence period" for senior labour immigrants to return to their hometowns. It is understood that this is mainly because children of this age have almost finished reading and come out to work. Therefore, the pressure on elder labour immigrants is reduced and most of them choose to return to their hometowns. A female worker said: "I used to go home once when I gave birth to a child. Then my child went to school, and my husband was not at home. I felt bored at home, so I went out to work again. Now the children are out to do things. I discussed with my husband. Maybe they have returned home in one or two years, anyway, there is a house in their hometown. How the children develop is their business…".

The above explanation is due to the dilemma of career development, the embarrassment of life prospects, and the helplessness in several critical periods of life. It also explains the wishes of labour immigrants who cannot stay in the city. However, the author found that there are also some labour immigrants have the conditions to stay in the city, they have a certain foundation in the city, their children go to school in the city, and even their elders also receive the city, so why do they choose to return to their hometown in a few years?

Although some scholars oppose the hometown and homeland complex as the reason for labour immigrants to return to their hometowns, and believe that people with high education or development prospects will not be willing to return to their hometowns, but this only shows that the homeland and homeland complex is not a cause of returning home. The main reason, in fact, the author found that this complex still plays a role that cannot be ignored today, but this role is not just as simple as "emotion". Many labour immigrants love their hometown more because they feel that they have no "roots" in the city and are always floating. This is mainly related to household registration and also related to the social circle.

Most of the senior labour immigrants' contacts come from their fellow villagers. They are a large number of friends who have walked out of the countryside to the city together. They have a natural intimacy of sympathy and sympathy. Many labour immigrants have expressed their trust in their fellow villagers. At the same time, while they stated that natives exclude foreigners, they also subconsciously exclude natives. This identification of "he" and "I" keeps their lives in the city always on guard and vigilance. Returning home is for them to choose a more relaxed and idle life for themselves. This is the reason why they love the hometown. The social circle of young labour immigrants has expanded a lot. In the survey, the author found many small groups of social contacts with both non-locals and local employees, which will be discussed in the following content.

In addition, the love home complex also contains the connotation of family responsibility. Many labour immigrants attach importance to the role of "house" just as they value "home". They involuntarily define the criterion of "stabilization" as whether they have "own house" or not. Of course, there are very few labour immigrants who can afford houses in cities, but it is more likely to build houses in their hometowns or to buy houses in towns near their hometowns. With this concept of "where the house is, home is", they are more willing to choose to return to their hometown. Can't it work without a house? Some labour immigrants said, "It feels like they have no roots under their feet". "With a house, we have fulfilled our parents' responsibilities. In the future, children will have their own homes. Otherwise, if the children will float in the future, we will also follow them…".

Just like the "push–pull" theory of labour immigrants' motivation to go out to work, labour immigrants are also considering push and pull when facing the future. The push is the embarrassment of career development in the city, the embarrassment of life prospects, and the different periods of life. The awkwardness of the pull is the more familiar and adaptable interpersonal environment in the hometown, the more relaxed and idle life, and the house as the "root". The interaction of the two forces made most of the senior labour immigrants make the choice to return to their hometown.

### *9.1.2 Possibility and Realization Ways of Returning Labour Immigrants*

What has been mentioned above is that most of the senior labour immigrants are willing to go back to their hometown and the reasons for that. Is there any possibility for them to go back? What are they going to do when they get back?

To the former question, most labour immigrants say "it is very possible, it is only a matter of time". The reason why it is very possible is objective conditions. Most senior labour immigrants think that they are originally outsiders in cities, who have no solicitude in cities, can choose to go back at any time, and their household registration is still at home, so they are most likely to be accepted by their hometown. In addition, many labour immigrants said they still have farmland at home, some are ploughing by their families, some are contracting by others, those who owned farmland are more likely to return home.

However, labour immigrants' decision making for return is related to many factors, one important factor is their children. It is easier to understand that if children are brought to the city to study, then labour immigrants may not be able or willing to return to their hometowns in the short term. If their children stay in their hometowns, their parents are responsible for making money for their education and will not easily return home. As many labour immigrants said that they might return home after their children came out for work, which means that when the burden of "supporting their families" for senior labour immigrants is lessened, they will choose to return to their hometown. This is different from urban people, but it fits the motivation of going out to work. They don't plan to retire at an old age, but choose to return to their hometown when they don't have to work hard to make money to support their families. This is one of the characteristics for the senior labour immigrant.

What are the plans of the labour immigrants returning home? About 30% of labour immigrants said that they have no mature ideas and would like to return and see. Only 5% choose to go to township enterprises to continue working, while about 45% choose to open a shop (business) or set up a small factory by themselves. 5% do nothing, and another 15% choose farming.

From the perspective of future plans, very few people choose to keep working as employee after returning to their hometowns, but most choose to do "business" or "farm" for themselves. When asked why, the respondent said that this kind of life is "freer", you can take it for yourself, do what you want, and rest if you don't want to do it. There is not much pressure and it is not that hard. "Business" seems to exist as a vague concept, and "opening a small shop" is a popular answer. A labour immigrant said: "My best hope is to open a small shop on the first floor of my own house, and live on the second floor above. I can come down to see the small shop during the day without renting someone else's place or looking at other people's face. So, I am building a house at home now, and it is will be finished in one or two years, I will return home and don't have to work so hard outside."

As for the place for return, not all labour immigrants choose to return to their countryside home, 50% of them choose to return to towns nearby, or cities in that region. Only those who own farmlands at home, more likely to return their home for farming. This shows that the return of labour immigrants is not simply a "regressive" return, but to choose a place they think is better for themselves, as a foothold for the future. A labour immigrant said: "I've been out for so many years myself, and I'm not used to it when I go back and continue farming, so I'd better choose the town of our county, which is more… how to say… it's more prosperous than my home. Well, in the future, my children will no longer come from my poor county. I have not worked in vain for so many years, and I am still a little better." Some labour immigrants also said that they chose towns because it is easier to do small businesses in towns.

Therefore, from the perspective of career plans and return locations, the labour immigrants who intend to return have chosen a more "free" and "easy" career for themselves on the one hand, rather than returning to farming, on the other hand they have chosen a compromise location for themselves It is more "prosperous" than the countryside, and has a similar interpersonal environment and regional culture to the hometown.

### *9.1.3 Return as a Subjective Choice for Labour Immigrants*

Academia often discusses the issue of "return" as a "way out" or even "retreat" for labour immigrants. Even at the beginning of the investigation, the author believes that "job-hopping" is adopted for labour immigrants in pursuit of a better new life strategy, while "returning" is the end of the old generation of labour immigrants' working life. However, in fact, we have found that whether it is "return" or "job-hopping" is their subjective choice for a new life opportunity.

Through the previous discussion, the author believes that, on the one hand, labour immigrants are more likely to make the choice to return to their hometown because of the push of the city and the pull of their hometown. On the other hand, their return to hometown is not a simple migration of population, but a renewal of their personal life. Sorting out and positioning.

From the perspective of life cycle, old labour immigrants are now about to enter the middle-aged period. They are entering a period where the pressure of survival is reduced, their life capital is more, and they are more mature about future choices. They are not only facing the need for a job career. The question of continuation is still the question of how to live in the future. This is different from urban people's retirement at age. They choose not to retire for themselves, but to truly have the ability to start a life they "long for".

The survey found that if labour immigrants initially chose to go out to work out of survival pressure, and choose a lifestyle that is easier to obtain economic benefits, then this generation of labour immigrants choose to return, which means that they have chosen themselves more "yearning" for themselves Life, this is more "yearning" includes several key elements: freedom, ease, self-esteem and so on.

"Freedom" is a term often used by labour immigrants. They think that working means being managed and oppressed by others. They have to check in to and from get off work every day, and their wages are time-keeping. Therefore, they have no freedom and think that they are like a machine. If you return to your hometown to do business and open a shop, you can choose whether to start work according to your own wishes. Although the term "easy" is often mentioned, the concept is very vague. Many labour immigrants just think that it is easier to be a boss, without further consideration. This shows that labour immigrants are also subject to some perceptual factors in their future choices. "Self-esteem" means that labour immigrants believe that they can never be accepted and respected in the city. Their marginal position in the city makes them eager to be recognized and respected by people around them.

Therefore, when they return to their hometowns, they not only someone who has some savings or someone who has met the world in the city. This perceptual cognition is very obvious labour immigrants. One labour immigrant said: "I can still live on the same. Why do you want to be a worker here? I am still envied by people when I return to my hometown. That's great. I don't want to embarrass myself. I am just forced to live for a while."

There are also some labour immigrants who say that they have to go home for "family responsibilities". On the one hand, they have to return to their parents. On the other hand, it is not a permanent solution for couples to stay outside for too long. On the other hand, they "leave a root for their children in their hometown". To leave some land and houses for their children, parents can be regarded as fulfilling some responsibilities. If they let their children drift outside with their parents, parents will feel uneasy.

It can be seen that simply using the "push–pull" theory does not seem to be able to fully explain the phenomenon of backflow, because the push and pull of cities exist at the same time, and the same is true for rural areas. In fact, pushing and pulling force acting on labour immigrants is a relatively complicated process, and labour immigrants must be calculated and weighed as the main body of choice. However, this trade-off is not entirely based on rationality, but also contains many emotional ingredients. But whether it is rationally or emotionally, in daily life or economically, choosing to return to their hometown is a way for labour immigrants to choose for themselves to spend their old age.

Therefore, "returning" as a new choice and positioning of enterprise labour immigrants in their future life is actually a new beginning of labour immigrants' lives. This choice highlights the nature of going out to work as a "stoppable measure". All in all, "return" is a subjective choice made by labour immigrants in accordance with their own survival logic. This choice is more in line with the needs of labour immigrants themselves. At the same time, this conclusion also shows that the possibility of labour immigrants in the older generation of enterprises staying in the city is not so, they are even less likely to become "vagrants" who endanger the future of the city's development. We can neither imprison them on the land in their hometown, nor simply follow our logic to discuss how they will integrate into city life in the future. In fact, only through their own efforts to transform from a labour type to an owner type, intellectual and technical type, can labour immigrants stay in the city and become a part of the urban new immigrants.

## 9.2 Research on "Job-Hopping" for Labour Immigrants

"Job-hopping" is a word that the author picked up through the interviews with labour immigrants, the other expressions such as "switch factories" or "job changing" is similar. In the achieve searching, the author found that the Chinese academia often use the expression of "transfer", the meaning is the same, but the author chooses to use the word "job-hopping", which is often use by the subject of study.

It can be seen that although in-depth research related to the academic circle is still rare, the society has begun to pay attention to the frequent job-hopping of labour immigrants. Frequent job-hopper not only have a great impact on the production of enterprises, but directly point to the improvement of treatment by enterprises. The requirements for improving the environment and assuming responsibility also reflect the group changes that are taking place in the labour immigrant. Therefore, it is of great significance to pay attention to the more and more frequent "job-hopping tide".

In the stationery factory, frequent job changes have also attracted our attention. We focused on the behaviour and attitudes of these young labour immigrants who frequently switch jobs or tend to change jobs. The research method is a combination of questionnaires and in-depth interviews to understand the frequency of labour immigrants changing jobs, the reasons for changing jobs, the general way of changing jobs, and the changes in life after changing jobs. The research theory is to explore how the "job-hopping" action produced by the main body of labour immigrants is transformed into a practical strategy from the perspective of the theory of action and practice.

### *9.2.1 "Job-Hopping": A Subjective Action for Labour Immigrants*

During the participant observation on the stationary factory, the author was serving in the human resources (HR) department, naturally comes into contact with the company's personnel changes. The author deeply feels that it is really easy to deal with the HR work of a stationery factory on the one hand, and it is really difficult on the other hand. Every morning, the staff in charge of personnel are faced with the resignation and entry work. They are dealing with similar matters almost all morning. There is almost no one resigning, and there are sporadic new employees joining almost every day. In the afternoon, the staff in charge of training will start the induction training for new employees. Since there is no fixed recruitment time, the induction training has become irregular. There will be one training for a few new employees, at least 3 to 4 days a week. There are trainings for new employees, but only a few people each time.

This happens almost every day, 365 days a year. After a year of accumulation, the number is staggering. We can see the number of employees leaving in 2007 provided by the HR department (see Table 9.1).

It can be seen that in a factory with only more than 1000 people, more than 4000 people left their jobs throughout the year. There are currently 767 people in the FA ring clip department, but 2199 people left their jobs throughout the year.

The astonishing resignation numbers raised the attention of the author to concern the job-hopping issue for labour immigrants. Why they leave? Why some of them stay? Why some of the senior labour immigrants return? How the frequent job-hopping affects young workers?

**Table 9.1** Total number of resignations in 2007 (Unit: person)

| Heshui factory resignations | | Hecheng factory resignations | |
|---|---|---|---|
| Department | Quantity | Department | Quantity |
| Fa ring clip | 2199 | T-card | 434 |
| La accessories | 94 | Manufactory | 59 |
| La assembly | 536 | Drawing | 26 |
| Engineering | 17 | Materials | 9 |
| Whiteboard | 7 | Quality control | 47 |
| Cold rolling | 47 | Injection moulding | 213 |
| Materials | 46 | Cold rolling | 323 |
| Quality control | 59 | Accounting | 3 |
| Powder application | 3 | Information | 1 |
| Training | 15 | Production | 2 |
| Accounting | 7 | General office | 1 |
| Human resource | 34 | Management | 1 |
| Production | 1 | Parts warehouse | 2 |
| Information | 2 | Human resource | 17 |
| Management | 2 | | |
| Total | 3069 | Total | 1138 |

*Note* Table 9.1 is provided by the HR Department

The author divides the worker into three surveyed groups to conducts investigations on the job-hopping issues: (1) in the HR department, simple interviews and questionnaire surveys are conducted on the employees who go through the resignation procedures to understand the reasons for their job-hopping. Some of them are in the investigation established a good friendship with the author. The author understands the changes in his life after the job-hopping through follow-up interviews; (2) in the training department, interviews with newly recruited employees to understand the general situation of their previous job-hopping, the reasons and reasons for coming to our factory expectations; (3) part of the questionnaire is also issued to general workers on the production line to find out whether they have a tendency or experience to change jobs.

The questionnaire consists of four main parts: (1) basic information, including age, number of years of working outside the home, and the number of job-hopping, etc.; (2) reasons for changing jobs, the purpose and needs of changing jobs, and what is most important when changing jobs; (3) problems that may arise in the process of changing jobs, such as whether they will seek the opinions of others, whose opinions will be solicited, whether they will encounter obstacles in the system and rules, etc.; (4) attitudes towards changing jobs and obtaining information about changing jobs, and pass information about job-hopping to others, etc.

There were 150 copies of the questionnaire distributed, 128 copies were returned, and 112 valid questionnaires were obtained. The contents of the last three parts of the questionnaire will be discussed later in conjunction with interviews. Here, we can first look at the basic situation of the job-hopper group.

First, in terms of age: 6% of the respondents are over 30 years old, 28% are 15–20 years old, 64% are 21–30 years old, and 2% are visitors and did not fill in this column.

Second, the number of years for labour immigrants: according to the results of the questionnaire, "old labour immigrants" over the age of 30 all went out to work in the late 1980s and early 1990s, and their working time were all over 5 years. During this period, some labour immigrants had a brief return. Young labour immigrants include labour immigrants between 15 and 30 years old, and the average length of their working time is 1–4 years. Judging from the total number, more than 90% of the respondents have gone out to work for more than one year.

Third, the frequency of changing jobs: 29% of labour immigrants over 30 choose "2 to 3 times", 71% choose "1 time"; labour immigrants aged 21 to 30 choose "4 times or more" 27%, 62% chose "2–3 times", 5% chose "1 time"; 2% chose "4 times or more" for labour immigrants aged 15–20, and "2–3 times" 36%, 37% choose "1 time".

Fourth, tracking statistics for resigned workers:

The author followed up and interviewed the resignation in some months. Take October 2007 as an example, a total of 133 people resigned from the factory, of which:

- 52% are 15–20 years old, 90.5% are 15–30 years old, and about 9.4% are 31–40 years old;
- Among the 133 people, 77% have junior high school education, 20% have high school education, and about 3% are undergraduates;
- The time spent working outside the home accounts for 28% in 1–2 years, 66.4% in 2–5 years, and 5% in 5–10 years;
- After resigning, 60% have a clear job destination, 15% return home temporarily, and another 25% still have to consider where to work;
- The time between resignation and the last resignation (the first resignation is calculated as the last time they left the country to work) is within 1 year for 34%, 1–2 years for 53%, and 2–5 years for 11%.

It can be seen from the survey that young labour immigrants account for the overwhelming majority of the job-hoppers, and most of these job-hopping workers have a low educational background; a considerable number of young labour immigrants are not changing jobs for the first time, but have more than twice. From the perspective of 25% of the unclear where to go, there is still a considerable number of labour immigrants who quit their jobs without certain future plans (rather than changing jobs).

The reason why labour immigrants change jobs is not just as simple as dissatisfaction with their wages. Although this is one of the most important reasons, factors such as fairness, freedom, and openness have also become important considerations for young labour immigrants.

According to previous statistics, 93% of the interviewees have changed jobs. In terms of "reasons for changing jobs", 42% chose "wage issues", 24% chose "working environment problems", and "no fairness, management restrictions, etc." accounted for 16%, "interpersonal relationship problems" accounted for 8%, "only because there are better choices" accounted for 3%, "other reasons" accounted for 7%, and another 4% default.

The author summarized the reasons for changing jobs mentioned in all interviews and questionnaires, and summarized the reasons for labour immigrants changing jobs into the following four aspects:

First, the job-hopping due to the lack of basic needs: this basic need is usually expressed by labour immigrants as "low wages", "delay payment", "working environment is too noisy, and the environment is not good". In the expressions of labour immigrants, this type of reason only accounts for about a quarter, but in the expressions of old labour immigrants, this element accounts for the vast majority. In any case, the dissatisfaction of wages and working environment is one of the reasons that trigger the job-hopping of labour immigrants. If the wages and social welfare provided by the factory to workers cannot meet the existing basic needs or when other companies provide more generous, workers will leave resolutely when they are paid.

Some people are willing to pursue a stable job and stable income, and are not willing to change jobs easily, but labour immigrants are not strong in this desire. A young labour immigrant said: "We just came out to break out and float out. We don't expect to stay in one place for a lifetime. I do it when I am happy or leave when I am unhappy. I want to go by myself, so there is nothing to be embarrassed about."

Moreover, labour immigrants also need good interpersonal relationships. Since people have the need for friendship and belonging, they hope to establish friendly relationships and feelings with others. Therefore, if labour immigrants cannot establish good interpersonal relationships with colleagues or superiors in the professional group, there are many contradictions, and under this pressure, they will leave whenever they have a chance. During the investigation in the factory, resignation often occurred due to the conflicts between new and old labour immigrants, the disharmonious relationship, and the different views on an issue. Young labour immigrants believed that there is no reason for "tolerance" when dealing with conflicts, the only way is to leave.

Second, the job-hopping of workers due to lack of fairness: most of the labour immigrants are currently engaged in manual labour, which is a "dirty, messy, and poor" working environment, but they still have their own expectations of remuneration in their hearts.

Since the stationery factory is a hardware industry and a labour-intensive industry, and the labor-intensive factories in the Pearl River Delta have been on a decline in recent years, the profit rate has been reduced, and the wages of labour immigrants in this field have only fallen but not increased. An important reason for employees to switch jobs.

The author found that the job needs of young labour immigrants are not as simple as making money to support their families. They have to compare with their partners who also go out of their hometown to work, wherever the salary is high and the environment is good, they will find ways to go there. In addition, the reason why some young labour immigrants changed jobs was due to the difference of wages in the factory. A clerk said before changing jobs: "I can stay here, but do you know, that girl got RMB2,800 a month. I don't understand why she is higher than me. Although she has done more years than me, I have a high degree than her, and now I have a higher position than her. I think this factory is not fair, salaries are given without a fair standard, so I don't want to stay here anymore."

Third, job-hopping caused by lack of happiness: the professional happiness of many labour immigrants not only comes from wages, but also from their desire for freedom of self-control. Many labour immigrants think that it is a better career or way out, which is to "be your own boss", and the reason is "freedom". The reason for their frequent job-hopping is also the desire for a "free" life.

What does "freedom" mean? To a large extent, labour immigrants are comparing a modern industrial world with their familiar hometown lifestyle. Farmers are not subject to "working hours", "work discipline", and "operating procedures" in agricultural life and production. "Check-in" and other new things brought by modern industrial civilization are restricted. Therefore, all kinds of restrictions after entering the factory have severely reduced the happiness of labour immigrants. Whether labour immigrants are "hopping jobs" or "returning", there are A deep desire for "freedom".

The development of modern society requires workers to have a modern industrial spirit, and the young and new generation of labour immigrants are also lacking in this regard. On the one hand, they yearn for urban modern civilization, on the other hand, they cannot adapt to the constraints of modern industry. The resulting psychological and behavioral contradictions have prompted them to change jobs frequently. This irrational job-hopping behavior not only did not help this group to achieve "self-realization", but hindered the process of self-realization, making it even more confused. Living conditions even lead to serious social anomie such as youth crime.[7]

Fourth, job-hopping caused by mass incidents: in factories, "collective evaporation" incidents occurred several times a year, and job-hopping caused by quarrels, conflicts, or group disagreements are common. Group job-hopping of workers has even become an action for labour immigrants to resist. In order to avoid too many mass incidents and too large scale, the director of the factory does not recruit people from the same province when recruiting, and tries not to recruit people who are aggressive in the place where the trouble occurred.

Then, how do labour immigrants make their job-hopping decision? From generating ideas to putting them into practice, what is process behind? After changing jobs, did the labour immigrants achieve their desired goals? What is the impact of job-hopping on labour immigrants?

First, the attitude toward the "job-hopping" may affect the subsequent behavior, not only the acquisition of information about job-hopping, but also the final decision

---

[7] Sun Jiuxia, The Ethnic Group Components of Foreign Enterprises in the Pearl River Delta, *Journal of Guangxi Minzu University (Philosophy and Social Science Edition)*, No. 4 (2001), pp. 39–46.

to it. According to the results of the questionnaire survey: none of labour immigrants over the age of 30 chose "job-hopping shall be advocated" and "just a personal choice". 71% chose "understandable but it shall not be advocated", 29% chose "job hopping is an incorrect behavior and should be restricted or prohibited". However, labour immigrants aged 21–30 choose the first two items accounted for 58%, and only 36% choose the latter two items. Those aged 15–20 choose the first two items for 64%, and those who choose the last two items account for 26%.

In the interviews, most young labour immigrants think that job hopping is "quite normal", and even regard those who job-hopping frequently as "courageous and thoughtful". An 18-year-old young female labour immigrant said: "The purpose of changing jobs is to find a better job. If I stay in this factory all my life, or stay for decades, I think it's weird. A normal person wouldn't do that. If I stay here, it's only because I haven't found a better place..." Young labour immigrants think that they are always ready to go to a better place. Few people think they should be in the same factory and doing the same job for a long time. When asked if it was because of this factory, a labour immigrant replied: "I think there are still few places that are really satisfied. Maybe even if I am in another factory, I will leave when I meet a better one." However, in the opinion of old labour immigrants, frequent job-hopping is the biggest problem of young labour immigrants, and it is also an extremely bad phenomenon. An old labour immigrant told the author: "I have said repeatedly to my daughter that we must work steadily in one place. It takes a while to learn something, so don't care too much about money, and don't be capricious" The old labour immigrant said: "Once before, workers would not leave so easily, because everyone still hopes to be steadily and familiar with each other. Young people nowadays just hopping around, it won't be any good for them going elsewhere."

Second, the majority of labour immigrants obtains information about new jobs from personal contacts. Because the young and middle-aged labour immigrants always take job-hopping as positive, they are always open to job-hopping information. During the survey, the author found that one of the main contents of the daily chat among young labour immigrants is where the job is better, where the salary is higher, who leaves, what kind of development, what are their future plans, etc. Regarding the composition of external sources of information, about 25% of young labour immigrants believe that the information about job-hopping comes from their fellow countrymen or relatives, and about 50% of the young labour immigrants believe that the information about job-hopping comes from friends. Friends here include former colleagues. Only about 20% of young labour immigrants think they can rely on job advertisements to obtain information, and they pay less attention to job advertisements. In other words, information about job-hopping basically relies on personal network, and there is no formal channel.

Third, the process of judging and weighting, from obtaining information to making a decision for job hopping. In this process, what are the major criteria for job-hopping decision? In the survey, the author interviewed dozens of young labour immigrants, and asked them to rate the criteria for job-hopping, the four most important factors in orders are "wage", "colleague relation", "superior relation" and "working environment", the listed criteria such as "skill obtain", "future development", "professional counterpart" is somehow irrelevant. Before making the final decision of job-hopping, 80% of young labour immigrants will consult their friends, but their friend's opinion is not a primary factor for final decision. One respondent told the author: "I don't take other's opinion that serious, because most people do encourage you to leave, unless they are really attached to you in this workplace, otherwise job-hopping is good." Besides, only 10% of the hopper will inspect the future working place before they make the final decision, most people think it is too much trouble to inspect, and usually wouldn't be given the chance to do it. In general, the decision-making process is mostly emotional and impulsive for young labour immigrants.

Fourth, the "push–pull" theory also applies the job-hopping situation. The pull factors for the "hop on" workplace, as the respondents told the author, are good wages, good interpersonal environment, and good working environment. The push factors are varying from low wages and some other incidents, these incidents include "disagreement with foreman", "generally below expectation", "tired from work". Some answered "quarrel with foreman", "be arranged to clean toilets", "broke up with boyfriend/girlfriend".

Fifth, the damage to the stationary factory caused by frequent job-hopping is obvious, but the factory does not have any legal means to restrain this situation. The director at the human resources department said: "According to the labour law, when we sign a contract, we can only stipulate the norms that employees should abide during their stay in the factory. In addition, we can neither collect deposit nor stipulate how long they must stay. They can leave at any time. The only loss is they may not get the salary of the month when they leave, so fewer people will quit by the end of the year, because everyone wants to get the year-end benefits before leaving. There are even some employees who left without even going through the formalities. I am not sure if this is illegal." According to the factory regulations, employees need to register when going through the resignation procedures and return the tools belonging to the company and the keys to the dormitory without paying any other price. Moreover, the human resources department has become accustomed to this situation, unless the quitter is an important person in the management, otherwise they will not conduct any persuasion or inquiries.

Sixth, is job-hopping really helped young labour immigrants to achieve their expectations and goals? As we know, the labour migration is more than a geographical move, it is also a professional move, the job-hopping, can be seen as a readjustment for labour immigrants' career in cities. Immigrants don't just move into a job at once and stay there for a long time, instead, they adjust their fit with their career and find a "better" place/job. Most immigrants consider their living condition improved after immigrated cities, however, there is no consensus on the impact of frequent job-hopping.

There are 70% of respondents said they are slightly improved after hopping, one clerk who just hopped said: "I think all-around considered, it's a bit of a novelty in the new environment, at least. The salary is the same as before, but it helps me to see more new things."

As for salary, most respondents said there is only slight raise, but for the subjective satisfactory, they considered improved a lot. That mean even those who hopped, aiming the target for higher salary, eventually missed their primary goal after hopping, however, they are more satisfied psychologically, which is obtained by comparison. When there is a new hopping opportunity, the change of satisfaction disappears, and the satisfaction is more about the psychological recognition of "hopping" itself, rather than the achieve the objective goals.

Seventh, most labour immigrants said they will share information of new job opportunities with their friends, the reason is: "Since people can introduce me to good opportunities, I should also introduce them to others, and it's better for friends to work together." What do you include in your message to others? The top four are "wage", "interpersonal environment", "working environment" and "living conditions such as dining hall and accommodation". The author also found that the relationship between foreman and young immigrants are more inclined to make a positive evaluation, and more inclined to convey the information to the others, in the interview, the young immigrants repeatedly mentioned "my foreman is nice", this kind of emotional evaluation to show their job satisfaction.

### 9.2.2 "Job-Hopping": A Subjective Strategy for Labour Immigrants

The job-hopping of labour immigrants is often considered normal in general opinion, the stability of job for labour immigrants is much lower than their urban resident counterparts. Low stability is caused by the nature of informal occupation for labour immigrants, urban residents usually engaged in the stable occupation such as civil service or skilled position, while the immigrants is facing the free market of labour force. Most of immigrants are engaged in the "temporary jobs", rather than being "formal employee", even the salary is different for formal and informal employees. Otherwise, the restrains are less effective for labour immigrants, since they are less attached to urban society, a piece of paper contract usually cannot restrain their mobility.

In recent years, job-hopping is becoming more frequent than before, which means the younger generation of labour immigrants are more likely to hop than their predecessors. During the survey, the author found out it is cause by the higher self-esteem and lower affiliate to workplace. In other words, younger labour immigrants require more salary, better working environment, and they are much harder to satisfied.

Higher self-esteem for younger labour immigrants is cause by their better education. Senior labour immigrants, even if they want to be promoted, are not qualified due to lack of education. A senior labour immigrant said: "I'm thinking of changing factories, but it's hard to find a job now, and I don't have any special skills, so it's not easy… I also want to help my sister to hop and stay with me, at least in the same city, but now it is not easy to find, not I do not want to, is my ability is limited, our factory is not always hiring."

For the younger generation of labour immigrants, "job-hopping" is not only a kind of dissatisfaction and resistance to the former employer, but also a practice of seeking higher goals. It can be said that it is an "involuntary strategy" for them to pursue group self-realisation. The following part will analyse the strategic significance of "job-hopping" as the strategy of labour immigrants from several aspects.

Job-hopping is a way for self-protection, it is always associated with "the secure the basic demands", "equality", "happiness in workplace", "comparable satisfactory", and "the secure the basic demands" among all above is most important. Some senior labour immigrants said: "Young people are not willing to bear hardships. When we came out to work, no matter how hard and tired we were, we could do any job as long as we could earn money. However, nowadays, young people are not the same." On contrast, some young labour immigrants said: "We are not afraid of hardship, because there is no need to take these hardships, those old people endure too much, a few months delay on paying salary, they still insist on working, it should not be that, we must go, sure I am here for work, but I won't let the bosses bully me." Indeed, many young labour immigrants took job-hopping as a mean for self-protection, while the old generation usually endure the inequality in workplace.

Frequent job-hopping is also a way to achieve goals, rather than stay and wait for promotion in the same place, young immigrants often aggressively seeking the opportunity for promotion. They improve their skill and keep learning, in order to get the qualification for a better place to work for. However, frequent job-hopping makes immigrants lack of accumulating skills. Job-hoppers often change their industry, and different industry have different technical requirement, most of them cannot keep the technical continuity. This situation makes job-hoppers weary to learn new skills, some experienced workers are not willing to take young labour immigrants as apprentice, for they are unable to learn a skill earnestly. Lack of professional skills prevent frequent job-hopper to promotion, which leading them to dissatisfaction, then cause resignation, a vicious circle.

In this chapter, we focus on the flows of labour immigrants, in our survey, two flows are significant, "return" and "job-hopping". As a new choice and orientation for the future life of the older generation of labour immigrants, "return" is actually a new beginning of their life. This choice highlights immigrants still sees working in cities as an "expedient measure". "Return" is their subjective choice according to their own survival logic, which is more in line with their own needs. At the same time, this conclusion also shows that the older generation of labour immigrants are less likely to stay in cities. We can neither take them for granted that they will be confined to the land of their hometown, nor simply follow our logic to discuss how they will integrate into cities in the future. In fact, only through their own efforts, from labour to entrepreneur, or through intellectual technology transformation, they can stay in cities. "Job-hopping" frequently for new generation of immigrants, there are not only a normal phenomenon, behind the phenomenon, hiding is a new generation of immigrants' definition of "success", and the pursuit and the choices of the strategies to achieve these goals, further also implies new generation immigrants in adapt to urban life adjustment in the way of transformation.

# Chapter 10
# Social Security for Urban New Immigrants

The idea of social security was first introduced by the Social Security Act by the U.S. Congress in 1935. The 1999 edition of *The Social Work Dictionary* defined social security as: "a program of old-age, unemployment, health, disability, and survivor's insurance maintained by government through compulsory payments by specific employer and employee groups." *In The New Encyclopedia Britannica*, social security is defined as "any of the measures established by legislation to maintain individual or family income or to provide income when some or all sources of income are disrupted or terminated or when exceptionally heavy expenditures have to be incurred." In China, social security generally refers to the social system that the state and society established, to protect the basic living of the members of the society defined by the law, through the distribution and redistribution of national income.[1] Generally speaking, China's social security consists of social insurance, social relief, social welfare, and special care and replacement. Social insurance is the core of social security. Social insurance is national compulsory means through legislation on the redistribution of national income, it is formed by special fund of insurance of the workers by age, disease, birth, disability, death, and lose working ability or unemployment, oneself and family lost revenue, the state and society shall provide the necessities and helps.[2] In the White Paper on China's Social Security and Its Policy issued by the State Council in 2004, social insurance is divided into pension insurance, unemployment insurance, medical insurance, work injury insurance and maternity insurance.

Before 1978, the planned economy has a "national insurance mode" to cover the few urban residents, the majority rural population was left uncovered. After 1978, with the reform from planned economy to market economy, the reform on the social security system began to change coordinately. The massive immigration

[1] Dong Keyong, *Thirty Years of China's Economic Reform*, Chongqing: Chongqing University Press, 2008.

[2] Zhang Hongtao, Zheng Gongcheng, *Insurance*, Beijing: China Renmin Unirersity Press, 2008.



D. Zhou, *Urban Migrants in China*,
https://doi.org/10.1007/978-981-99-3114-9_10

was emerged at the same time, different area deal with immigration with different social security exploration. Due to the great differences between different parts of China, the practical modes explored are also different. The academia classifies these different explorations into the Guangdong model, Shanghai model, Beijing model, Chengdu model and Shenzhen model, and etc. In other words, classified into rural model, integrated model, independent model, complete urban model, differentiated urban model, elastic model, and etc. As a form of national income redistribution, social security is realised through a certain system. China's current social security system is a plan for which the central government and local governments at all levels are jointly responsible. But for now, the national policy has not issued a unified standard. It is necessary and important to carry out the social security of urban new immigrants.

First of all, with the development of China's modernisation and urbanisation process, labour mobility and settlement between urban and rural areas and cities will become the norm, and the number of new immigrants will continue to increase. How to solve the social security problems of these new immigrants is a major issue related to social fairness, social stability and development.

Second, new immigrants often engage in dirty and aversive work that some urban residents do not want to engage. Their jobs are risky, their working environment is poor and their social status is low. Among the immigrant, there are still a considerable number of women and children, and there is still much work to be done to protect the basic rights and interests of these relatively vulnerable groups in society.

Finally, integrating urban new immigrants into the social security system is necessary for both social fairness and efficiency, and is an important prerequisite for ensuring people to live and work in peace and contentment.

From the perspective of institutional economics, the social security for new immigrants in China is facing the obstacles caused by the administrative segmentation of the labour market. Labour market segmentation is generally divided into two situations. One is vertical labor market segmentation, also known as technical segmentation, which refers to the objective boundary of labour's occupational level. This segmentation comes from the gap between the education and training of workers. The other is horizontal segmentation, which refers to professional segmentation, industrial segmentation, urban–rural segmentation and regional segmentation. Among them, the horizontal segmentation is mostly expressed as household registration segmentation. It is embodied in the segmentation of labour market by household registration system and by means of ownership of means of production.[3] New immigrants of different types are employed in different levels and types of urban labour market, so they are faced with different types of administrative segmentation and social security with institutional characteristics.

[3] Liu Chuanjiang, Xu Jianling, *Research on the Urbanization Process of Rural Migrant in China*, Beijing: People's Publishing House, 2008.

## 10.1 Quantitative Study on Social Security for Urban New Immigrants

Table 10.1 below shows the intellectual immigrants is most concern with housing, as well as junior labour immigrants and entrepreneurial immigrants. Entrepreneurial immigrants and senior labour immigrants are more concern with education, and new generation labour immigrants take recreational welfare more than other groups.

Figure 10.1 below shows the equality of welfares between immigrants and locals, housing the most inequal welfare, intellectual immigrants are slightly better all-around. Besides, the public insurance coverage on intellectual immigrants are better than the other immigrants, entrepreneurial immigrants have the less insurance coverage (commercial insurance excepted) among all immigrants (see Fig. 10.2).

Table 10.2 below shows Zhengzhou is the less institutional segmented city among all surveyed cities, different types of immigrants do not have significant gaps.

**Table 10.1** Weighting on public welfare for new immigrants (Unit: %)

| Items | Intellectual immigrants | Junior labour immigrants | Senior labour immigrants | Entrepreneurial immigrants |
|---|---|---|---|---|
| Education | 14.94 | 25.49 | 34.69 | 36.69 |
| Health | 17.68 | 13.83 | 16.80 | 11.69 |
| Recreation | 3.76 | 10.57 | 4.07 | 3.47 |
| Housing | 61.89 | 46.51 | 34.15 | 43.52 |
| Total (N) | 984 | 918 | 369 | 864 |

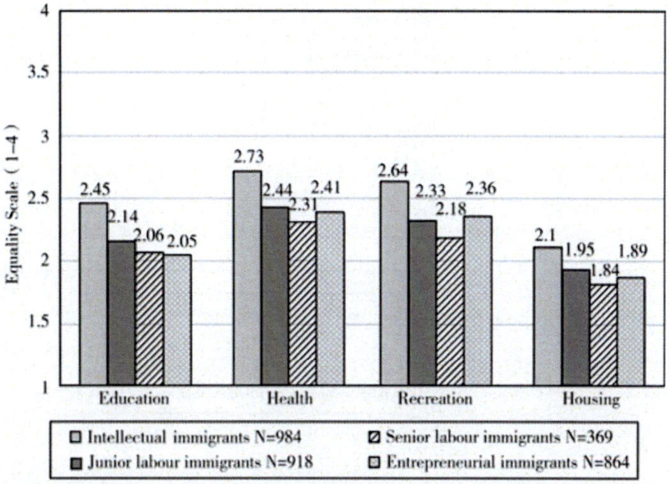


**Fig. 10.1** Equality on welfares for immigrants and locals

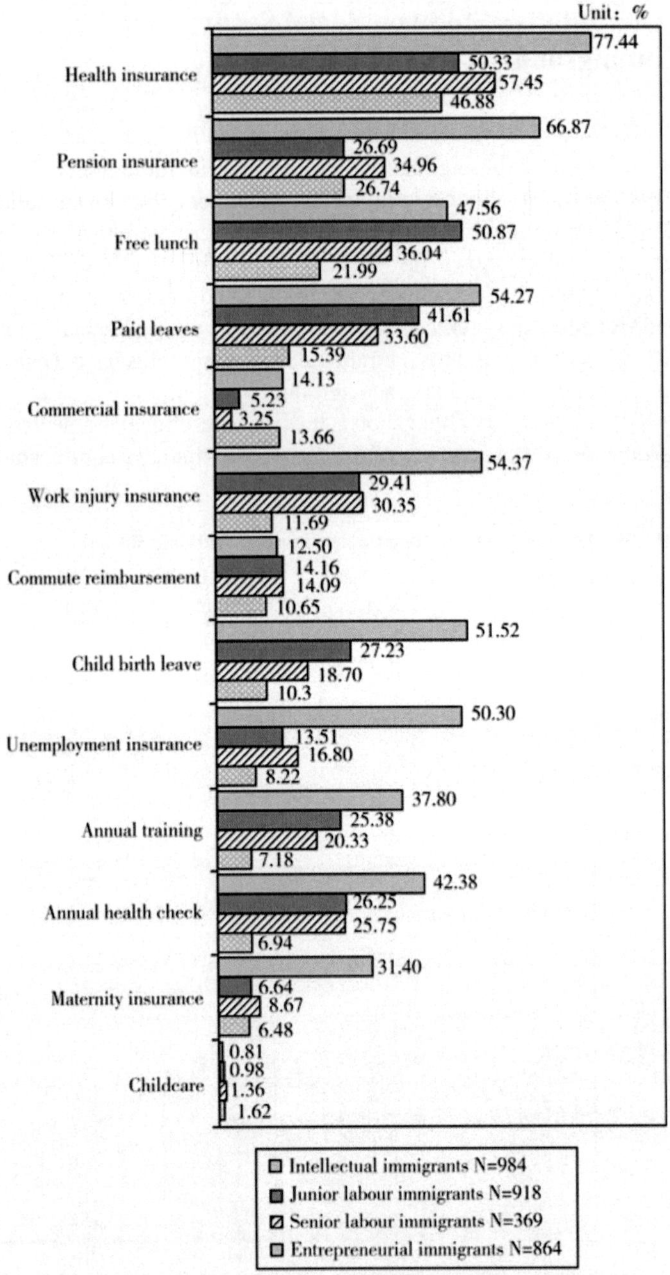


**Fig. 10.2** The coverages on welfares for new immigrants

**Table 10.2** Equality on welfares for immigrants and locals in Zhengzhou

| Items | Intellectual immigrants | Labour immigrants | Entrepreneurial immigrants |
|---|---|---|---|
| Education | 2.25 | 2.02 | 2.09 |
| Health | 2.20 | 1.98 | 2.26 |
| Recreation | 2.21 | 1.98 | 2.18 |
| Housing | 1.69 | 1.63 | 1.82 |
| Total (N) | 152 | 236 | 130 |

The above data reflect the lagging supply of urban public services (such as housing, education, infrastructure, etc.), and the unguaranteed benefits such as housing, medical insurance and social security that are the concerns of new immigrants. Among them, the proportion of intelligent immigrants cover by all kinds of insurance and welfare provided by workplace is higher than that of other new immigrants, and the coverage rate of entrepreneurial immigrants is the lowest. In addition, it is found that in the case of regional system division, both intelligent immigrants and labour immigrants are at a disadvantage in securing security. Next, we will further analyses the social security problems of urban new immigrants through the way of case analysis.

## 10.2 Social Security for Labour Immigrants

### *10.2.1 Discontinuous Jobs and Fractured Security*

Institutionally, China's social security doesn't exclude labour immigrants who engaged in formal employment. Immigrants are protected by the Labour Law and other administrative regulations, they are qualified to basic public insurance, but in fact, the coverage on immigrants are low everywhere.

However, labour immigrants who engaged in informal employment, are almost excluded from the urban social security, and they remain outsider status in urban lifestyle, health and employment. Take unemployment insurance as an example, according to Regulations on Unemployment Insurance, all employee must have the unemployment insurance, when the contract expires or suspend, a compensation will be given to the employee according to the length of work year. But most of employers do not pay the unemployment insurance for labour immigrants, and labour immigrants usually do not appeal to for this. At the same time, the labour immigrants who are temporary hired and low income, some of them work less than a year in one workplace, which is unqualified for the minimum insurance time length. Most of labour immigrants rather take cash for salary than any means of insurance.

Case: Shenyang – Labour – 001

The foreign language school Mr. Lu worked for last year fully paid for his pension insurance, health insurance and housing fund, but only for 4 months, with his discharge, all the security was suspended, Mr. Lu has no intention to continue his insurance, mostly because he is short for money.

Case: Shenyang – Labour – 002

Mr. Zhang has been a construction worker, but he never has work injury insurance. He thinks he is earning too little, and have a family to support, there is no money left. Even the workplace willing to paid for his insurance, it is his earning after all, so he would rather cash it.

Academia had made five macroscopic suggestions as the countermeasure for labour immigrant's social security. First, establish the legal status for labour immigrants; second, strengthen macro-control and micro-management for immigrants; third, make full use of the existing system to promote the social security coverage for labour immigrants; fourth, promote legislation for social security; fifth, actively carry out publicity activities, adopt various forms, make full use of the power of public supervision, and promote the publicity of social security.[4] As a matter of fact, the public insurance of immigrants is mainly affected by the institutional arrangement, payment amount and method.[5] Some provinces and cities where labour migrants are concentrated have been developing and adjusting relevant social security policies to bring more new immigrants into the social security system.

Take the reform of health insurance in Guangzhou as an example, the regulations for labour immigrants issued by Guangzhou authority in 2009, apply to all full-time employee in the region, and it is mandatory insurance, make sure full coverage. Employer must pay for their employee's basic health insurance, and selectively join the urban employee health fund. Labour immigrants do not pay, the employer must pay the insurance monthly by the amount of 1.2% of the average monthly salary of general employee in Guangzhou last year, now the payment is RMB40 per month per person. Since the regulations issued, social security for labour immigrants who engaged in formal employment was largely improved. Some observations consider, the future development shall aim for, firstly, the expansion of pension insurance, and promote the coverage and participate population, study on how to transfer pension fund from region to another. Secondly, the coverage of health insurance, focus on the serious illness of immigrants. Thirdly, focus on implementation of the policy for work injury and unemployment insurance. Coverage and policy implementation are related to the nature, industry and scale of employment of labour immigrants. It is an important step for the implementation of policies and the realisation of social security coverage to carry out social security work in every type of enterprises.

However, the facts are, neither province in China achieve public security funds mobility, nor for across provinces. That means no matter for urban employee or

---

[4] Zheng, Hangsheng, Hong Dayong, Emphasizing and Developing the Social Security for Urban Immigrant Workers: An Importance in the Process of Social Transformation, *Academic Exchange*, Vol. 5, 1994.

[5] Zheng Gongcheng, Some Reflections on Pushing Forward the Reform of China's Social Security System, *China Soft Science*, No. 4 (2001), pp. 19–22,107.

immigrants, it is quite impossible for them to move any type of their public security funds across country. Some immigrants chose to return, and when they return, the public security they accumulate in cities cannot transfer to rural regions.

At individual level, the awareness and pay capacity for public insurance affects labour immigrants' behavior on join the social security program. At the enterprise level, from the aspect of saving expense, they natural avoid responsibility to the labour immigrants they hired. Some employer uses the measure such as "conceal actual employee", "settle things in private", "blockade information" to avoid the responsibility of buying insurance and deal with the aftermath. Therefore, the social insurance of immigrants often becomes a game field between the government, enterprises and workers, and the workers are obviously in a weak position.

Case: Zhengzhou – Labour – 004

He does not have pension insurance, but have medical insurance, his wife pays for him at home, it seems to be a year pay RMB80, reimburse 60%, not particularly clear. Work injury insurance have never been bought, nor asked, the working site is too complex, a big project has dozens of subcontractors, the subcontractors have their sub subcontractors. When contractors and the subcontractors sign a contract, the injury and accident responsibility also being subcontracted, let these small contractor bear responsibility, this is not a fair deal, because they don't pay injury insurance for labour immigrants, small contractor usually don't have the ability to bear labour immigrants compensation, therefore, in case of accident, the lives of labour immigrants are the least valuable.

### *10.2.2 Struggling to Settle Down*

Household registration system reforming is an important subject to the immigration, the reform taken place now, settles three types of immigrants' household registration. First, those who qualified for "dependent" but have no quota; second, those who have household registration in urban village, their household registration can transfer from villager to urban resident; third, investment and elite. There are different types of household registration resettle policies in different regions, but all basically draw a bottom line at "domicile, stable income, high profession". Only a few of labour immigrants who have special contribution, were granted the urban household registration and settle down in cities, for the majority of labour immigrants, it is still quite impossible for them to settle down in cities. Among the ten labour immigrants we interviewed in Shenyang, only one was settled in Shenyang (Shenyang-Labour-001). He was granted the household registration of Shenyang by investment, buying his own apartment in Shenyang. This path of accumulating wealth and processes urban real estate, it is a non-replicable path, for the majority is impossible to gain such wealth.

Case: Hangzhou – Labour – 001

Real estate here in Hangzhou is really expensive, it is not easy for me to buy a house here. Fortunately, our company provides dormitory. Otherwise, it is not cheap to rent a house, let alone buy a house. To become a city resident that at least to have my own house, at least to have the ability to afford a house, Hangzhou housing price is equal to say a blockade to immigrants.

Case: Zhengzhou – Labour – 003

Never have the thought to settle down here, nor bringing my children to here, it is impossible financially. Very few immigrants are able to settle down in cities, I don't even imagine that, just do what I can do, day by day, for my children, when they are capable to earn their own money, then I can rest, return home and plough my farmland, just enough for my daily meal.

### *10.2.3 Utilized Personal Archive*

Personal archive in China, is a record kept by the officials such as school, government, state-own enterprise and organisation, it consists personal information and family information. Labour immigrants, mostly engaged in private sector, and frequently change their job, personal archive has no particular means for them. Some of our respondents even ask us "what is a personal archive?", because they don't have, and don't need for personal archive. However, when social mobility happens, archive is needed for middle to higher level positions. For immigrants who change jobs frequently, the temporary residency permit is more important.

Case: Shenyang – Labour – 003

Mrs. Wu and her husband's archive were keeping in their hometown, Anshan. This couple think they are old, so it doesn't matter anymore. She is expecting the state will compensate her in the future, for they are laid off by the state's command. She hopes her daughter's personal archive will be keeping in Shenyang, it is easier for her future.

Case: Shenyang – Labour – 004

Mr. Lin's archive and household registration were keeping in his hometown, he has no idea what these are use for, and his employer doesn't require those files. In his opinion, the state's policy is quite positive for rural, and rural farmer have some beneficial advantage. Mr. Lin said: "I have half acre of farmland at home, annually RMB300 of subsidy. I was told there will be more in the future, I planned to return at the end of this year, so household registration and archive does bother me."

### *10.2.4 Moral Struggle and Physical Struggle: Strive for Unpaid Wages*

The arrears of wage are the most urge problem for labour immigrants. In 2010, labour dispute arbitration institutions at all levels across the country established a "green channel" to facilitate the appeal of immigrants' labour disputes. They adopted simple procedures to handle labour dispute cases in a rapid manner, and made final rulings and early implementation of small labor remuneration dispute cases. According to statistics, the number of disputes involving immigrants' wages handled by labour dispute arbitration agencies at all levels nationwide dropped by 15.1% in 2010. However, the establishment of national system is often unable to connect with the social reality of labour disputes, the launch of this "green channel" is based on the declaration of labour disputes. In practice, labour disputes are often resolved outside formal channels. Our interviews with some of the labour immigrant who worked as contractors reflect the problem of wage arrears from the perspective of the state and its resolution. In this process, contractors that directly have labour-capital relations with labour immigrants stand in the united front with immigrants, and jointly have a dialogue with commissioning bosses. Sometimes we also see that it is the contractors and labour immigrants have disputes. No matter what kind of dispute, labour immigrants are always in a disadvantage and passive position.

Labour contract is a formal written agreement to determine the employment relationship, and it is also an important proof to protect the rights and interests of labour immigrants. In the case of Zhengzhou – Labour – 004, the construction worker, Mr. Lu, has came to a costly conclusion, that before starting to work, must sign the contract. Contract is a weapon for the disadvantaged party to strive for their wages, in their strive, moral struggle and physical struggle is their strategy.

Case: Zhengzhou – Entrepreneurial – 008

Boss Zhang said, now we are not dare to delay worker's pay. The state has very strict order on this, workers' wages must be paid after disbursement of the project funds. Before we are permit to start the project, a percentage of wages were hold at the funds of construction office, if someone dare to delay worker's pay, the office will shut down the site by force. Work injury is easier, if someone injured, the foreman usually settles the matter by negotiation.

## 10.3 Social Security for Entrepreneurial Immigrants

Entrepreneurial immigrants usually engaging in small business, operate independently, and be responsible for their own profits and losses. They are often drifting from the formal employment and social security coverage. Most of our respondents are not covered by any public insurance, some show no idea regarding household registration and personal archive. Wholistic speaking, restricted by the financial system and overall regional policies, policy-based social security is not fully implemented for small business owner. Some transform from formal employee to individual business, the responsibility of social insurance is transfers to oneself entirely.

Case: Shenyang – Entrepreneurial – 004

When Mr. Li was clerk at an electronic commerce company, the company handles all 5 social insurances for him, when he resigned from the company, he has to deal with the insurances by himself, now he opened his own shop, so the insurance was paid by himself.

Case: Zhengzhou – Entrepreneurial – 002

In 2006, when he returns to Zhengzhou and work for a real-estate company, the company pay 3 social insurances for him, now he is doing business by himself, but he continues those insurances, by putting his papers under the company, and pay the insurances through the company. He values those insurance very much.

The entrepreneurial immigrant who never engaged in formal employment, usually take commercial insurance as their way for future security, the sense of planning ahead is crucial for the decision of buy insurance, therefore, it varies on different individuals. In fact, most entrepreneurial immigrants are insured in some way, those whose paper still in their rural home, often join the New Rural Cooperative Medical Insurance System[6] and New Rural Social Endowment Insurance.[7]

Case: Shenyang – Entrepreneurial – 005

She doesn't have any insurance, "if I pay for insurance, there will be no money left for me, I will rather return to my rural home."

Case: Zhengzhou – Entrepreneurial – 003

He doesn't have any insurance, his own words: "It is only a matter of money, if I have the money, who cares about insurance." he put all his money on his business, and have no intention to buy any insurance.

Case: Zhengzhou – Entrepreneurial – 001

When Mr. Liu was asked whether join any insurance, he said when he was in his hometown, there is a scheme, one pays RMB250 for a year, pays 15 years, then after age 60, one can collect money from the government monthly (The NRSEI). He already saw some old people collect that money, he cannot recall the name of that insurance, the government officials came to his village to advertise it, he thinks it is a good scheme, and he joins and pay for his whole family. I asked about the NRCMS, he don't remember the name of the scheme, I explain how it works, and he said he already join the scheme for 4 years, RMB25 a year, first year excluded, after the first year, if one goes to hospital, then one can claim a portion of reimbursement, he saw officials advertise these in his village, and some people already gained benefit from these, peasants don't know the names, they see people benefit from it, then they are willing to join. He said in his village, most people join both insurances, and all his family had joined.

In the fore mentioned case, Mr. Liu's village is in Luoshan County, in Henan Province, is the pilot county on New Rural Social Endowment Insurance. All rural residents age above 20, after paying fees for 15 years and reaching the age of 60, eligible to receive pension insurance by the month. The county government shall subsidize The NRSEI with 3% of its fiscal revenue every year. From the perspective of individuals, entrepreneurial immigrants get rural pension insurance as farmers, and their life after 60 is guaranteed to a certain extent. On the social level, it is beneficial to social stability and social equity. However, those who immigrated into cities, even their papers are still in their rural home, it is very hard for them to return completely to rural environment, most of them have to continue their lives in cities, and the problem of social security in cities cannot be avoid by solving the problem in rural.

[6] New Rural Cooperative Medical Insurance System, is a new basic health social security system, started in 2003, that combines insurance and social assistance, and targets all farmers and rural residents. Coverage in 2009 amounted to 833 million people. Participation is voluntary but with 94% of the target population covered in 2009. -Tr.

[7] New Rural Social Endowment Insurance, launched in 2009, the new scheme is voluntary and includes a combination of personal payments and collective and governmental subsidies. It is open for rural residents aged 16 and above (excluding full-time students) who do not participate in the Basic Pension Scheme for Urban Employees (BPSUE). -Tr.

## 10.4 Social Security for Intellectual Immigrants

Comparatively speaking, the social security status of intellectual immigrants is relatively better among the three types of new immigrants. This is closely related to the employment status of intellectual immigrants. Intellectual immigrants are usually engaged in formal employment, and most of these sectors are in the taxation and supervision system of the government, their occupational stability, income and welfare benefits and other aspects can be guaranteed. Normally, when a college student graduate, a three-party agreement will be signed by the college, the graduate, and the employer. After the agreement was signed, the graduate will report to the employer with an assignment certificate issued by the college, and one's formal employment as "cadre status" was started from the day one reports to the workplace.

The "cadre status", along with other status such as "farmer status" and "worker status", is a status confirmed by the status system, which is a framework for people's identity before the reform in 1978. There were 14 system arrangement upon the status system, there are the household registration system, the grain supply system, the fuel and subsidiary supply system, education system, employment system, medical system, pension system, human resource system, conscript system, marriage system, and maternity system. After the reform, some personal identity binding frameworks were abolished. However, the cadre status of the college graduate, still a useful social capital in terms of social mobility and identity.

Case: Shenyang – Intellectual – 002

Ms. Miao's company has a good profit. For three years, the company has been paying her "five insurances" (including endowment insurance, medical insurance, unemployment insurance, work injury insurance and maternity insurance). Three months ago, the company has begun to pay for her housing accumulation fund.

Case: Hangzhou – Intellectual – 015

I think the social security system is good, one pay for the insurance, and save a lot of money in case of accident. My colleague's brother, once was injured in workplace, the medical bill was around RMB3,000, the insurance pay for RMB2,000, and the company compensated around RMB4,000. With the insurance, health care does cost so much, those who without insurance, pays a lot when sick.

### *10.4.1 Personal Archive and Household Registration*

In China, only the state organs, the state-owned enterprises and public institutions, have the authority to keep personal archives, if someone employed by a public institution, then one's archive and household registration is received by one's employer. At present, private companies do not have the authority to keep archives, those who employed by private sector, can only send their archives to the state-run human resource centre. When the employment is temporary, the archive and household registration may be returned to one's hometown.

Case: Shenyang – Intellectual – 002

Ms. Miao's household registration was transferred to her university in Shenyang. After graduation, Ms. Miao put her archive in the human resource centre, and her household registration was moved back to her hometown, until she bought an apartment and finally settled in Shenyang.

Case: Shenyang – Intellectual – 003

When Ms. Gao went to college in Shenyang, her household registration and archives were moved with her to her college. After graduation, she moved her household registration to her grandmother's household registration book in Shenyang. However, her employment is temporary and she has nowhere to keep archives, so archives has been a big problem for her. Until now, the archive is still keeping in her home.

Case: Shenyang – Intellectual – 006

His archive and household registration are kept in the human resource centre of Shenzhen, and his current company headquarters is in Shenzhen, so he left his archive and household registration in the human resource centre of Shenzhen after he signed the contract. He can leave all his papers in the human resource centre; he only needs to pay the keep fee upon his transfer to another place. Because his household registration is not in Shenyang, he met some obstacles when he obtained the qualification of assistant engineer, he had to go to his household registration place to get the qualification. It is not so much trouble after all, he doesn't need to get a TRP, because there is no need for TRP to get insurance in Shenyang.

For those who did not engage in formal employment after graduation from college, there are three ways to put those personal archives, first is to transfer to their hometown, the paper will be kept by the personnel bureau of a prefecture-level city; second is left the papers to college, the papers will be kept in college for two years after graduation; third is transfer the papers to human resource centre.

Under the circumstance of massive migration, the role of personal archive has been diluted, it is a social progress for the mobility of people. Quite a few cities in China have promised that elite talents will not be restricted by their archives, which on the one hand shows that talent has become an important strategic resource for local social development, and the curtain of the talent war has been slowly opened. On the other hand, the function of personal archive system is slowly weakening, although it is no more the decisive paper, still it has some importance for individuals, in China's current system, archive affects such as pension insurance, medical insurance, unemployment insurance and other state-mandated welfare policies, professional title evaluation, various political audit, and etc. There are some effects try to make a breakthrough in this respect, for example, Shanghai already lifted the limitation of archive for social insurance. However, cross-region insurance reimbursement still cannot be processed, and archives still have an impact on the implementation of social security.

Case: Shenyang – Intellectual – 006

He has medical, work injury, pension, unemployment and commercial insurance. "Hundreds of yuan deducted from my salary every month." However, due to the regional limitation of medical insurance, his medical insurance in Shenzhen is not available in Shenyang, so he buys medicine and goes to the hospital at his own expense. The location limitation of insurance is still very strict.

### 10.4.2 Settling Household Registration and Urban Resident Identity

Household registration system has a significant impact on the integration for new immigrants, especially on the social security and social psychological aspect. Taking the employment types into consideration, the employment can be further divided into three types, state employment, collective employment and private employment. Those immigrants who engaged in private employment, including self-employment and foreign-invested enterprises, are far less covered by the social security in compare with state employment, and the state employment is closely related to one's household registration status.

Some provinces and municipalities have began trying to reform the household registration system. Guangdong province, started to register household registration base on the actual residency, and administrate the household registration system as a whole, instead of separate rural and urban residency. Shenyang city, relaxed the restriction on move in household registration for those who possessed legal residence in the city, whose family and dependent may move their household registration into Shenyang all together. Some intellectual immigrants who move their household registration in to Shenyang, may have the social security as well as the locals, which enhanced their social psychological identity as urban resident. Some people regard state employment as a more important indicator for urban resident identity, only those who engaged in formal employment, are truly integrated into the city. This kind of social mentality reflects the cognition of the structure and type of formal and informal employment.

Case: Shenyang – Intellectual – 002

After graduated from secondary school, Ms. Wang did not engage in formal employment, and she has to put her papers back to her hometown. She only engaged in temporary employment since then, so there is no need for her to move her personal archive, but she take the household registration very serious, she desperately wants to be a true urban resident, and only through moving her household registration into Shenyang, she can affirm being an urban resident. To achieve her goal, she finally purchased an apartment and settle her household registration in the city.

Case: Shenyang – Intellectual – 010

His household registration was moved to Shenyang after he purchased his own apartment, in the future, Mr. Xu may put his child's household registration in Shenyang with him, then the child may attend the local school. Although he and his wife are engaged in private employment, they are senior staffs in the office, so the employer paid for all the insurances as the formal employment would do, there is nothing to worry about.

The attitude of intellectual immigrants towards insurance shows different tendency and even opposite attitude. Some employer and employee each bear a part of the insurance payment, some still think that the money in their own hands is safer. Although some employers do not pay for public insurances, some of their employee will buy commercial insurance to seek security. The existence of a variety of circumstances is also one of the characteristics of the diversity of the social value orientation of the intellectual immigrant.

Case: Zhengzhou – Intellectual – 001

His employer only bears a small part of the insurance payment, most of the money is from his own pocket, according to his income, RMB1,500 was deducted from his salary each month for the insurances. He is working for a private company, and his sense for stability is weak, if he does not work for this company any longer, the insurances payment will be stop unless he pays all by himself. He is not sure about the future, and he thinks it is only safe for him to have the money in his own pocket.

After the reform and opening up in 1978, to accommodate the market economy reform, the social security system started to change, with the emergence of massive urban immigrants, each city started to find a comparative way of social security for its immigration situation. However, quantitative and qualitative data both reflect that the supply of public goods (such as housing, education, infrastructure, etc.) in the surveyed cities is lagging behind, and the household registration, personal archive, insurance, etc. have all set tiers of restrictions, which cannot satisfy urban new immigrants. The new immigrants were excluded in cities' social security system by institutional barriers. Household registration is the most important barrier for labour immigrants, their living condition, employment and health care is not covered by the existing social security system, they are treated equally despite they are living in cities, this inequality strike hard on their basic living and motivation to settle down in cities. For entrepreneurial immigrants, due to their independent management and self-financing informal employment characteristics, they often drift outside of the state's formal employment and public social security systems. However, for some business immigrants with future awareness and considerable income, purchasing commercial insurance has become a way for them to obtain future security. In contrast, the proportion of intellectual immigrants who covered by all kinds of public insurance and welfare is higher than that of other new immigrants, which has become an important reason for their high sense of belonging and strong urban identity.

# Chapter 11
# Summary and Suggestions

## 11.1 Summary

At present, the term of "urban new immigrants" is seldom seen in the statement of academic and authority of China. The "migration" officially recognized by the state is mainly refers to the involuntary migration, such as ecological migration, reservoir migration and so on. Studies on "urban new immigrants" were often take the narrative of "floating population" or "rural labour". Since the emergence of the first wave of "rural labour" in the late 1980s, China's population migration between urban and rural areas and between regions has entered a new stage, showing new characteristics. In terms of occupational structure, the urban survival and living conditions of the immigrant groups appear obvious occupational differentiation and social stratification. In terms of their intention divergence, some immigrant groups show strong urban migration and settlement intention and tend to step into the stage of family migration. In terms of age structure, "the new generation of immigrants" have increasingly become an important component of today's urban new immigrants, and their identification were inclined with the city more than that with the countryside.

Although the source and occupation distribution are different, the immigration trend today shows that the urban new immigrants mostly migrate from rural to cities as well as from towns to metropolises. Urban new immigrants not only work and live in cities, but also has the desire to settle down in cities. Therefore, we classify them as "urban new immigrants" and we define the term with the following elements: the place of birth and the original household registration were not in cities, and they lived in cities for 2–5 years; willingness to settle in cities; having a legal residence; having legal income. In our research, urban new immigration was divided into 3 types, say labour immigration, intellectual immigration and entrepreneurial immigration.

The tension between the current institutional and structural constraints on immigrants and the strong willingness and behaviour of immigration is difficult to be calmed in a short time, and the social consequences caused by "move but not migrate or migrate but not settle" even affect the stability of the institution of China's society


D. Zhou, *Urban Migrants in China*,
https://doi.org/10.1007/978-981-99-3114-9_11

to a certain extent. Therefore, the study of urban new immigration and its countermeasures shows great significance to China's economic development and social stability. First of all, with the development of China's modernization and urbanization process, labour mobility from rural to urban, as well as from one city to another city became the norm, and the population of new immigrant will continue to increase. The integration of urban new immigrants and locals is a major issue related to social equity and social stability as well as development. Secondly, in the initial stage, urban new immigrants are often engaged in aversive and hardship works that most locals do not want to engage, these occupations often associate with high risks, poor working environment and low social status. Besides the urban new immigrants includes a considerable number of women and children, and there is still much work to be done to protect the basic rights and interests of these relatively vulnerable groups in society. Finally, bringing urban new immigrants into the scope of urban social construction and resident management is the need for both social fairness and efficiency, an important prerequisite for ensuring people's living and contentment, and a manifestation of China's socialism. The study subject of urban new immigrants focuses on the classification of urban new immigrants and the characteristics of each group, as well as the social differentiation of this group and its impact on the urban social structure, and then discusses the adaptation and integration of new immigrants, as well as the settlement and the transformation of urban resident identity.

We analysed the migration and settling process of three types of urban new immigrants, and found that intellectual immigrants have higher education background, relatively clear urbanization willingness and entry strategy, and intellectual capital has become the most prominent advantage of intellectual immigrants. Generally speaking, with their higher education and intellectual capital, the intellectual immigrants will gradually find a suitable development platform in cities. Labour immigrants are active participants in economic construction, but under the current civil and social policies and economy conditions, it is difficult for them to obtain the social equality of resident, and they are severely isolated in urban society. Although they are willing to settle down in the city, the instability of their jobs and the low income make "going back home" their ultimate plan after depleting their youth, and they may still be "passenger" who make contributions to cities but unable to settle. A considerable proportion of the entrepreneurial immigrants were transformed from the labour immigrants, who started their own business after the early accumulation of capital. Compared with labour immigrants, their willingness to settle down in cities is stronger, and the trend of family orientation is obvious. But given the consideration of living cost for urban households and the benefits of rural land ownership, they want to benefit from both and swinging willingness to move to cities.

The fore-mentioned three types of immigrants have differentiated social network and capacity of mobilising the source of which. Source gathering based on social network is the most important adaptation strategy for entrepreneurial immigrants, it is effective for social statues promotion of urban new immigrants. For those entrepreneurial immigrants from Teochew language speaking regions, the support provided by the strong regional association is the key to business success the young immigrants. Labour immigrants usually only networking with their relatives and countrymen, but still a considerable number of labour immigrant trying to breach this social norm and make a broader network. This effort of broader social contact, also means the process of social adaptation in urban social context, it helps greatly for labour immigrants to settle down in cities. Intellectual immigrants have access to more social resources than other immigrants, which means they are more likely to settle down. However, the gap between locals and immigrants cannot be easily bridged, the structural and primary social interaction and integration still needs a long process.

This book shows the self-identity and community participation for urban new immigrants have already affect their integration and social construction. To be more specific, new immigrants scattered in urban communities, usually rent but not own residence, is at the bottom of the society, they are not self-organised but disciplined by state governance. The identity of new immigrants tends to be "outsider", and the degree of belongingness is low, they have the obvious "community apathy" complex, and their community participation is very low. The lifestyle of new immigrants is almost the same as that of urban residents, but new immigrants are not taken into account in community construction led by governments at all levels.

As for social insurance, since 1978, China's social insurance system started to sharp along with the development of market economy. The spurt growth of urban residents brought great challenge for the governance and public sector supply. Our study shows most of the city authorities cannot provide sufficient public service to catch up with the growing pace of urban population, therefore, the authorities put restrictions on household registration, personal portfolio, and etc. to exclude a certain number of immigrants from public service such as housing, education and infrastructure. The exclusion mostly based on household registration, is an institutional inequality for immigrants, especially for labour immigrants, which has caused great damage to both their basic life and their willingness to settle down in cities. As for entrepreneurial immigrants, due to their characteristics of informal employment, self-management and self-financing, they are often excluded from the basic social insurance system of the state, however, some of the entrepreneurial immigrant who have a relatively high income and future awareness, turn to seek the coverage of commercial insurance as their way to obtain social security. In contrast, the intellectual immigrants are less likely to be excluded in cities' social insurance system than labour and entrepreneurial immigrants, which gave the intellectual immigrants a good reason to have some degree of belongingness in the city.

Return home, as the choice and orientation for the old generation of labour immigrants, is de facto a new beginning of life, and this choice highlights the nature of migrate as an expedient measure. Return home is their subjective choice according to their own survival logic, which is more in line with their own needs. At the same time, this conclusion also shows that the older generation of labour immigrants are less likely to stay in the city. We can neither take them for granted that they will be confined to the land of their home country, nor simply follow our logic to discuss how they will integrate into the city life in the future. In fact, only through their own efforts, take the transformation from labour to entrepreneurial and intellectual, and lead to the settle down in cities. The fact that immigrants frequently change jobs is not only as a normal phenomenon exists, behind this phenomenon, there is a new generation of immigrant groups search for the way of "success", the pursuit and the choices of the strategies to achieve the goals, further also implies new generation immigrant groups in adapting to city life on the way of adjustment.

## 11.2 Policy Suggestions

### *11.2.1 Differentiate the Immigration Demarcation, Rather Than Treat the Immigration as One Plate*

At present, the policies towards rural–urban immigrants are mainly aimed at labour immigrants. Since the emergence of labour immigrants in 1980s, which caught the city authorities unprepared, the authorities always adopted the policy of restricting the population flow and stabilizing the urban population. The management of urban migrant population started from coping with the influx of rural migrants. This policy orientation has not changed substantially to this day. There are many presuppositions in the formulation and implementation of labour-oriented policies: first, it leads people to pay attention to the labour immigrants in cities, while ignoring other types immigrants in cities. Especially in today's increasingly complex immigration situation, such policy guidance is difficult to cover all the immigrant groups that need to be served and managed, such as the intellectual immigrants with higher education and the entrepreneurial immigrants who invest in many industries in cities, and there are many deviations considering the group positioning of policies. Second, the labour-oriented policies have formed some stereotypes on immigrants. The emergence of domestic immigration is considered as the result of the urban–rural dual social structure. On the one hand, the influx of immigrants is considered as a potential danger to cities, which may bring negative impact on the normal order of the city, especially in social security, urban environment, distribution of public resources and other aspects. On the other hand, they are considered as disadvantaged and marginal, they live in rented houses, have low income, and find it difficult to adapt to the urban lifestyle and so on. These two stereotypes on immigrants lead to two kinds of policies. One is to promote the effective management on immigrants to prevent them from bringing negative effects to urban society; The other is to advocate policy support for immigrants and recognize their contribution to urban social construction. Both kinds of policies are necessary for effective urban management, but the common weakness is the lack of flexibility to cope with the changes of the immigration situation. Third, due to the large scale of immigration in some cities, the government appears extremely cautious in issuing relevant improvement policies. If immigrants are taken as the one whole target group of the policy, cities need to pay a heavy price when they want to solve the problem of the coordinated development of immigrant and cities. Every policy aims to improve the living and development conditions of immigrants will have a far-reaching impact on the economic and social development of cities. Meanwhile, with public opinion calling for equal treatment on immigrants, and central and provincial governments guiding policies on building a harmonious society, local governments are in an awkward position and must be cautious with every step of reform.

For all of these reasons, we suggest the labour-oriented policies, in a general sense targeting all immigrants as a whole group, need to be adjusted. To coordinated the development of immigration and governance, the authorities need to differentiate the division of the immigration, investigation and grasp the features of different immigrants' groups, assess the new demand and development trend of the different immigrants' groups, tailor-made management policies to accommodate every different type of immigrants.

### *11.2.2 Focus on Fulfilling the Demands of Urban New Immigrants Who Have Willingness to Settle Down in Cities*

In the recent three decades of massive domestic migration, the most significant demarcation is the formation of a group of immigrants that work and live in cities for a long time. In our research, we studied immigrants who stay in the surveyed 6 cities for 2–5 years. This group of immigrants are differentiating than the immigrants in a general sense, and their demographic features, interest appeal and urban identity is not identical with the general immigrants.

According to the our research, we suggested that the new immigrants with the willingness and ability to settle down in cities should be taken as primary, and policies should be formulated to gradually meet the needs of this group of new immigrants, realize their integration into the urban society and enjoy the social public welfare fairly. The priority is to improve children's educational welfare, including compulsory education for children, local participation in the high school entrance examination and college entrance examination. In terms of social insurance, the authority shall gradually expand the coverage of public insurance based on existing basic needs guarantees. In terms of household registration entry policy, the authority should formulate and implement a point-based household registration entry policy as soon as possible for new immigrant groups. In terms of social services, they shall facilitate the living convenience of new immigrants.

First, the intellectual immigrant is a younger demographic group compare to the labour and entrepreneurial immigrants, and they have a strong willingness to settle down in cities. Most of the intellectual immigrants are unmarried, after shifting in cities for a few years, the majority of this group is highly likely find a stable job and settle down in cities, own residence and finally integrate into urban population. However, this path is not an easy one, they are stimulated to burst their motivations at most, to keep up with their expectations for future life. To meet the needs of this group of immigrants, the research team believes that relevant policies shall focus on housing system, public recreational welfare and work insurance and some other aspects. As for housing, the intellectual immigrant generally share residence with their schoolmates and colleagues, scattered in downtown residential area in rented houses or urban villages. We suggest the development of public low-rent housing shall be taken as priority, and it shall cover the intellectual immigrants who have higher education, continuous residence in cities for a certain number of years and legal income. As for recreational welfare, we suggested to make full use of the power of network such as Communist Youth League, enterprises and communities to enrich the spare time life of intellectual immigrants. In terms of work insurance and job protection, our survey shows frequent job changes of intellectual immigrants, for some of the intellectual immigrants who are engaging in the private sector, violations of the labour law such as unpaid overtime, embezzle salary, is not uncommon. We suggest the protection shall referring to the relevant policies of labour immigrants, formulate a series of policy to cover the interests of intellectual immigrants.

Second, among the labour immigrant who shows willingness to settle down in cities, only a small number of them believe they are capable to stay. According to our survey, the respondents of labour immigrants, despite many of them have live in cities for a long time, and some of their families already migrated to cities, they still did not wish to finally settle down in cities, and they wish to return home some day in the future. However, their children think otherwise, the new generation of those whose parents are labour immigrants grew up in cities, they do not have any rural living experience, and they consider themselves more as urban residents rather than rural ones. We suggest the policy towards this group of immigrants shall take filial education as the priority, and expand the social insurance coverage of them.

Finally, the entrepreneurial immigrants, who are generally older than labour and intellectual immigrants. Our survey shows many of them start from labour immigrants, gradually accumulate capital, and finally own their business. The entrepreneurial immigrants usually doing business base on the network of kinship and countrymen, and shows a strong character of regional and industrial clusters. They are the important node in the chain migration of labour force, which not only provides the initial shelter and workplace for the new immigrants, but also provides them with a good platform for entrepreneurship and integration into the environment. Based on the relationship between relatives and countrymen, it is an effective civil mechanism to realize the adaptation and social mobility of new immigrants. We suggest in the future policy adjustment, this group shall be considered to systematically manage new immigrants with various types of informal employment, and new policies shall accommodate a better environment for the survival and development of this group in cities, especially to solve the problems of their filial education and family planning.

### 11.2.3 Keep Reinforcing the Disciplinary Management on General Migration

For a long time to come, the cities will still be the concentrated areas of all kinds of new immigrants. The research group believes that it is necessary to further improve the disciplinary management of the migrant population, especially the disciplinary management of the general migration. The emphasis of strengthening disciplinary management should be put on the collection, management and sharing of general migration information, which is the foundation of realizing disciplinary management. We suggest the authority shall improve in the following aspects.

First, expand the coverage of population information management, especially the coverage of the informal employment of casual labourer. At present, in the disciplinary management system established with the starting point of rental housing and work, the immigrants of other living types and informal employment cannot be included into this management system. Such as sheds, dormitories, small hotels and other immigrants concentrated living hubs. A large number of immigrants are engaged in informal occupations such as apprenticeships, waste collection, delivering and unlicensed vendors. This group has greater mobility, higher degree of concealment, it is difficult for the authority to inflict measures of public security, family planning, health management to these people. We suggest, on the one hand, employers in informal employment, especially the entrepreneurial immigrants, shall be used to help grasp the information of informal employees, so as to encourage the entrepreneurial immigrants to join the management of informal employee, and take it as one of the social responsibilities for the entrepreneurial immigrants to integrate into cities. On the other hand, it is necessary to thoroughly check all kinds of illegal construction and illegal employment places in the city, so as to grasp the information of informal employees.

The second is to strengthen the information exchange between the origins and the destinations of migration. Some cities have gradually established the population management information system, but the construction of the information system in the destination of the population is often unable to connect with the information system in the origin, and the information system has not fully played its due function, leading to the loopholes in the management of the migration. At the same time, the collection content of population information can be appropriately expanded, including the unchanged information such as individual gender, age, date of birth, fingerprint and blood type, as well as variable information such as domicile and current residence, workplace and type of work, marital status, time of leaving a certain place, and social insurance.

The third is to coordinate the population information with other department information, strengthen the cooperation of all departments, and establish a complete population dynamic management information base. At present, the management of migrate population based on rental housing and work is still in the stage of information collection and other management driven by information collection. The important premise of migration management is to establish an information system that can share resources, which is the key to save management cost. Moreover, the population information at all levels also needs to be effectively connected with the information of other departments at all levels so as to truly provide an effective information basis for the work of each department. Under the premise of information sharing between governments and departments at all levels, the management functions at all levels should be clarified and the management centre should be moved down to districts and streets as far as possible. The municipal government shall formulate policies, build management models and databases, and implement the daily management of the grassroots level in the form of blocks, squares and grids.

### 11.2.4 The City Identity of New Immigrants Should Be Enhanced and the Harmonious Culture of Diversity Should Be Advocated

New immigrants live in cities, but most of them still consider themselves as outsider, which is not helping their integration in the urban society. We believe the city identity of new immigrants should be enhanced and the harmonious culture of diversity should be advocated.

First, promote the volunteer service for immigrants, and promote the participation of immigrants in cities' social live. Take Guangzhou as an example, the volunteer service in Guangzhou is the most active one in China, its branch reaches every corner of communities and workplaces. Guangzhou Volunteers Association operates the internet platform, with the slogan of "be volunteer, be happy", attracted a great number of volunteers and did a brilliant job on promoting the belongingness and identity for general public. We found out in our survey that some immigrants already participated in the volunteer service and gained a lot through helping others and promoting oneself. We suggest this platform shall be put to full extend of use, it shall corporates with the youth league, designs some tailor-made program, includes the new immigrant into its serving and served object, makes the new immigrant builds identity to cities.

Second, actively cooperate with public organizations, promote the pilot work of government purchasing public services, and establish social work networks for new immigrants in communities and enterprises. Government purchase of public services means that some services for social development and public welfare once held by the government directly will be transfer to social organizations, and the government will assess and purchase those services provided by social organizations according to its quantity and quality. It is a way of "the government bear, commissions, contract management, assessment to cash" to provide public services. Since the government purchase of public services have not been practiced in many Chinese cities, we suggest some branch of the government shall practice as a pilot, on the basis of careful investigation, put forward to promote a batch of new immigrant identity has a direct role of specific projects, public bidding to the qualified social organizations, by the government to provide a certain fixed number of year (e.g., a three years period) of funds to support.

In this regard, Hong Kong SAR. has very mature experience to learn from. Take Hong Kong's Grand Del Sol real-estate compound as an example, it provides accommodation and psychological counselling for women who are victims of domestic violence. The services are provided by the Boys' and Girls' Clubs Association of Hong Kong for new immigrant families are subsidized by the government for a limited period of time. Hong Kong SAR. has been very successful in transferring some of its public service functions to communities and civil organizations and actively participate in community affairs and being an effective provider of social services, there are more than 17,000 community non-governmental organisations (NGOs) in Hong Kong SAR, making a significant contribution to Hong Kong SAR. Most of Hong Kong's thousands of welfare services are run by community and civil society organisations. Community and civil society organisations are engaged in community services, elderly services, family and child services, rehabilitation services and youth development. They provide welfare services for about 2 million people and play an important role in Hong Kong SAR. In terms of the delivery of public services, Hong Kong SAR. has successfully established a "partnership" between the government and NGOs, defining a model in which public services are jointly undertaken by the government and NGOs. The funds are mainly provided by the government and supplemented by private fund-raising, Services are mainly provided by the private sector, supplemented by government services. Hong Kong's community and civil society organizations are thriving. Government purchase of services is the core of public management reform around the world. It can not only enable new immigrants to enjoy diversified and high-quality services, but also reduce the burden of overstaffed government agencies and overburdened services, so as to truly realize a small government and a big society.

Thirdly, in terms of cultural and entertainment, government services should focus less on improving the people's cultural life and protecting their cultural rights and interests. The construction standards and service standards are formulated according to the characteristics of the construction of grassroots public cultural positions, and the targets and tasks of the construction of grassroots cultural positions are refined and quantified to facilitate the implementation and implementation. We should expand the scope of public finance to cover the construction of cultural fronts at the grassroots level, and constantly increase the proportion of cultural funds spent at the grassroots level. We will strive for the allocation of social funds to the field of cultural construction at the grass-roots level, and gradually change the situation of insufficient venues, equipment, equipment and other facilities for people's cultural activities. We will support the development of cultural facilities and cultural activities at the community level and gradually improve the basic framework of the system of cultural services at the community level. In this regard, Chengdu's approach is noteworthy. Across the country in 2006, Chengdu first proposed the concept of "cultural supermarket" construction, covering the whole city of "recreation within 15 min" and "fitness within 10 min", which means the walking distant from any community resident to cultural facility is less than 15 min, and to fitness facility less than 10 min.

### 11.2.5 Promote Participatory Urban Management and Increase the Social Participation of New Immigrants

In view of the low social participation of new immigrants in cities, the research group believes that it is necessary to promote participatory urban management to attract new immigrants to participate in community management. Participatory community management is a new set of project implementation concepts and methods in the field of international development projects since the 1990s. Participatory emphasized on the basis of the analysis of various interest groups, each stakeholder groups, especially in the past neglected opportunity for ordinary people and disadvantaged groups in project design, implementation, management and supervision, have the opportunity to express themselves in the process of sound, thus improve project implementation of the public recognition rate and its mutual eligibility with local culture. The research group believes that urban new immigrants participate in the management of urban society from the following two aspects.

First, as a stakeholder of urban public service and public supply policy, the new immigrants and the general public and business groups shall be taken equally in the setting of public policies, they shall have the right to fully understand the particular way and process of urban public service, to participate in the urban public services and public goods planning practical process, sufficient feedback on their own interests, wishes and requirements, the supervision of the public service as one of the stakeholder groups, recognition and benefit sharing. In particular, the introduction of public policies which are closely related to the new immigrants especially needs the participation of urban new immigrants. The participation of new immigrants in urban public policies can also help train the economic ability and social bearing capacity of new immigrant organizations and individuals, enhance the social self-organization function, reduce the resistance in the implementation process of the government, and make the effectiveness of urban management more easily exert.

Second, explore the potential of urban new immigrants and their self-organization, and participate in the management of urban communities and migrant population. The research group found that self-organization of new immigrants is relatively common among the new immigrant groups, with relatives, fellow townsmen, classmates and friends, peers and so on as the link showing a certain degree of settlement. Using these new immigrants to organize themselves for the management of community and migrant population can relieve the pressure of government public management and promote new immigrants to participate in urban community management. May consider by the local government guidance, hires the new immigrant personnel in the quality is higher, establishes the alien personnel service center, strengthens the alien personnel management in this way. We can also try to set up party branches, youth league branches, family planning associations, women's associations, legal rights aiding stations, training centre and so on. Of course, the first thing you need to raise awareness of the community participation of new immigrants, is to make them realize that must rely on our own can actively participate in effectively improving the management of the community and their more actively participate in community service management work, the more realize their own interests, on the contrary, if the passive state, all the work in the community will stagnate. Secondly, the local government should pay attention to the reasonable guidance of the self-organization of new immigrants, prevent the emergence of cliquing phenomenon, so as to realize the purpose of the self-organization of new immigrants to serve the immigrant population.

### 11.2.6 Guiding the New Immigrant to Establish Harmonious Social Relations with Local Residents

Social environment is a decisive factor for new immigrants to settle down in cities. Our survey shows more friendly immigrants felt in the city, more likely they will settle down. Whether the new immigrants successfully establish their social network, is crucial for their decision making to stay. This social network will provide supports for new immigrants that government, employer and others cannot provide, and it is important than income, working environment, social welfare and living condition, as a key factor for new immigrants to settle down.

The relation of new immigrants and local residents, provide affluent social resources for immigrants and directly shows the integration level of immigrants. However, numerous researches showing that as the majority of immigrants, the labour immigrants' will to associate with local residents, was prevented by the segregation of household registration, lifestyle, culture, personal traits and social consciousness. Which eventually keeping the labour immigrants out of local social network, and leading the immigrants forming homogeneous communication circles. Compare to labour immigrants, the intellectual immigrants who are well-educated, usually having a stable job and decent income, are more likely to establish social contact with local residents. Our survey shows that most of the intellectual immigrants have local contacts, among which, the instrumental support and social companionship overwhelming the emotional support relationship. As for features, the intellectual immigrants who are female, long stayed in cities, working and living among locals, more likely to have social relationships with local residents. The immigrants who have local household registration, tend to have more instrumental support from local residents, but the higher the income for immigrants, the less the emotional support relationship they have with locals. Labour immigrants are very different from intellectual in the matter of social contact, labour immigrants' primary social network is built on countrymen, relatives and schoolmate circles, then on colleagues and neighbors, follows by employers and bosses, local residents are at the bottom. Therefore, the research group suggests that on the basis of making good use of the network of blood and geographical relations, the government should guide the new immigrants and local residents to form a civil mechanism to effectively realize the adaptation and social mobility of new immigrants.

### *11.2.7 Gradient Resettlement for New Immigrants*

China has long been a developing country, and it will remain an important and special developing country for a long time. Urbanise and modernise is two themes in the developing path, during the urbanising and modernising transformation, urban immigration is a major topic in our era. Migration is an important component in the urbanising progress, the increase of urban population and the decrease of rural population is a crucial factor for urbanising. There are two causes leading population concentrate in urban area, one is the industrial development increases job in cities and decreases job in countryside, and the convenience of life in cities also attracts some rural population to immigrate; the other one is the population natural growth inclined in rural area, provide surplus labour force to emigrate.

After the founding of People's Republic of China in 1949, China was tightly griping by the state's politic machine, and an urban stratum was reinforced by the state's installation. One of the important features is the combination of the hierarchical installation of political facilities and the unbalanced investment under the centralised state power. The higher the political level of the city, the more investment the city receives and the faster its economic and social scale grows. At present, most of the cities with prominent new immigration issues are municipalities directly under the central government, provincial capitals or cities specifically designated in the state plan. In this urban differential pattern, villages have the lowest political status. Many counties are larger in population size and economic strength than the local city, but they are administratively affiliated to the city, so the county naturally becomes the blood transfusion bank of the city. On the other hand, the policy system of urban and rural division further enlarges the difference between urban and rural areas, and brings about the new wave of urban immigrants which has been pushing rural labour force to cities, especially the major cities.

From the experience and lessons of the urbanization development in the third world countries, the population gathered from the countryside to the city without restriction, which caused the rapid expansion of the city. Rapid urbanization has brought a great deal of social problems to the country. Unemployment, poverty, crime, environmental pollution and so on have all happened and are happening. Therefore, one-sided development of cities, excessive population growth and over-concentration are lessons worth learning.

Thus, urbanization does not simply refer to an increasing number of people living in cities and towns, but rather to an increasing process of interaction and interconnectedness between urban and rural areas of a country. Urbanisation should include the development of original cities and the local urbanization of rural areas and modern transformation. In fact, China's urbanization process from the rural urbanization and urban internationalization at the same time. In terms of the genealogy from rural to urban areas, there are several stages and types of village town, town, town urbanization, big city of county and small city and internationalization of large and medium-sized cities. In our view, urbanization includes five aspects: first, the polarization of population structure, with more people engaged in non-agriculture;

second, the diversification of economic structure, the proportion of secondary and tertiary industries gradually increased, agricultural management mode from traditional agriculture to export-oriented, commercialization and modernization of agriculture; third, the urbanization of life style, people's daily life and leisure life to urban life change; fourth is the popularization of mass media. With the improvement of rural living standards, mass media has increasingly penetrated into rural society and become one of the driving forces of rural social changes; fifth, the modernization of people's thinking and concepts. People's ideas and concepts have changed from conservative, regressive a to open, advanced and progressive. People's cultural level has improved and their overall quality has been provided. The changes in these five aspects are the correct direction and train of thought for the development of modernization and urbanization in China, and keep the internal consistency with the scientific development concept.

Apply this line of thinking on the solution of the urban new immigration, rural labour force shall seek employment nearby, combining with the migration for employment, on the one hand, as for a view of whole China, gradient resettle immigrants to large, medium, small cities and towns; on the other hand, hierarchical classify new immigrants in each city, tailor-made policies to settle their different needs, deal with the stratum relations and the market relations, rationalising the employment structure and the disciplining the spatial settlement for new immigrants.

# References

Abraham H.Maslow, Xu Jinsheng(trans), *Motivation and Personality*, Beijing: Huaxia Publishing House, 1987.

Alan B. Simmons, Ramiro Ramirez Cardona, "Rural-Urban Migration, Who Comes, Who Stays, Who Returns? The Case of Bogotá, Columbia, 1929–1968", *International Migration Review*, Vol. 6, No. 2, Internal Migration in Latin America (Summer, 1972), pp. 166–181.

Alden Speare, Jr., "A Cost-Benefit Model of Rural to Urban Migration in Taiwan", *Population Studies*, Vol. 25, No. 1 (March, 1971), pp. 117–130.

Arnold Rose, Leon Warshay, "The Adjustment of Migrants to Cities", *Social Force*, (October, 1957), pp. 72–76.

Ben Chenglong, Going to the City or Going Home? Analysis of the Phenomenon of Migrant Workers and Farmers Returning to Cities under the New Situation, *Rural Economy*, Vol. 11, 2007.

Bian Yanjie, Zhang Wenhong, Economic Systems, Social Networks and Occupational Mobility, *Social Science in China*, Vol. 2, 2001.

Blaine E. Mercer, "Rural Migration to Urban Settings, Educational and Welfare Problems", *International Migration Digest*, Vol. 2, No. 1 (Spring, 1965), pp. 52–62.

Cai Fang, China's Future Demographic Dividend: Digging New Source of Economic Growth, *Chinese Journal of Population Science*, Vol. 1, 2009.

Cai He, ed., *Rural Migrant in the Process of Urbanisation: A Study from the Pearl River Delta*, Beijing: Social Sciences Academic Press (China), 2009.

Cai He, Wang Jin, Study on the Willingness of Permanent Migration of Rural Migrants, *Sociological Studies*, Vol. 6, 2007.

Calvin Goldscheider, "Migration and Social Structure, Analytic Issues and Comparative Perspectives in Developing Nations", *Sociological Forum*, Vol. 2, No. 4, Special Issue, Demography as an Interdiscipline (Autumn, 1987), pp. 674–696.

Charles Hirschman, "Problems and Prospects of Studying Immigrant Adaptation from the 1990 Population Census, From Generational Comparisons to the Process of 'Becoming American'", *International Migration Review*, Vol. 28, No. 4, Special Issue, The New Second Generation (Winter, 1994), pp. 690–713.

Chen Wenzhe, Zhu Yu, Changes and Internal Differentiation of the Migrant Population's Willingness to Settlement: Evidence from Four Cities in Fujian Province, *South China Population*, Vol. 2, 2008.

Chen Yingfang, Focus on Urban Immigrant (A Public Speech), *Jiefang Daily*, August 22, 2004.

Chen Yingfang, Rural migrant: System and Identity, *Sociological Studies*, Vol. 3, 2005.

Cheng Aihua, "Research on the Dynamic Mechanism of Migrant Workers' Return in China", *Journal of Anhui Agricultural Sciences*, Vol. 2, 2007.



D. Zhou, *Urban Migrants in China*,
https://doi.org/10.1007/978-981-99-3114-9

Zheng Hangsheng, Hong Dayong, Emphasizing and Developing the Social Security for Urban Immigrant Workers: An Importance in the Process of Social Transformation, *Academic Exchange*, Vol. 5, 1994.

Douglas S. Massey, Joaquin Arango, Graeme Hugo, Ali Kouaouci, Adela Pellegrino, J. Edward Taylor. "Theories of International Migration: A Review and Appraisal." *Population and Development Review*, Vol. 19, No. 3 (1993), pp. 431–66.

Du Weiming, *Modern Spirit and Confucian Tradition*, Beijing: SDX Joint Publishing Company, 1997.

Du Ying, *Out of the Village, Empirical Research of Rural Labour Migration in China*, Beijing: Economic Science Press, 1997.

E. G. Ravenstein, "The Birthplace of the People and the Laws of Migration", *The Geographical Magazine*, Vol. 3 (1876), pp. 173–177, 201–206, 229– 233.

E. G. Ravenstein, "The Laws of Migration", *Journal of the Royal Statistical Society*, Vol. 52, No. 2. (June, 1889), pp. 241–305.

E. G. Ravenstein, "The Laws of Migration", *Journal of the Statistical Society of London*, Vol. 48, No. 2. (June, 1885), pp.167–235.

Editorial, Let Rural Migrant Play a Role in the Construction Industry, *Economic Daily*, June 7, 1984.

Everett S.Lee, "A Theory of Migration", *Demography*, Vol. 3, No. 1. (1996), pp.47–57.

Fei Xiaotong, Chinese Modernisation: Rethinking the Rebuilding of Urban Communities, *Jiangsu Social Sciences*, Vol. 1, 2001a.

Fei Xiaotong, *Peasant Life in China*, Beijing: The Commercial Press, 2001.

Feng Xiaotian, Taking Roots: Social Adaptation of Rural Migrants in the Three Gorges Area], *Sociological Studies*, Vol. 5, 2004.

Gao Jun, Promotion of Employment and Citizenization of Rural Migrants, *Theory Monthly*, Vol. 10, 2008.

Gao Qiang, Jia Haiming, Analysis on the Causes and Effects of Migrant Workers' Backflow, *Management of Agricultural Science and Technology*, Vol. 4, 2007.

Gong Yong, Recruit Permanent Labour, Replace Rural Labour, Nanjing Nitrogen Fertilizer Factory Tap the Potential of Labour, *People's Daily*, August 17, 1988.

Gu Shengzu, A Study on Population Migration, *Wuhan University Journal(Social Science Edition)*, Vol. 2 1989;

Gu Shengzu, Yi Shance, Zheng Lingyun, Sustainable Development of Industrialization and Urbanization in China: An Approach Based on the Characteristics of Rural Migrant Workers, *Population Research*, Vol. 5, 2006.

Guo Xinghua, Chu Huijuan, From Rural Areas to Urban Areas, Integration and Segregation, *Jianghai Academic Journal*, Vol. 3, 2004.

Guo Zhenglin, Zhou Daming, Going Out for Work and the Cultivation of Modernity of the Farmers, *Journal of Sun Yat-sen University(Social Science Edition)*, Vol. 5, 1996.

Hagen Koo, "Rural-Urban Migration and Social Mobility in Third World Metropolises: A Cross-National Study", *The Sociological Quarterly*, Vol. 19, No. 2 (Spring, 1978), pp. 292–303.

Harley L.Browning, Waltraut Feindt, "Selectivity of Migrants to a Metropolis in a Developing Country: A Mexican Case Study", *Demography*, Vol. 6, No. 4 (November, 1969), pp. 347–357;

He Pichan, Guo Jinfeng, An Analysis of Rural Migrants' Tendency to Immigration under Institutional Constraints, *Chinese Rural Economy*, Vol. 10, 2004.

Hu Chunjuan, Urbanization of Migrant Workers Should Pay Attention to Classified Guidance, *Guangming Daily*, November 13, 2009.

Huang Ping, ed., *Seeking Survival*, Kunming: Yunnan People's Publishing House, 1997.

Huang Qian, An Empirical Analysis on the Factors Affecting the Rural Migrants' Willingness to Settle in Cities: Based on the Survey of Five Cities, *Journal of Shanxi University of Finance and Economics*, Vol. 4, 2008.

Huang Xiaoyan, The Social Integration Dimensions and Methods of New Urban Citizens, *Social Scientist*, Vol. 3, 2010.

James Coleman, "Social Capital in the Creation of Human Capital", *American Journal of Sociology*, Supplement 94, 1988, pp. 95-120.

Jorge Balan, "Migrant-native Socioeconomic Differences In Latin American Cities, A Structural Analysis", *Latin American Research Review* Vol. 4, No. 1, (Spring, 1969), pp. 3–29.

Josh DeWind, Philip Kasinitz, "Everything Old is New Again? Processes and Theories of Immigrant Incorporation", *International Migration Review*, Vol. 31, No. 4, Special Issue, Immigrant Adaptation and Native-Born Responses in the Making of Americans (Winter, 1997), pp. 1096–1111.

Kim Hyung-Kook, "Social Factors of Migration from Rural to Urban Areas with Special Reference to Developing Countries: The Case of Korea", *Social Indicators Research*, Vol. 10, No. 1 (January 1982), pp. 29-74.

Li Debin, The Characteristics and Causes of Migration in Contemporary China, *Sociological Studies*, Vol. 4, 1993.

Li Ke, Liu E, The Process of Urbanisation Urgently Needs Migrant Workers to Integrate into the City, *Theory Front*, Vol. 23, 2009.

Li Minghuan, An Overview of the Western Theories of International Migration in the 20th Century, *Journal of Xiamen University*, Vol. 4, 2000.

Li Peilin, Chen Guangjin, China Entering New Stage of Development, Analysis and Forecast of Social Development for 2009–2010: Analysis and Forecast of Social Situation Institute, Research Group of Chinese Academy of Social Sciences, *Society of China Analysis and Forecast (2010)*, Beijing: Social Sciences Academic Press(China), 2009, pp. 9–10.

Li Peilin, Modernity and Chinese Experience, *Chinese Journal of Sociology*, Vol. 3, 2008.

Li Peilin, Rural Migrant's Social Network and Status, *Sociological Studies*, Vol. 4, 1996a.

Li Peilin, Social Networks and Social Status of Rural Migrants, *Sociological Studies*, Vol. 4, 1996b.

Li Peilin, Tremendous Changes, the End of Villages: A Study of Villages in the Center of Guangzhou City, *Social Sciences in China*, Vol. 1, 2002.

Li Qiang, Labour Market of Urban Migrant Workers in China, *Journal of Dalian Minzu University*, Vol. 7, 2000.

Li Qiang, Tang Zhuang, The Rural-Urban Workers in Cities and the Irregular Employment in Cities, *Sociological Studies*, Vol. 6, 2002.

Li Qiang, The Occupational Mobility of Rural Migrant in Cities of China's Mainland, *Sociological Studies*, Vol. 3, 1999;

Li Ruojian, "Urban White-collar" Group: Analysis on Guangzhou Population Data, *Chinese Journal of Population Science*, Vol. 2, 2009.

Li Ruojian, A Preliminary Analysis of the Settlement and Mobility of the Migrant Population in Guangdong Province, *Population Research*, Vol. 6, 2007.

Lu Ruojian, The Opportunities and Obstacles for Status Attainment: An Analysis of Occupational Structure in Communities with Dense Floating Population, *Chinese Journal of Population Science*, Vol. 5, 2006.

Lu Wei, *Residential Urbanization: A Perspective on Sciences of Human Settlements*, Beijing: Higher Education Press, 2005, p. 140.

Li Weidong, The Social Fusion of Rural Migrants in View of Social Distance, *Social Sciences of Beijing*, Vol. 6, 2007.

Li Yaqin, Li Kairong, Research on the Effectiveness of Settlement- oriented Migration of Rural Migrants, *Rural Economy*, Vol. 12, 2007.

Liao Zhenghong, *Population Migration*, Taipei: San Min Book Co., Ltd, 1985.

Lin Nan, V. M. Ensel, J. C. Vaughan, Zheng Lu, eds. Trans, Social Resources and Strength of Ties, Structural Factors in Occupational Status Attainment, *Social Science Abroad*, Vol. 3, 2001.

Lin Peng, Research on Young Rural Migrants Marching towards the "Post-survival Era", China Rural Studies Website, retrieved at November 3, 2003, http,//www.cucc.org.cn/html/74-3/3087.htm.

Lin Yifu, Rural Modernization and Urban Development, *China Financial and Economic News*, August 17, 2001.

Liu Chuanjiang, A Study of Urbanization of Rural Migrant in China, *Theory Monthly*, Vol. 10, 2006.

Liu Huiwu, Cultural Capital and Integration into Cities of Rural Migrants, *Rural Economy*, Vol. 7, 2007.

Liu Jian'e, A Study on Rural—Urban Migrants' Practical Strategies for Social Inclusion: Towards a Community Inclusion Approach, *Chinese Journal of Sociology*, Vol. 1, 2006.

Liu Xin, Class Habitus and Taste: Bourdieu's Theory of Social Class, *Sociological Research*, Vol. 6, 2003.

Liu Zheng, The Paradox of Unlimited Supply of Labour Force: An Analysis on the Causes and Effects of Migrant Workers' Backflow, *Journal of Tsinghua University (Philosophy and Social Science)*, Vol. 3, 2006.

Lu Xueli, China's Population Facing Situation of Trisection, retrieved at May 21, 2020, http,//politics.people.com.cn/GB/1027/9129924.html.

M. S. Granovetter, "The Strength of Weak Ties",*The American Journal of Sociology*, 1973, Vol. 78, No. 6, pp. 1360–1380.

Ma Guoqing, The Foundation of a Family-Oriented Civil Society: Family Theory and Continued Vertical Society, *Academic Research*, Vol. 8, 2007.

Manoela Guidorizzi Borges, *Citizenship for the Urban Poor? Inclusion trough Housing Program in Rio De Janeiro*, Brazil, Doctoral Dissertation, The University of Colorado, 2005.

Mi Qingcheng, Analysis on the Sense of Belonging of Rural Migrants, *Youth Studies*, Vol. 3, 2004.

Michael P. Todaro, "A Model of Labour Migration and Urban Unemployment in Less Developed Countries", *The American Economic Review*, Vol. 59, No. 1 (1969), pp. 138-148.

Milton M.Gordon, "Assimilation in America: Theory and Reality, Daedalus", Vol. 90, No. 2, *Ethnic Groups in American Life* (Spring, 1961), pp. 263–285.

Pan Yi, Chinese Women Workers: the Call of the Emerging Working Class, Hong Kong: Ming Bao Publishing House Co., Ltd, 2007.

Peng Xizhe, *The Changing Population of China*, Wiley-Blackwell, 2000, p. 180.

Peter M.Blau, Otis Dudley Duncan, *The American Occupational Structure*, New York: Wiley, 1967.

Qu Jingdong, The Strength of Relations in the Life World, the Life Trajectory of Rural Migrants, Ke Lanjun, Li Hanlin, *Villagers in the City, Migrant Population in Big Cities in China,* Beijing: Central Compilation & Translation Press, 2001.

Ren Yuan, Dai Xingyi, Model Analysis of the Tendency of the Migrant Population to Settle in Cities for a Long Time, *South China Population*, Vol. 4, 2003.

Ren Yuan, Wu Minle, Social Integration of Floating Population in Urban China: A Literature Review, *Population Research*, Vol. 3, 2006a.

Ren Yuan,"Gradually Settlement" and "Residence Determines Residence": Analysis of the Residential Mode of the Immigrants in Shanghai, *Chinese Journal of Population Science*, Vol. 3, 2006b.

Research Group of the State Council Research Office, *Research Report on Chinese Rural Migrants*, Beijing: China Yan Shi Press, 2006.

Richard Alba, Victor Nee, "Rethinking Assimilation Theory for a New Era of Immigration", *International Migration Review*, Vol. 31, No. 4, Special Issue, Immigrant Adaptation and Native-Born Responses in the Making of Americans (Winter, 1997), pp. 826–874.

Robert V.Kemper, "Rural-Urban Migration in Latin America: A Framework for the Comparative Analysis of Geographical and Temporal Patterns", *International Migration Review*, Vol. 5, No. 1 (Spring, 1971), pp. 36–47.

Ronald S.Burt, *Structural Holes: The Social Structure of Competition,* Harvard University Press, 1992.

Rural Labour Research Group, The Status of Rural Labour in Pearl River Delta, *Social Science in China*, Vol. 4, 1995.

Seymour M.Lipset, Reinhard Bendix, *Social Mobility in Industrial Society*, Berkeley: University of California Press, 1959, pp. 204–216.

Stanley Lieberson, "Generational Differences among Blacks in the North", *American Journal of Sociology*, Vol. 79, No. 3 (November, 1973), pp. 550–565.

Tong Xing, Ma Xiheng, "The Others in Harmonious Community" and "Reconciliation": Community Integration of Urban Immigrants, *Social Science Research*, Vol. 1, 2008.

W. Arthur Lewis, "Economic Development with Unlimited Supplies of Labour", *Manchester School of Economic and Social Studies*, Vol. 22, No. 2. (1954), pp. 139–191.

Wang Chunguang, A Study of Floating People's "Semi-urbanization", *Sociological Studies*, Vol. 5, 2006.

Wang Chunguang, Social Identity of the New Generation of Rural Migrant and Merger of Urban and Rural, *Sociological Studies*, Vol. 3, 2001.

Wang Chunlan, Ding Jinhong, Analysis of Influencing Factors of the Migrant Population's Willingness to Settle in the City, *South China Population*, Vol. 1, 2007.

Wang Dong, Qin Wei, A Study on Intergenerational Difference of Rural Migrant, Comparative Stratification of Rural Migrant in Chengdu, *Population Research*, Vol. 5, 2002.

Wang Guangling, Yang Chen, Analysis on the Difficulties of Social Integration of Rural Migrants under the Background of Financial Crisis, *Theory Journal*, Vol. 3, 2010.

Wang Miao, Employment of College Students: Unable to Escape the "Black Hole" of Big Cities, *China Reform Daily*, June 24, 2010.

Wang Xiaozhang, From "Existence" to "Recognition": Issues of Peasant Workers in the View of Citizenship, *Sociological Studies*, Vol. 1, 2009.

Wang Yanhua, A Sociological Study on Urbanization of New Generation Rural Migrant, *China Youth Study*, Vol. 5, 2007.

Wang Yijie, Analysis of Influencing Factors Affecting Rural Migrants' Willingness to Settle in Cities, *Jiangsu Social Sciences*, Vol. 5, 2005.

Wen Jun, Are They Migrant or Resident? A Study on Shanghai's New Labour Migrant Since 1980s, The Pablicity Department of CPC Shanghai Municipal Committee, ed., *Modern Awareness and Urban Studies*, Shanghai: Shanghai People's Publishing House, 2006, pp. 32-67.

Wen Jun, The System Composition and Behavior Preference of Urban New Immigrant in China, *Nanjing Journal of Social Sciences*, Vol. 1, 2005.

William P.Mangin, "Latin American Squatter Settlements: A Problem and a Solution", *Latin American Research Review*, Vol. 2, No. 3(Summer, 1967), pp. 65–98.

Wu Xiaogang, Plunging into the Private Business: Self-employment and Social Stratification in the Transformation of China's Urban and Rural Labour Market (1978–1996), *Sociological Studies*, Vol. 6, 2006.

Wu Xinglu, An Empirical Study on the Influencing Factors of Rural Migrants' Decision-making on Settlement-oriented Migration, *Population & Economics*, Vol. 1, 2005.

Wu Xinhui, Paying Attention to the Social Integration of Children of Migrants, *Chinese Journal of Society*, Vol. 9, 2004.

Xia Fei, A Rational and Objective View on University Enrolment Expansion, *Guangming Daily*, 10, December, 2008, retrieved at May 21, 2020, https,//epaper. gmw.cn/ gmrb/html/2008–12/10/nw. D110000gmrb_20081210_1–05.htm?div=-1.

Xiang Biao, *Communities Crossing the Boundary: The Life history of "Zhejiang Village" in Beijing*, Beijing: SDX Joint Publishing Company, 2000.

Xiong Bo, Shi Renbing, An Analysis on the Factors Affecting the Rural Migrants' Willingness to Settle Permanently in the City, *South China Population*, Vol. 2, 2007.

Xiong Caiyun, Push-pull Model on the Factor of Transfer Decision-making for Rural Migrants' Settling Down in the City and Empirical Analysis, *Issues in Agricultural Economy*, Vol. 3, 2007.

Yang Dongping, Monitoring Educational Equity and Conducting Background Checks on Students' Families, retrieved at March 19, 2009, http,//blog. tianya.cn/blogger/post_show.asp?BlogID=296700&PostID=16802752.

Yang Hongshan, An Investigation on Pearl River Delta's "Rural Labour Rush", *Demography Studies*, Vol. 2, 1995.

Yang Junhua, From Isolation and Selection to Integration: Theoretical Contemplating on Integration of Immigrant, *Population Research*, Vol. 1, 2009.

Yang Min, Community as State Governance Unit: A Case Study on Residents' Community Participation and Cognition in the Process of Community Building Campaign, *Sociological Studies*, Vol. 4, 2007.

Yao Jun, Survey on the Willingness to Settle in Cities of Rural Migrants: An Empirical Analysis in Three Cities of Jiangsu Province, *Urban Problems*, Vol. 9, 2009.

Yi Qingchuan, Yi Yuanhua, Tan Yangjin, A Study on the Backflow of Migrant Workers' Elite Group to Construct New Countryside, *Rural Economy*, Vol. 9, 2007.

Yin Xiaoqing, The Obtain and the Maintain of Mobile Employment, *Journal of Nanjing Normal University(Social Science Edition)*, Vol. 5, 2001.

Yu Xianzhong, The Floating Position of Chinese Population, *Journal of University of Jinan (Social Science Edition)*, Vol. 6, 2004.

Zeng Xuhui, Qin Wei, Analysis on the Influencing Factors of Rural Migrants' Tendency to Stay in the City, *Population & Economics*, Vol. 3, 2003.

Zhai Zhenwu, Hou Jiawei, Migrant Population Concentrated Communities in Beijing: Patterns and Trends, *Population Research*, Vol. 1, 2010.

Zhang Cheng, Analysis of the Phenomenon of Migrant Workers' Return in China, *Journal of Nanjing Forestry University (Humanities and Social Sciences Edition)*, Vol. 3, 2007.

Zhang Jijiao, *Adaptation to the City*, Beijing: The Commercial Press, 2004.

Zhang Jijiao, The "Differential Mode of Association" from Rural to Urban Version: With Examples of the Migrant Population's Employment in Cities, *Ethno-National Studies*, Vol. 6, 2004b.

Zhang Wenhong, Lei Kaichun, Structural Equation Model of the Social Identifications of the New Urban Immigrant, *Sociological Studies*, Vol. 4, 2009a.

Zhang Wenhong, Lei Kaichun, Structural Model of Social Identity of Urban Immigrants, *Sociological Studies*, Vol. 4, 2009b.

Zhang Wenhong, Lei Kaichun, The Urban New Immigrants' Social Inclusion: Internal Structure, Present Situation and Influential Factors, *Sociological Studies*, Vol. 5, 2008.

Zhang Wenhong, Social Capital: Theoretical Discussion and Empirical Studies, *Sociological Studies*, Vol. 4, 2003.

Zhang Zheng, Settling in the City or Returning to the Hometown for Development?: Analysis of the Life Cycle of Rural Migrants' Decision on Migration, *Chinese Rural Economy*, Vol. 7, 2006.

Zhang Zhiyong, Intergenerational Urbanization of Rural Migrant: A Comparative Study Based on Multiple Employment of Rural Migrant, *Jianghan Tribune*, Vol. 11, 2009.

Zhao Yandong, Wang Fenyu, The Acquisition and Determinants of Economic Status of the Migrant Population in Urban and Rural Areas, *Chinese Journal of Population Science*, Vol. 4, 2002.

Zheng Chuangui, Lu Xiaohui, The Causes, Harms and Countermeasures of Social Groups Segregation of Cities in China Today, *Urban Problems*, Vol. 6, 2003.

Zheng Gongcheng, Wong Linda, et al. ed., *Immigrant Workers' Problems and Social Protection in China*, Beijing: People's Publishing House, 2007.

Zhong Dajun, *Examining the Inequality of National Treatment: China under the Dual Structure*, Beijing: China Workers Publishing House, 2002.

Zhou Daming, *"Free" Urban Marginal People: A Study on Casual Workers in China's Southeast Coast*, Guangzhou: Sun Yat-sen University Press, 2007.

Zhou Daming, A Study on Migrant Casual Labour in Guangzhou, *Sociological Studies*, Vol. 4, 1994.

Zhou Daming, Cheng Luxiao, A Study of Migrant Workers' Occupational Diversification and Their Children's Education: Taking Youxian Hunan Province as an Example, *Journal of South China Normal University (Social Science Edition)*, Vol. 6, 2009.

Zhou Daming, Immigration Culture: A Hypothesis, *Jiangsu Social Sciences*, Vol. 5, 2005a.

Zhou Daming, *Longing for Survival: An Anthropology Observation on Rural Labour's Mobility*, Guangzhou: Sun Yat-sen University Press, 2005.

Zhou Daming, The Eternal Pendulum, Study on the Mobility of Labour Immigrants, Ke Lanjun, Li Hanlin, *Urban Villagers, The Floating Population in China's Big Cities*, Beijing: Central Compilation & Translation Press, 2001.

Zhou Daming, Zhou Jianxin, Liu Zhijun, *"Free" Urban Marginal People: Study on Casual Labourer in Southeast Coast of China*, Guangzhou: Sun Yat: sen University Press, 2007, pp. 3–11.

Zhou Tianyong, The Hesitation, Missteps and Consequences of Urbanization in China, *Urban Insight*, Vol. 1, 2009.

Zhou Yu'e, Ruan Zhengyu, The Status Quo and Trend of Theoretical Research on Contemporary International Immigration, *Jinan Journal* (Philosophy & Social Sciences), Vol. 2, 2003.

Zhu Kaojin, Liu Ruiqing, Research on the Social Support Network and Urban Integration of Young Rural Migrants, *Youth Studies*, Vol. 8, 2007.

Zhu Li, Adaptation to Urban Life: Taking the Spiritual and Cultural Life of Migrating Population for Example, *Journal of Hehai University*, Vol. 3, 2005.

Zhu Li, Characteristics and Social Status of the Peasant Worker Class, *Journal of Nanjing University (Philosophy, Humanities and Social Sciences)*, Vol. 6, 2003.

Zhu Li, Debate on the Urban Adaptability of the Rural Migrants Strata, *Jianghai Academic Journal*, Vol. 6, 2002.

Zhu Li, Group Prejudice and Discrimination, *Jianghai Academic Journal*, Vol. 6, 2001.

Zhu Li, How to Recognize Rural Migrant, A Preface, Zhu Li, Chen Ru, ed. *New Urban Immigrant: Report on Immigrant of Nanjing*, Nanjing: Nanjing University Press, 2003, p. 22.

Zhu Limin, Pros and Cons of Massive Rural Labour Recruit by Enterprises, *Shanghai Enterprise*, Vol. 8, 1988.

Zhu Yu, Households Registration System Reform and the Dwelling Willing of Floating Population in the Areas They are Dwelling and Its Restrictive Mechanism, *South China Population*, Vol. 3, 2004.

Zhu Yu, Overseas Research on Non-permanent Migration and Its Implications for the Issues of Floating Population in China, *Population Research*, Vol. 5, 2004.